Universitext

For other titles published in this series, go to
http://www.springer.com/series/223

Helge Holden
Bernt Øksendal
Jan Ubøe
Tusheng Zhang

Stochastic Partial Differential Equations

A Modeling, White Noise Functional Approach

Second Edition

 Springer

Helge Holden
Department of Mathematical Sciences
Norwegian University of Science
   and Technology
7491 Trondheim
Norway

and

Center of Mathematics and Applications
University of Oslo
0316 Oslo
Norway
holden@math.ntnu.no

Bernt Øksendal
Department of Mathematics
University of Oslo
0316 Oslo
Blindern
Norway
oksendal@math.uio.no

Jan Ubøe
Department of Economics
Norwegian School of Economics
   and Business Administration
5045 Bergen
Norway
jan.uboe@nhh.no

Tusheng Zhang
University of Manchester
School of Mathematics
Manchester M13 9PL
United Kingdom
tzhang@maths.man.ac.uk

ISBN 978-0-387-89487-4   e-ISBN 978-0-387-89488-1
DOI 10.1007/978-0-387-89488-1
Springer New York Dordrecht Heidelberg London

Library of Congress Control Number: 2009938826

Mathematics Subject Classification (2000): 60H15, 35R60, 60H40, 60J60, 60J75

To Eva, Qinghua, and Tatiana

At vide hvad
  man ikke véd,
er dog en slags
  alvidenhed.

Knowing what
  thou knowest not
is in a sense
  omniscience.
                    Piet Hein

Preface to the Second Edition

Since the first edition of this book appeared in 1996, white noise theory and its applications have expanded to several areas. Important examples are

(i) White noise theory for fractional Brownian motion. See, e.g., Biagini et al. (2008) and the references therein.
(ii) White noise theory as a tool for Hida–Malliavin calculus and anticipative stochastic calculus, with applications to finance. See, e.g., Di Nunno et al. (2009) and the references therein.
(iii) White noise theory for Lévy processes and Lévy random fields, with applications to SPDEs.

The last area (iii) fits well into the scope of this book, and it is natural to include an account of this interesting development in this second edition. See the new Chapter 5. Moreover, we have added a remarkable, new result of Lanconelli and Proske (2004), who use white noise theory to obtain a striking general solution formula for stochastic differential equations. See the new Section 3.7. In the new Chapter 5 we provide an introduction to the more general theory of white noise based on Lévy processes and Lévy random fields, and we apply this theory to the study SPDEs driven by this type of noise. This is an active area of current research.

We show that the white noise machinery developed in the previous chapters is robust enough to be adapted, after some basic modifications, to the new type of noise. In particular, we obtain the corresponding Wick product, generalized Skorohod integration and Hermite transform in the Lévy case, and we get the same general solution procedure for SPDEs. The method is illustrated by a study of the (stochastic) Poisson equation, the wave equation and the heat equation involving space or space-time Lévy white noise.

In this second edition we have also improved the presentation at some points and corrected misprints. We are grateful to the readers for their positive responses and constructive remarks. In particular we would like to thank (in alphabetical order) Atle Gyllensten, Jørgen Haug, Frank Proske, Mikael Signahl, and Gjermund Våge for many interesting and useful comments.

Trondheim
Oslo
Bergen
Manchester

January 2009

Helge Holden
Bernt Øksendal
Jan Ubøe
Tusheng Zhang

Preface to the First Edition

This book is based on research that, to a large extent, started around 1990, when a research project on fluid flow in stochastic reservoirs was initiated by a group including some of us with the support of VISTA, a research cooperation between the Norwegian Academy of Science and Letters and Den norske stats oljeselskap A.S. (Statoil). The purpose of the project was to use stochastic partial differential equations (SPDEs) to describe the flow of fluid in a medium where some of the parameters, e.g., the permeability, were stochastic or "noisy". We soon realized that the theory of SPDEs at the time was insufficient to handle such equations. Therefore it became our aim to develop a new mathematically rigorous theory that satisfied the following conditions.

1) The theory should be *physically meaningful* and realistic, and the corresponding solutions should make sense physically and should be useful in applications.
2) The theory should be *general* enough to handle many of the interesting SPDEs that occur in reservoir theory and related areas.
3) The theory should be *strong* and *efficient* enough to allow us to solve these SPDEs explicitly, or at least provide algorithms or approximations for the solutions.

We gradually discovered that the theory that we had developed in an effort to satisfy these three conditions also was applicable to a number of SPDEs other than those related to fluid flow. Moreover, this theory led to a new and useful way of looking at stochastic *ordinary* differential equations as well. We therefore feel that this approach to SPDEs is of general interest and deserves to be better known. This is our main motivation for writing this book, which gives a detailed presentation of the theory, as well as its application to ordinary and partial stochastic differential equations.

We emphasize that our presentation does not make any attempts to give a comprehensive account of the theory of SPDEs in general. There are a number of important contributions that we do not mention at all. We also emphasize that our approach rests on the fundamental work by K. Itô on stochastic calculus, T. Hida's work on white noise analysis, and J. Walsh's early papers on SPDEs. Moreover, our work would not have been possible without the inspiration and wisdom of our colleagues, and we are grateful to them all.

In particular, we would like to thank Fred Espen Benth, Jon Gjerde, Håkon Gjessing, Harald Hanche-Olsen, Vagn Lundsgaard Hansen, Takeyuki Hida, Yaozhong Hu, Yuri Kondratiev, Tom Lindstrøm, Paul-André Meyer, Suleyman Üstünel and Gjermund Våge for helpful discussions and comments.

And, most of all, we want to express our gratitude to *Jürgen Potthoff*. He helped us to get started, taught us patiently about the white noise calculus and kept us going thanks to his continuous encouragement and inspiration.

Finally, we would like to thank Tove Christine Møller and Dina Haraldsson for their excellent typing. Bernt Øksendal would like to thank the University of Botswana for its hospitality during parts of this project. Helge Holden acknowledges partial support from Norges forskningsråd. We are all grateful to VISTA for their generous support of this project.

April 1996

Helge Holden
Bernt Øksendal
Jan Ubøe
Tusheng Zhang

How to use this book

This book may be used as a source book for researchers in white noise theory and SPDEs. It can also be used as a textbook for a graduate course or a research seminar on these topics. Depending on the available time, a course outline could, for instance, be as follows:

Sections 2.1–2.8, Section 3.1, Section 4.1, and a selection of Sections 4.2–4.8, Sections 5.1–5.7.

Contents

Chapter 1
Introduction

1.1 Modeling by Stochastic Differential Equations

The modeling of systems by differential equations usually requires that the parameters involved be completely known. Such models often originate from problems in physics or economics where we have insufficient information on parameter values. For example, the values can vary in time or space due to unknown conditions of the surroundings or of the medium. In some cases the parameter values may depend in a complicated way on the microscopic properties of the medium. In addition, the parameter values may fluctuate due to some external or internal "noise", which is random – or at least appears so to us.

A common way of dealing with this situation has been to replace the true values of these parameters by some kind of average and hope that the corresponding system will give a good approximation to the original one. This approach is not always satisfactory, however, for several reasons.

First, even if we assume that we obtain a reasonable model by replacing the true parameter values by their averages, we might still want to know what effect the small fluctuations in the parameter values actually have on the solution. For example, is there a "noise threshold", such that if the size of the noise in the system exceeds this value, then the averaged model is unacceptable?

Second, it may be that the actual fluctuations of the parameter values effect the solution in such a fundamental way that the averaged model is not even near to a description of what is actually happening.

The following example may serve as an illustration of this.

Suppose that fluid is injected into a dry, porous (heterogeneous but isotropic) rock at the injection rate $f(t, x)$ (at time t and at the point $x \in \mathbb{R}^3$). Then the corresponding fluid flow in the rock may be described mathematically as follows:

Let $p_t(x)$ and $\theta(t, x)$ denote the pressure and the saturation, respectively, of the fluid at (t, x). Assume that either the point x is dry at time t, i.e.,

H. Holden et al., *Stochastic Partial Differential Equations*, 2nd ed., Universitext,
DOI 10.1007/978-0-387-89488-1_1, © Springer Science+Business Media, LLC 2010

$\theta(t, x) = 0$, or we have complete saturation $\theta_0(x) > 0$ at time t. Define *the wet region at time t, D_t,* by

$$D_t = \{x;\ \theta(t, x) = \theta_0(x)\}. \tag{1.1.1}$$

Then by combining Darcy's law and the continuity equation (see Chapter 4) we end up with the following *moving boundary problem* for the unknowns $p_t(x)$ and D_t (when viscosity and density are set equal to 1):

$$div(k(x)\nabla p_t(x)) = -f_t(x); x \in D_t \tag{1.1.2}$$

$$p_t(x) = 0; x \in \partial D_t \tag{1.1.3}$$

$$\theta_0(x) \cdot \frac{d}{dt}(\partial D_t) = -k(x)N^T(x)\nabla p_t; x \in \partial D_t, \tag{1.1.4}$$

where $N(x)$ is the outer unit normal of ∂D_t at x.

(The divergence and gradients are taken with respect to x.)

We assume that the initial wet region D_0 is known and that $\mathrm{supp} f_t \subset D_t$ for all t. For the precise meaning of (1.1.4) and a weak interpretation of the whole system, see Chapter 4. Here $k(x) \geq 0$ is the *permeability* of the rock at point x. This is defined as the constant of proportionality in Darcy's law

$$q_t(x) = -k(x)\nabla p_t(x), \tag{1.1.5}$$

where $q_t(x)$ is the (seepage) velocity of the fluid. Hence $k(x)$ may be regarded as a measure of how freely the fluid is flowing through the rock at point x. In a typical porous rock, $k(x)$ is fluctuating in an irregular, unpredictable way. See for example, Figure 1.1, which shows an actual measurement of $k(x)$ for a cylindrical sample taken from the porous rock underneath the North Sea.

In view of the difficulty of solving (1.1.2)–(1.1.4) for such a permeability function $k(x)$, one may be tempted to replace $k(x)$ by its x-average $\bar{k}$ (constant) and solve this system instead. This, however, turns out to give

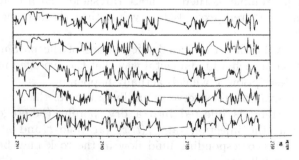

Fig. 1.1 Measurements of permeability in a porous rock. (A linear interpolation is used on intervals where the values are missing). Courtesy of Statoil, Norway.

a solution that does not describe what actually happens! For example, if we let $f_t(x) = \delta_0(x)$ be a point source at the origin and choose D_0 to be an open unit ball centered at 0, then it is easy to see (by symmetry) that the system (1.1.2)–(1.1.4) with $k(x) \equiv \bar{k}$ will give the solution $\{D_t\}_{t \geq 0}$ consisting of an increasing family of open balls centered at 0. See Figure 1.2.

Such a solution is far from what actual experiments with fluid flow in porous rocks show; see, for example, Figure 1.3.

In fact, it has been conjectured that ∂D_t is a fractal with Hausdorff dimension 2.5 in $\mathbb{R}^3$ and 1.7 for the corresponding 2-dimensional flow. In the 2-dimensional case this conjecture is supported by physical experiments (see Oxaal et al. (1987) and the references therein). Moreover, in both two and three dimensions these conjectures are related to the corresponding conjectures for the (apparently) related fractals appearing in DLA processes (see, e.g., Måløy et al. (1985)).

We conclude from the above that it is necessary to take into account the fluctuations of $k(x)$ in order to get a good mathematical description of the flow. But how can we take these fluctuations into account when we do not know exactly what they are?

We propose the following: The lack of information about $k(x)$ makes it natural to represent $k(x)$ as a *stochastic* quantity. From a mathematical point of view it is irrelevant if the uncertainty about $k(x)$ (or some other parameter) comes from "real" randomness (whatever that is) or just from our lack of information about a non-random, but complicated, quantity.

If we accept this, then the right mathematical model for such situations would be partial differential equations involving stochastic or "noisy" parameters – *stochastic partial differential equations* (SPDEs) for short.

In order to develop a theory – and constructive solution methods – for SPDEs, it is natural to take as a starting point the well-developed and highly successful theory and methods from stochastic ordinary differential equations (SDEs). The extension from the 1-parameter case (SDEs) to the multiparameter case (SPDEs) is in some respects straightforward, but in other respects surprisingly difficult. We now explain this in more detail.

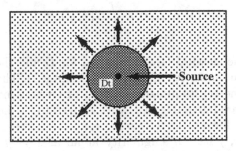

Fig. 1.2 A constant permeability $k(x) \equiv \bar{k}$ leads to solutions of the moving boundary problem consisting of expanding balls centered at the injection hole.

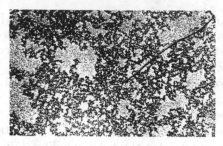

Fig. 1.3 A physical experiment showing the fractal nature of the wet region (dark area). Courtesy of Knut Jørgen Måløy.

In SDE theory the fundamental concepts are the *white noise* $W_t(\omega)$ (where t denotes time and ω some random element), the Itô integral $\int_0^t \phi(s,\omega)dB_s(\omega)$, or the Stratonovich integral $\int_0^t f(s,\omega) \circ dB_s$, where $B_s = B_s(\omega)$ denotes n-dimensional Brownian motion, $s \geq 0$. There is a canonical extension of 1-parameter white noise $W_t(\omega)$ to multiparameter white noise $W_{x_1,x_2,\ldots,x_n}(\omega)$. Similarly, the 1-parameter Itô and Stratonovich integrals have canonical extensions to the multiparameter case, involving multiparameter Brownian motion $B_{x_1,x_2,\ldots,x_n}(\omega)$ (sometimes called the , see Chapter 2). From that point of view the extension from SDE to SPDE appears straightforward, at least in principle.

The following example, however, related to an example discussed by Walsh (1986), shows that there are unexpected difficulties with such an extension.

Consider the following model for the temperature $u(x)$ at point x in a bounded domain D in $\mathbb{R}^d$. When the temperature at the boundary ∂D of D is kept equal to 0, and there is a random heat source in D modeled by *white noise* $W(x) = W(x_1, x_2, \ldots, x_n, \omega)$, then

$$\begin{cases} \Delta u(x) = -W(x); \ x = (x_1,\ldots,x_n) \in D \\ u(x) = 0; \ x \in \partial D. \end{cases} \tag{1.1.6}$$

It is natural to guess that the solution must be

$$u(x) = u(x,\omega) = \int_D G(x,y)dB(y), \tag{1.1.7}$$

where $G(x,y)$ is the classical Green function for D and the integral on the right is the multiparameter Itô integral (see Chapter 2). The problem is that for the (multiparameter) Itô integral in (1.1.7) to make sense, it is necessary that $G(x,\cdot)$ be square integrable in D with respect to the Lebesque measure. But this only holds for $d \leq 3$.

For $d = 1$, $G(x,\cdot)$ has no singularity at all at $x = y$. For $d = 2$, the singularity of $G(x,y)$ at $x = y$ is $\log 1/|x - y|$, which belongs to $L^p_{\text{loc}}(\mathbb{R}^2)$ for all $p < \infty$. In $\mathbb{R}^d$ for $d \geq 3$, the singularity is $|x - y|^{-d+2}$, and using polar coordinates we see that $G(x, \cdot) \in L^p_{\text{loc}}(\mathbb{R}^d)$ if and only if $p < d/d - 2$ for $d \geq 3$. It was shown in Walsh (1986) that for any d there exists a unique distribution valued stochastic process $u(x, \omega)$ solving equation (1.1.6). More precisely, there exists a Sobolev space $H^{-n}(\mathbb{R}^d)$ and an $H^{-n}(\mathbb{R}^d)$-valued stochastic process

$$u = u(\omega): \quad \Omega \to H^{-n}(\mathbb{R}^d) \quad \text{(with } n \in \mathbb{N} \text{ large enough)}$$

such that (1.1.6) holds, in the sense of distributions, for almost all ω. Therefore, for SPDEs one can no longer expect to have solutions represented as ordinary (multiparameter) stochastic processes, unless the dimension is sufficiently low.

This fact makes it necessary to reconsider the concept of a solution of an SPDE in terms of some kind of generalized process. The Walsh construction, albeit elegant, has the disadvantage that it leads to the problem of defining the multiplication of (Sobolev or Schwartz) distributions when one considers SPDEs where the noise appears multiplicatively. Also, it does not seem to allow extension to types of noise other than white noise. To be able to model, for example, our problem (1.1.2)–(1.1.4) as an SPDE, it will be necessary to represent the permeability $k(x)$ as a *positive* noise. Moreover, there the noise appears multiplicatively.

Because of this difficulty we choose a different approach: Rather than considering the distribution–valued stochastic processes

$$\omega \to u(\cdot, \omega) \in H^\alpha(\mathbb{R}^d); \quad \omega \in \Omega,$$

we consider functions

$$x \to u(x, \cdot) \in (\mathcal{S})_{-1}; \, x \in \mathbb{R}^d,$$

where $(\mathcal{S})_{-1}$ is a suitable space of stochastic distributions (called the Kondratiev space). See Chapter 2.

This approach has several advantages:

(a) SPDEs can now be interpreted in the usual strong sense with respect to t and x. There is no need for a weak distribution interpretation with respect to time or space.

(b) The space $(\mathcal{S})_{-1}$ is equipped with a multiplication, the *Wick product* denoted by a $\diamond$. This gives a natural interpretation of SPDEs where the noise or other terms appear multiplicatively.

(c) The Wick product is already implicit in the Itô and the Skorohod-integrals. The reason for this is the remarkable fact that if $Y(t) = Y(t, \omega)$ is Skorohod-integrable, then

$$\int_0^T Y(t)\delta B(t) = \int_0^T Y(t) \diamond W(t)dt; \, T \geq 0, \tag{1.1.8}$$

where the integral on the left is the Skorohod-integral and the integral on the right is the Pettis integral in $(\mathcal{S})_{-1}$. If $Y(t, \omega)$ is *adapted*, then the Skorohod-integral coincides with the Itô integral, and (1.1.8) becomes

$$\int\limits_0^T Y(t)dB(t) = \int\limits_0^T Y(t) \diamond W(t)dt. \qquad (1.1.9)$$

(See Chapter 2.)

(d) When products are interpreted as Wick products, there is a powerful solution technique via the Hermite transform (or the related $\mathcal{S}$-transform) (see Chapter 2).

(e) When applied to ordinary stochastic differential equations (where products are interpreted in the Wick sense), our approach gives the same result as the classical one, when applicable. In fact, as demonstrated in Chapter 3, our approach is often easier in this case. This is because of the useful fact that Wick calculus with *ordinary* calculus rules, e.g.,

$$(u \diamond v)' = u' \diamond v + u \diamond v' \qquad (1.1.10)$$

is equivalent to Itô calculus governed by the Itô formula.

(f) It is irrelevant for our approach whether the quantities involved in an SDE are anticipating or not, the Wick calculus is the same. Therefore, this approach makes it possible to handle (Skorohod-interpreted) SDEs where the coefficients or initial conditions are anticipating. See Chapter 3.

We emphasize that although the Kondratiev space $(\mathcal{S})_{-1}$ of stochastic distributions may seem abstract, it does allow a relatively concrete interpretation. Indeed, $(\mathcal{S})_{-1}$ is analogous to the classical space $\mathcal{S}'$ of tempered distributions, the difference being that the *test function space* for $(\mathcal{S})_{-1}$ is a space of "smooth" *random variables* (denoted by $(\mathcal{S})_1$). Thus, if we interpret the random element ω as a specific "experiment" or "realization" of our system, then generic elements $F \in (\mathcal{S})_{-1}$ do not have point values $F(\omega)$ for each ω, but only average values $\langle F, \eta \rangle$ with respect to smooth random variables $\eta = \eta(\omega)$. In other words, knowing the solution

$$x \rightarrow u(x, \cdot) \in (\mathcal{S})_{-1}$$

of an SPDE does not tell us what the outcome of a specific realization ω would be, but rather what the average over a set of realizations would be. This seems to be appropriate for most applications, because (in most cases) each specific singleton $\{\omega\}$ has zero probability anyway.

For example, if $u = u(x, \cdot) \in (\mathcal{S})_{-1}$ is applied to the constant (random) test function $1 \in (\mathcal{S})_1$, we get the *generalized expectation* of $u(x, \cdot)$, $\langle u(x, \cdot), 1 \rangle \in \mathbb{R}$. This number may be regarded as *the best ω-constant approximation* to $u(x, \cdot)$.

This corresponds to the first term (zeroth order term) in the *generalized Wiener–Itô chaos expansion* of $u(x, \cdot)$ (Chapter 2). Similarly, the next term (the first order term) of this expansion gives in a sense the *best Gaussian approximation* to $u(x, \cdot)$ and so on. This will be discussed in detail in Chapter 2.

Finally we mention that for several specific applications it may be more appropriate to consider the *smoothed* form of noise rather than the idealized, singular noise usually applied. For example, for the idealized 1-parameter white noise $W(t) = W(t, \omega)$ it is usually required that $\{W(t, \omega)\}_{t \in \mathbb{R}}$ is a (generalized) stochastic process that is stationary, has mean zero, and satisfies the requirement

$$t_1 \neq t_2 \Rightarrow W(t_1, \cdot) \quad \text{and} \quad W(t_2, \cdot) \quad \text{are independent.} \qquad (1.1.11)$$

In most applications the specific form of noise we encounter does not satisfy (1.1.11), because usually $W(t_1, \cdot)$ and $W(t_2, \cdot)$ are *not* independent if t_1 and t_2 are close enough. In these cases it is natural to modify the noise to the *smoothed* white noise

$$W_\phi(t) := W_\phi(t, \omega) := \langle \omega, \phi_t \rangle = \int \phi_t(s) dB_s(\omega) \qquad (1.1.12)$$

(the Itô integral of ϕ_t with respect to 1-parameter Brownian motion B_t), where ϕ is a suitable (deterministic) test function (e.g., $\phi \in \mathcal{S}(\mathbb{R})$), and ϕ_t is the *t-shift* of ϕ defined by

$$\phi_t(s) = \phi(s - t) \quad \text{for} \quad s, t \in \mathbb{R}. \qquad (1.1.13)$$

Then $\{W_\phi(t)\}_{t \in \mathbb{R}}$ will be an ordinary (not generalized) stochastic process. It is stationary as $\{W(t)\}$, and the mean value of $W_\phi(t)$ is still zero, for all t. However, we have independence of $W_\phi(t_1)$ and $W_\phi(t_2)$ (if and) only if the condition $\operatorname{supp} \phi_{t_1} \cap \operatorname{supp} \phi_{t_2} = \emptyset$ is satisfied.

Therefore, in specific applications it is natural to choose ϕ such that the size of the support of ϕ gives the maximal distance within which $W_\phi(t_1)$ and $W_\phi(t_2)$ are correlated. Thus ϕ will have a physical or modeling significance, in addition to being a technical convenience, in that it replaces the singular process $\{W(t)\}$ by the ordinary process $\{W_\phi(t)\}$.

In this case the solution u of the corresponding SDE will depend on ϕ also, so we may consider the solution as a function

$$u(\phi, t) : \mathcal{S}(\mathbb{R}) \times \mathbb{R} \to (\mathcal{S})_{-1}.$$

Similarly, for the multiparameter equation the solution u of the corresponding SPDE will be a function

$$u(\phi, x) : \mathcal{S}(\mathbb{R}^d) \times \mathbb{R}^d \to (\mathcal{S})_{-1}.$$

Such processes are called *functional* processes. We stress that from a modeling point of view these processes are of interest in their own right, not just as technically convenient "approximations". In fact, there may be cases where it is not even physically relevant to ask what happens if $\phi \to \delta_0$ (the Dirac measure at 0). Nevertheless, such questions may be mathematically interesting, both from the point of view of approximations and in connection with numerical methods. We will not deal with these questions in depth in this book, but give some examples in Chapter 3.

Finally, in Chapter 4 we apply the techniques developed in Chapter 2 to stochastic partial differential equations. Our general strategy is the following: Consider a stochastic partial differential equation where the stochastic element may be a random variable in the equation or in the initial and boundary data, or both. In general, the solution will be a (stochastic) distribution, and we have to interpret possible products that occur in the equation, as one cannot in general take the product of two distributions. In our approach, products are considered to be Wick products. Subsequently, we take the Hermite transform of the resulting equation and obtain an equation that we try to solve, where the random variables have been replaced by complex–valued functions of infinitely many complex variables. Finally, we use the inverse Hermite transform to obtain a solution of the regularized, original equation.

The equations we solve here are mostly equations where we obtain the final solution on a closed form expressed as an expectation over a function of an auxiliary Brownian motion. There are also methods for solving equations where the solution cannot be obtained in a closed form, see, e.g., Benth (1996) and Våge (1995a).

Our first example is the stochastic Poisson equation

$$\Delta U = -W$$

on a domain D in $\mathbb{R}^d$ with vanishing Dirichlet boundary data, where W is singular white noise. First taking the Hermite transform (no Wick products are required in this equation), we obtain the equation

$$\Delta \widetilde{U} = -\widetilde{W}$$

on D with the same boundary condition, which leads to the solution (Theorem 4.2.1)

$$U(x) = \int_{\mathbb{R}^d} G(x, y) W(y) dy$$

in $(\mathcal{S})^*$ with G being the corresponding Green function of the deterministic Poisson equation. If we instead first regularize white noise, i.e., replace W by the smooth white noise W_ψ for some test function $\psi \in \mathcal{S}(\mathbb{R}^d)$, we find, see equation (4.2.10), correspondingly the solution $U_\psi(x) = \int_{\mathbb{R}^d} G(x, y) W_\psi(y) dy \in L^p(\mu)$ for all finite p. If ψ approaches Dirac's delta-function, then the solution U_ψ will converge to U in $(\mathcal{S})^*$. The stochastic Poisson equation has been studied in Walsh (1986) using different techniques, and his solution differs from ours in the sense that his solution takes x-averages for almost all realizations ω, while our approach considers ω-averages for each point x in space.

The next equation that is analyzed is the linear transport equation, Gjerde (1996b),

$$\frac{\partial U}{\partial t} = \frac{1}{2}\sigma^2 \Delta U + V \cdot \nabla U + KU + g$$

with initial data given by $U(x,0) = f(x)$. Here all functions V, K, g, and f are elements in $(\mathcal{S})_{-1}$, and are assumed to satisfy regularity conditions, see Theorem 4.3.1. We first insert Wick products, obtaining

$$\frac{\partial U}{\partial t} = \frac{1}{2}\sigma^2 \Delta U + V \diamond \nabla U + K \diamond U + g$$

before we make the Hermite transform to yield (4.3.7). The resulting equation can be solved, and we find the solution U in $(\mathcal{S})_{-1}$ given by equation (4.3.5). If we specialize to $V = g = 0$, we find the solution of the heat equation with stochastic potential (Corollary 4.3.2).

Closely related to the previous equation is the stationary Schrödinger equation with a stochastic potential, Holden et al., (1993b), Gjerde (1996b)

$$\frac{1}{2}\Delta U + V \diamond U = -f$$

on a domain D in $\mathbb{R}^d$ and with vanishing Dirichlet data on the boundary of D. We analyze the case where the potential V is the Wick exponential of white noise, i.e.,

$$V(x) = \rho \exp^\diamond[W(x)],$$

where ρ is a constant. The function f is assumed to be a stochastic distribution process. Under certain regularity conditions, we obtain the solution in closed form; see Theorem 4.4.1 and equation (4.4.7). If we replace singular white noise by regularized smoothed white noise, we obtain a solution that is in $L^1(\mu)$. This is the content of Theorem 4.4.2.

Our prime example of a nonlinear stochastic partial differential equation is the celebrated viscous Burgers equation (Burgers (1940), (1974)), which has been studied extensively. The key insight in all approaches to this equation is the Cole–Hopf transformation which effectively linearizes the Burgers equation. This transformation turns the Burgers equation into the linear heat equation. If we modify the Burgers equation by an additive (stochastic) source term, the Cole–Hopf transformation yields the linear heat equation with a multiplicative potential. We are able to solve this equation by the methods described above, and what remains is to apply the Cole–Hopf transformation in our stochastic setting where the Wick product replaces the ordinary product. This turns out to be possible, and we obtain a solution in $(\mathcal{S})_{-1}$ of Burgers' equation

$$\frac{\partial U}{\partial t} + \lambda U \diamond \frac{\partial U}{\partial x} = \nu \frac{\partial^2 U}{\partial^2 x} + F,$$

where we assume that the stochastic source F is a gradient, i.e., $F = -\partial N/\partial x$. The solution U is unique among solutions of the form $U = -\partial Z/\partial x$. The analysis is easily generalized to a Burgers system of equations (see Theorem 4.5.4), where the scalar U is replaced by a vector U in $\mathbb{R}^d$.

An important equation in the modeling of porous media is the *stochastic pressure equation* given by

$$\mathrm{div}(K(x) \diamond \nabla p(x)) = -f(x)$$

on a domain D in $\mathbb{R}^d$ and with vanishing Dirichlet data on the boundary of D. An important case is the case where K is has a log-normal distribution. A natural interpretation is then to consider

$$K(x) = \exp^{\diamond}[W(x)]$$

or the smoothed version $K(x) = \exp[W_\phi(x)]$. For a source term f in $(\mathcal{S})_{-1}$, we obtain a solution in closed form; see Theorem 4.6.3 and Theorem 4.6.1 and equations (4.6.37) and (4.6.6), respectively. We also describe a method for computing the actual solution based on approximations using the chaos expansion. An alternative method based on finite differences is described in Holden and Hu (1996). The 1-dimensional case is computed in detail in Theorem 4.6.2.

One may combine the stochastic heat equation with the pressure equation to obtain a heat equation in a stochastic, anisotropic medium, namely an equation of the form

$$\frac{\partial U}{\partial t} = \mathrm{div}(\mathbf{K} \diamond \nabla U) + g(x).$$

Here $\mathbf{K}$ is taken to be a positive noise matrix with components that are the Wick exponentials of singular white noise. The initial data $U(x,0)$ is a given element in $(\mathcal{S})_{-1}$, and the solution is in the same space.

If we consider the more general class of quasilinear parabolic stochastic differential equations given by

$$\frac{\partial U}{\partial t} = L(t, x, \nabla U) + \sigma(x) U \diamond W(t),$$

we obtain an equation with a solution in $L^p(\mu)$ when we assume a related deterministic SDE has a unique solution; see Theorem 4.8.1.

So far analysis has been exclusively with Gaussian white noise, start-ing with the Bochner–Minlos theorem. One could, however, replace the right-hand side of (2.1.3) by other positive definite functionals, thereby obtaining a different measure. An important case is the case of Poisson noise. Most of the analysis can be carried out in this case. A brief presentation of this, based on Benth and Gjerde (1998a), is given in Section 4.9, culminating in a solution of the viscous Burgers equation with the Gaussian noise replaced by Poisson noise.

In the new Chapter 5 we do this in the more general setting of *Lévy white noise*, based on the multiparameter Lévy process (Lévy random field). We establish a white noise theory in this setting and we use it to solve stochastic partial differential equations driven by such noise.

Chapter 2
Framework

In this chapter we develop the general framework to be used in this book. The starting point for the discussion will be the standard white noise structures and how constructions of this kind can be given a rigorous treatment. White noise analysis can be addressed in several different ways. The presentation here is to a large extent influenced by ideas and methods used by the authors. In particular, we emphasize the use of multidimensional structures, i.e., the white noise we are about to consider will in general take on values in a multidimensional space and will also be indexed by a multidimensional parameter set.

2.1 White Noise

2.1.1 The 1-Dimensional, d-Parameter Smoothed White Noise

Two fundamental concepts in stochastic analysis are *white noise* and *Brownian motion*. The idea of white noise analysis, due to Hida (1980), is to consider white noise rather than Brownian motion as the fundamental object. Within this framework, Brownian motion will be expressed in terms of white noise.

We start by recalling some of the basic definitions and properties of the 1-dimensional white noise probability space. In the following d will denote a fixed positive integer, interpreted as either the time-, space- or time–space dimension of the system we consider. More generally, we will call d the *parameter dimension*. Let $\mathcal{S}(\mathbb{R}^d)$ be the Schwartz space of rapidly decreasing smooth (C^∞) real-valued functions on $\mathbb{R}^d$. A general reference for properties of this space is Rudin (1973). $\mathcal{S}(\mathbb{R}^d)$ is a Fréchet space under the family of seminorms

$$\|f\|_{k,\alpha} := \sup_{x \in \mathbb{R}^d} \{(1 + |x|^k)|\partial^\alpha f(x)|\}, \tag{2.1.1}$$

H. Holden et al., *Stochastic Partial Differential Equations*, 2nd ed., Universitext, DOI 10.1007/978-0-387-89488-1_2, © Springer Science+Business Media, LLC 2010

where k is a non-negative integer, $\alpha = (\alpha_1, \ldots, \alpha_d)$ is a multi-index of non-negative integers $\alpha_1, \ldots, \alpha_d$ and

$$\partial^\alpha f = \frac{\partial^{|\alpha|}}{\partial x_1^{\alpha_1} \cdots \partial x_d^{\alpha_d}} f \quad \text{where} \quad |\alpha| := \alpha_1 + \cdots + \alpha_d. \tag{2.1.2}$$

The dual $\mathcal{S}' = \mathcal{S}'(\mathbb{R}^d)$ of $\mathcal{S}(\mathbb{R}^d)$, equipped with the weak-star topology, is the space of *tempered distributions*. This space is the one we will use as our basic probability space. As events we will use the family $\mathcal{B}(\mathcal{S}'(\mathbb{R}^d))$ of Borel subsets of $\mathcal{S}'(\mathbb{R}^d)$, and our probability measure is given by the following result.

Theorem 2.1.1. (The Bochner–Minlos theorem) *There exists a unique probability measure μ_1 on $\mathcal{B}(\mathcal{S}'(\mathbb{R}^d))$ with the following property:*

$$E[e^{i\langle \cdot, \phi \rangle}] := \int_{\mathcal{S}'} e^{i\langle \omega, \phi \rangle} d\mu_1(\omega) = e^{-\frac{1}{2}\|\phi\|^2} \tag{2.1.3}$$

for all $\phi \in \mathcal{S}(\mathbb{R}^d)$, where $\|\phi\|^2 = \|\phi\|^2_{L^2(\mathbb{R}^d)}$, $\langle \omega, \phi \rangle = \omega(\phi)$ is the action of $\omega \in \mathcal{S}'(\mathbb{R}^d)$ on $\phi \in \mathcal{S}(\mathbb{R}^d)$ and $E = E_{\mu_1}$ denotes the expectation with respect to μ_1.

See Appendix A for a proof. We will call the triplet $(\mathcal{S}'(\mathbb{R}^d), \mathcal{B}(\mathcal{S}'(\mathbb{R}^d)), \mu_1)$ the *1-dimensional white noise probability space*, and μ_1 is called the *white noise measure*.

The measure μ_1 is also often called the (normalized) *Gaussian measure* on $\mathcal{S}'(\mathbb{R}^d)$. The reason for this can be seen from the following result.

Lemma 2.1.2. *Let $\xi_1, \ldots, \xi_n$ be functions in $\mathcal{S}(\mathbb{R}^d)$ that are orthonormal in $L^2(\mathbb{R}^d)$. Let λ_n be the normalized Gaussian measure on $\mathbb{R}^n$, i.e.,*

$$d\lambda_n(x) = (2\pi)^{-\frac{n}{2}} e^{-\frac{1}{2}|x|^2} dx_1 \cdots dx_n; \; x = (x_1, \ldots, x_n) \in \mathbb{R}^n. \tag{2.1.4}$$

Then the random variable

$$\omega \to (\langle \omega, \xi_1 \rangle, \langle \omega, \xi_2 \rangle, \ldots, \langle \omega, \xi_n \rangle) \tag{2.1.5}$$

has distribution λ_n. Equivalently,

$$E[f(\langle \cdot, \xi_1 \rangle, \ldots, \langle \cdot, \xi_n \rangle)] = \int_{\mathbb{R}^n} f(x) d\lambda_n(x) \text{ for all } f \in L^1(\lambda_n). \tag{2.1.6}$$

Proof It suffices to prove this for $f \in C_0^\infty(\mathbb{R}^n)$; the general case then follows by taking the limit in $L^1(\lambda_n)$. If $f \in C_0^\infty(\mathbb{R}^n)$, then f is the inverse Fourier transform of its Fourier transform $\hat{f}$:

$$f(x) = (2\pi)^{-\frac{n}{2}} \int \hat{f}(y) e^{i(x,y)} dy$$

where

$$\hat{f}(y) = (2\pi)^{-\frac{n}{2}} \int f(x) e^{-i(x,y)} dx,$$

where (x, y) denotes the usual inner product in $\mathbb{R}^d$. Then (2.1.3) gives

$$E[f(\langle \cdot, \xi_1 \rangle, \ldots, \langle \cdot, \xi_n \rangle)] = (2\pi)^{-\frac{n}{2}} \int_{\mathbb{R}^n} \hat{f}(y) E[e^{i\langle \cdot, \sum_j y_j \xi_j \rangle}] dy$$

$$= (2\pi)^{-\frac{n}{2}} \int_{\mathbb{R}^n} \hat{f}(y) e^{-\frac{1}{2}|y|^2} dy$$

$$= (2\pi)^{-n} \int_{\mathbb{R}^n} \left(\int_{\mathbb{R}^n} f(x) e^{-i(x,y)} dx \right) e^{-\frac{1}{2}|y|^2} dy$$

$$= (2\pi)^{-n} \int_{\mathbb{R}^n} f(x) \left(\int_{\mathbb{R}^n} e^{-i(x,y) - \frac{1}{2}|y|^2} dy \right) dx$$

$$= (2\pi)^{-\frac{n}{2}} \int_{\mathbb{R}^n} f(x) e^{-\frac{1}{2}|x|^2} dx$$

$$= \int_{\mathbb{R}^n} f(x) d\lambda_n(x),$$

where we have used the well-known formula

$$\int_{\mathbb{R}} e^{i\alpha t - \beta t^2} dt = \left(\frac{\pi}{\beta} \right)^{\frac{1}{2}} e^{-\frac{\alpha^2}{4\beta}}. \tag{2.1.7}$$

(See Exercise 2.4.) □

For an alternative proof of Lemma 2.1.2, see Exercise 2.5.

Remark Note that (2.1.6) applies in particular to polynomials

$$(x_1, \ldots, x_n) = \sum_{|\alpha| \le N} c_\alpha x^\alpha, \quad N = 1, 2, \ldots;$$

where the sum is taken over all n-dimensional multi-indices $\alpha = (\alpha_1, \ldots, \alpha_n)$ and $x^\alpha = x_1^{\alpha_1} x_2^{\alpha_2} \cdots x_n^{\alpha_n}$. Let $\mathcal{P}$ denote the family of all functions $p : \mathcal{S}'(\mathbb{R}^d) \to \mathbb{R}$ of the form

$$p(\omega) = f(\langle \omega, \xi_1 \rangle, \ldots, \langle \omega, \xi_n \rangle)$$

for some polynomial f. We call such functions p *stochastic polynomials*. Similarly, we let $\mathcal{E}$ denote the family of all linear combinations of functions $f : \mathcal{S}'(\mathbb{R}^d) \to \mathbb{R}$ of the form

$$f(\omega) = \exp[i\langle \omega, \phi \rangle] \quad \text{where} \quad \phi \in \mathcal{S}(\mathbb{R}^d).$$

Such functions are called *stochastic exponentials*. The following result is useful.

Theorem 2.1.3. $\mathcal{P}$ and $\mathcal{E}$ are dense in $L^p(\mu_1)$, for all $p \in [1, \infty)$.

Proof See Theorem 1.9, p. 7, in Hida et al. (1993). □

Definition 2.1.4. The *1-dimensional (d-parameter) smoothed white noise* is the map

$$w : \mathcal{S}(\mathbb{R}^d) \times \mathcal{S}'(\mathbb{R}^d) \to \mathbb{R}$$

given by

$$w(\phi) = w(\phi, \omega) = \langle \omega, \phi \rangle; \quad \omega \in \mathcal{S}'(\mathbb{R}^d), \phi \in \mathcal{S}(\mathbb{R}^d). \qquad (2.1.8)$$

Remark In Section 2.3 we will define the singular white noise $W(x, \omega)$. We may regard $w(\phi)$ as obtained by smoothing $W(x, \omega)$ by ϕ.

Using Lemma 2.1.2 it is not difficult to prove that if $\phi \in L^2(\mathbb{R}^d)$ and we choose $\phi_n \in \mathcal{S}(\mathbb{R}^d)$ such that $\phi_n \to \phi$ in $L^2(\mathbb{R}^d)$, then

$$\langle \omega, \phi \rangle := \lim_{n \to \infty} \langle \omega, \phi_n \rangle \quad \text{exists in} \quad L^2(\mu_1) \qquad (2.1.9)$$

and is independent of the choice of $\{\phi_n\}$ (Exercise 2.6). In particular, if we define

$$\widetilde{B}(x) := \widetilde{B}(x_1, \ldots, x_d, \omega) = \langle \omega, \chi_{[0,x_1] \times \cdots \times [0,x_d]} \rangle; x = (x_1, \ldots, x_d) \in \mathbb{R}^d, \qquad (2.1.10)$$

where $[0, x_i]$ is interpreted as $[x_i, 0]$ if $x_i < 0$, then $\widetilde{B}(x, \omega)$ has an x-continuous version $B(x, \omega)$, which becomes a d-parameter Brownian motion.

By a *d-parameter Brownian motion* we mean a family $\{X(x, \cdot)\}_{x \in \mathbb{R}^d}$ of random variables on a probability space $(\Omega, \mathcal{F}, P)$ such that

$$X(0, \cdot) = 0 \quad \text{almost surely with respect to } P, \qquad (2.1.11)$$

$\{X(x, \omega)\}$ is a Gaussian stochastic process (i.e., $Y = (X(x^{(1)}, \cdot), \ldots,$ $X(x^{(n)}, \cdot))$ has a multinormal distribution with mean zero for all

$x^{(1)}, \ldots, x^{(n)} \in \mathbb{R}^d$ and all $n \in \mathbb{N}$) and, further, for all $x = (x_1, \ldots, x_d), y = (y_1, \ldots, y_d) \in \mathbb{R}_+^d$, that $X(x, \cdot) X(y, \cdot)$ have the covariance $\prod_{i=1}^d x_i \wedge y_i$. For general $x, y \in \mathbb{R}^d$ the covariance is $\prod_{i=1}^d \int_{\mathbb{R}} \theta_{x_i}(s) \theta_{y_i}(s) ds$, where $\theta_x(t_1, \ldots, t_d) = \theta_{x_1}(t_1) \cdots \theta_{x_d}(t_d)$, with

$$(2.1.12)$$

$$\theta_{x_j}(s) = \begin{cases} 1 & \text{if } 0 < s \leq x_j \\ -1 & \text{if } x_j < s \leq 0 \\ 0 & \text{otherwise} \end{cases}$$

We also require that

$$X(x, \omega) \text{ has continuous paths, i.e., that } x \to X(x, \omega)$$
$$\text{is continuous for almost all } \omega \text{ with respect to } P. \qquad (2.1.13)$$

We have to verify that $\widetilde{B}(x, \omega)$ defined by (2.1.10) satisfies (2.1.11) and (2.1.12) and that $\widetilde{B}$ has a continuous version. Property (2.1.11) is evident. To prove (2.1.12), we choose $x^{(1)}, \ldots, x^{(n)} \in \mathbb{R}_+^d, c_1, \ldots, c_n \in \mathbb{R}$ and put

$$\chi^{(j)}(t) = \chi_{[0,x_1^{(j)}] \times \cdots \times [0,x_d^{(j)}]}(t) \; ; \; t \in \mathbb{R}^d$$

where $x^{(j)} = (x_1^{(j)}, \ldots, x_d^{(j)})$, and compute

$$E\left[\exp\left(i \sum_{j=1}^n c_j \widetilde{B}(x^{(j)}) \right) \right] = E\left[\exp\left(i \left\langle \cdot, \sum_{j=1}^n c_j \chi^{(j)} \right\rangle \right) \right]$$

$$= \exp\left(-\frac{1}{2} \left\| \sum_{j=1}^n c_j \chi^{(j)} \right\|^2 \right)$$

$$= \exp\left(-\frac{1}{2} \int_{\mathbb{R}^d} \left(\sum_{j=1}^n c_j \chi^{(j)}(t) \right)^2 dt \right)$$

$$= \exp\left(-\frac{1}{2} \sum_{i,j=1}^n c_i c_j \int_{\mathbb{R}^d} \chi^{(i)}(t) \chi^{(j)}(t) dt \right)$$

$$= \exp\left(-\frac{1}{2} c^T V c \right),$$

where $c = (c_1, \ldots, c_n)$ and $V = [V_{ij}] \in \mathbb{R}^{n \times n}$ is the symmetric non-negative definite matrix with entries

$$V_{ij} = \int_{\mathbb{R}^d} \chi^{(i)}(y) \chi^{(j)}(y) dy = (\chi^{(i)}, \chi^{(j)}).$$

This proves that $Y = (\widetilde{B}(x^{(1)}, \cdot), \ldots, \widetilde{B}(x^{(n)}, \cdot))$ is Gaussian with mean zero and covariance matrix $V = [V_{ij}]$. For $x = (x_1, \ldots, x_d) \in \mathbb{R}^d$ let $\chi_x(t) = \chi_{[0,x_1] \times \cdots \times [0,x_d]}(t); \ t \in \mathbb{R}^d$. Then for $y = (y_1, \ldots, y_d) \in \mathbb{R}^d$, we have

$$D^2 := \|\chi_x - \chi_y\|^2 = \prod_{i=1}^{d} \|\chi_{[0,x_i]} - \chi_{[0,y_i]}\|^2 = \prod_{i=1}^{d} |x_i - y_i|.$$

Hence by (2.1.6)

$$E[|\widetilde{B}(x) - \widetilde{B}(y)|^2] = E[\langle \cdot, \chi_x - \chi_y \rangle^2]$$

$$= D^2 E\left[\left\langle \cdot, \frac{\chi_x - \chi_y}{D} \right\rangle^2\right] = D^2 \int_{\mathbb{R}} t^2 d\lambda_1(t) = D^2,$$

which proves that $\widetilde{B}$ satisfies (2.1.12).

Finally, using Kolmogorov's continuity theorem (see, e.g., Stroock and Varadhan (1979), Theorem 2.1.6) we obtain that $\widetilde{B}(x)$ has a continuous version $B(x)$, which then becomes a d-parameter Brownian motion. See Exercise 2.7.

We remark that for $d = 1$ we get the classical (1-parameter) Brownian motion $B(t)$ if we restrict ourselves to $t \geq 0$. For $d = 2$ we get what is often called *the Brownian sheet*.

With this definition of Brownian motion it is natural to define the d-parameter Wiener–Itô integral of $\phi \in L^2(\mathbb{R}^d)$ by

$$\int_{\mathbb{R}^d} \phi(x) dB(x, \omega) := \langle \omega, \phi \rangle; \quad \omega \in \mathcal{S}'(\mathbb{R}^d). \tag{2.1.14}$$

We see that by appealing to the Bochner–Minlos theorem we have obtained not only a simple description of white noise, but also an easy construction of Brownian motion. The relation between these two fundamental concepts can also be expressed as follows.

Using integration by parts for Wiener–Itô integrals (Appendix B), we get

$$\int_{\mathbb{R}^d} \phi(x) dB(x) = (-1)^d \int_{\mathbb{R}^d} \frac{\partial^d \phi}{\partial x_1 \cdots \partial x_d}(x) B(x) dx. \tag{2.1.15}$$

Hence

$$w(\phi) = \int_{\mathbb{R}^d} \phi(x) dB(x) = \left((-1)^d \frac{\partial^d \phi}{\partial x_1 \cdots \partial x_d}, B \right)$$

$$= \left(\phi, \frac{\partial^d B}{\partial x_1 \cdots \partial x_d} \right), \tag{2.1.16}$$

where $(\cdot, \cdot)$ denotes the usual inner product in $L^2(\mathbb{R}^d)$. In other words, in the sense of distributions we have, for almost all ω,

$$w = \frac{\partial^d B}{\partial x_1 \cdots \partial x_d}. \qquad (2.1.17)$$

We will give other formulations of this connection between Brownian motion and white noise in Section 2.5.

Using $w(\phi, \omega)$ we can construct a stochastic process, called the *smoothed white noise process* $W_\phi(x, \omega)$, as follows: Set

$$W_\phi(x, \omega) := w(\phi_x, \omega), \quad x \in \mathbb{R}^d, \ \omega \in \mathcal{S}'(\mathbb{R}^d), \qquad (2.1.18)$$

where

$$\phi_x(y) = \phi(y - x) \qquad (2.1.19)$$

is the x-shift of ϕ; $x, y \in \mathbb{R}^d$.

Note that $\{W_\phi(x, \cdot)\}_{x \in \mathbb{R}^d}$ has the following three properties:

If supp $\phi_{x_1} \cap supp\phi_{x_2} = \emptyset$, then $W_\phi(x_1, \cdot)$ and $W_\phi(x_2, \cdot)$ are independent.
$$\qquad (2.1.20)$$

$\{W_\phi(x, \cdot)\}_{x \in \mathbb{R}^d}$ is a stationary process, i.e., for all $n \in \mathbb{N}$ and for all $x^{(1)}, \ldots, x^{(n)}$ and $h \in \mathbb{R}^d$, the joint distribution of $\qquad (2.1.21)$

$$(W_\phi(x^{(1)} + h, \cdot), \ldots, W_\phi(x^{(n)} + h, \cdot))$$

is independent of h.

For each $x \in \mathbb{R}^d$, the random variable $W_\phi(x, \cdot)$ is normally distributed with mean 0 and variance $\|\phi\|^2$. $\qquad (2.1.22)$

So $\{W_\phi(x, \omega)\}_{x \in \mathbb{R}^d}$ is indeed a mathematical model for what one usually intuitively thinks of as white noise. In explicit applications the test function or "window" ϕ can be chosen such that the diameter of supp ϕ is the maximal distance within which $W_\phi(x_1, \cdot)$ and $W_\phi(x_2, \cdot)$ might be correlated.

Figure 2.1 shows computer simulations of the 2-parameter white noise process $W_\phi(x, \omega)$ where $\phi(y) = \chi_{[0,h] \times [0,h]}(y); y \in \mathbb{R}^2$ for $h = 1/50$ (left) and for $h = 1/20$ (right).

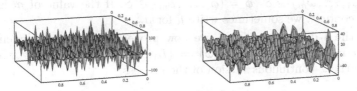

Fig. 2.1 Two sample paths of white noise ($h = 1/50, h = 1/20$).

2.1.2 The (Smoothed) White Noise Vector

We now proceed to define the multidimensional case. If m is a natural number, we define

$$\mathcal{S} := \prod_{i=1}^{m} \mathcal{S}(\mathbb{R}^d), \quad \mathcal{S}' := \prod_{i=1}^{m} \mathcal{S}'(\mathbb{R}^d), \quad \mathcal{B} := \prod_{i=1}^{m} \mathcal{B}(\mathcal{S}'(\mathbb{R}^d)) \qquad (2.1.23)$$

and equip $\mathcal{S}'$ with the product measure

$$\mu_m = \mu_1 \times \mu_1 \times \cdots \times \mu_1, \qquad (2.1.24)$$

where μ_1 is the 1-dimensional white noise probability measure. It is then easy to see that we have the following property:

$$\int_{\mathcal{S}'} e^{i\langle \omega, \phi \rangle} d\mu_m(\omega) = e^{-\frac{1}{2}\|\phi\|^2} \quad \text{for all} \quad \phi \in \mathcal{S}. \qquad (2.1.25)$$

Here $\langle \omega, \phi \rangle = \langle \omega_1, \phi_1 \rangle + \cdots + \langle \omega_m, \phi_m \rangle$ is the action of $\omega = (\omega_1, \ldots, \omega_m) \in \mathcal{S}'$ on $\phi = (\phi_1, \ldots, \phi_m) \in \mathcal{S}$, where $\langle \omega_k, \phi_k \rangle$ is the action of $\omega_k \in \mathcal{S}'(\mathbb{R}^d)$ on $\phi_k \in \mathcal{S}(\mathbb{R}^d)$; $k = 1, 2, \ldots, m$.

Furthermore,

$$\|\phi\| = \|\phi\|_{\mathcal{K}} = \left(\sum_{k=1}^{m} \|\phi_k\|^2 \right)^{\frac{1}{2}} = \left(\sum_{k=1}^{m} \int_{\mathbb{R}^d} \phi_k^2(x) dx \right)^{\frac{1}{2}} \qquad (2.1.26)$$

is the norm of ϕ in the Hilbert space $\mathcal{K}$ defined as the orthogonal sum of m identical copies of $L^2(\mathbb{R}^d)$, viz. $\mathcal{K} = \bigoplus_{k=1}^{m} L^2(\mathbb{R}^d)$.

We will call the triplet $(\mathcal{S}', \mathcal{B}, \mu_m)$ the *d-parameter multidimensional white noise probability space*. The parameter m is called the *white noise dimension*. The *m-dimensional smoothed white noise*

$$\mathbf{w} : \mathcal{S} \times \mathcal{S}' \to \mathbb{R}^m$$

is then defined by

$$\mathbf{w}(\phi) = \mathbf{w}(\phi, \omega) = (\langle \omega_1, \phi_1 \rangle, \ldots, \langle \omega_m, \phi_m \rangle) \in \mathbb{R}^m \qquad (2.1.27)$$

if $\omega = (\omega_1, \ldots, \omega_m) \in \mathcal{S}', \phi = (\phi_1, \ldots, \phi_m) \in \mathcal{S}$. If the value of m is clear from the context, we sometimes write μ for μ_m.

As in the 1-dimensional case, we now proceed to define m-dimensional Brownian motion $\mathbf{B}(x) = \mathbf{B}(x, \omega) = (B_1(x, \omega), \ldots, B_m(x, \omega))$; $x \in \mathbb{R}^d$, $\omega \in \mathcal{S}'$ as the x-continuous version of the process

$$\widetilde{\mathbf{B}}(x, \omega) = (\langle \omega_1, \chi_{[0,x_1] \times \cdots \times [0,x_d]} \rangle, \ldots, \langle \omega_m, \chi_{[0,x_1] \times \cdots \times [0,x_d]} \rangle). \qquad (2.1.28)$$

From this we see that $\mathbf{B}(x)$ consists of m independent copies of 1-dimensional Brownian motion. Combining (2.1.27) and (2.1.14) we get

$$\mathbf{w}(\phi) = \left(\int \phi_1(x)dB_1(x), \ldots, \int \phi_m(x)dB_m(x) \right). \qquad (2.1.29)$$

Using $\mathbf{w}(\phi, \omega)$, we can construct m-dimensional smoothed white noise process $\mathbf{W}_\phi(x, \omega)$ as follows:

$$\mathbf{W}_\phi(x, \omega) := \mathbf{w}(\phi_x, \omega) \qquad (2.1.30)$$

for $\phi = (\phi_1, \ldots, \phi_m) \in \mathcal{S}$, $\omega = (\omega_1, \ldots, \omega_m) \in \mathcal{S}'$, where

$$\phi_x(y) = (\phi_1(y - x), \ldots, \phi_m(y - x)); \quad x, y \in \mathbb{R}^d. \qquad (2.1.31)$$

2.2 The Wiener–Itô Chaos Expansion

There are (at least) two ways of constructing the classical Wiener–Itô chaos expansion:

(A) by Hermite polynomials,
(B) by multiple Itô integrals.

Both approaches are important, and it is useful to know them both and to know the relationship between them. For us the first construction will play the major role. We will therefore introduce this method in detail first, then sketch the other construction, and finally compare the two.

2.2.1 Chaos Expansion in Terms of Hermite Polynomials

The *Hermite polynomials* $h_n(x)$ are defined by

$$h_n(x) = (-1)^n e^{\frac{1}{2}x^2} \frac{d^n}{dx^n}(e^{-\frac{1}{2}x^2}); \quad n = 0, 1, 2, \ldots. \qquad (2.2.1)$$

Thus the first Hermite polynomials are

$$h_0(x) = 1, \ h_1(x) = x, \ h_2(x) = x^2 - 1, \ h_3(x) = x^3 - 3x,$$
$$h_4(x) = x^4 - 6x^2 + 3, \ h_5(x) = x^5 - 10x^3 + 15x, \ldots.$$

The *Hermite functions* $\xi_n(x)$ are defined by

$$\xi_n(x) = \pi^{-\frac{1}{4}}((n-1)!)^{-\frac{1}{2}} e^{-\frac{1}{2}x^2} h_{n-1}(\sqrt{2}x); \quad n = 1, 2, \ldots. \qquad (2.2.2)$$

The most important properties of h_n and ξ_n are given in Appendix C. Some properties we will often use follow.

$$\xi_n \in \mathcal{S}(\mathbb{R}) \quad \text{for all } n. \tag{2.2.3}$$

The collection $\{\xi_n\}_{n=1}^{\infty}$ constitutes an orthonormal basis for $L^2(\mathbb{R})$.

$$\tag{2.2.4}$$

$$\sup_{x \in \mathbb{R}} |\xi_n(x)| = O(n^{-\frac{1}{12}}). \tag{2.2.5}$$

The statement (2.2.3) follows from the fact that h_n is a polynomial of degree n. Proofs of statements (2.2.4) and (2.2.5) can be found in Hille and Phillips (1957), Chapter 21.

We will use these functions to define an orthogonal basis for $L^2(\mu_m)$, where $\mu_m = \mu_1 \times \cdots \times \mu_1$ as before. Since the 1-dimensional case is simpler and also the case we will use most, we first do the construction in this case.

Case 1 ($m = 1$). In the following, we let $\delta = (\delta_1, \ldots, \delta_d)$ denote d-dimensional multi-indices with $\delta_1, \ldots, \delta_d \in \mathbb{N}$. By (2.2.4) it follows that the family of tensor products

$$\xi_\delta := \xi_{(\delta_1, \ldots, \delta_d)} := \xi_{\delta_1} \otimes \cdots \otimes \xi_{\delta_d}; \quad \delta \in \mathbb{N}^d \tag{2.2.6}$$

forms an orthonormal basis for $L^2(\mathbb{R}^d)$. Let $\delta^{(j)} = (\delta_1^{(j)}, \delta_2^{(j)}, \ldots, \delta_d^{(j)})$ be the jth multi-index number in some fixed ordering of all d-dimensional multi-indices $\delta = (\delta_1, \ldots, \delta_d) \in \mathbb{N}^d$. We can, and will, assume that this ordering has the property that

$$i < j \Rightarrow \delta_1^{(i)} + \delta_2^{(i)} + \cdots + \delta_d^{(i)} \leq \delta_1^{(j)} + \delta_2^{(j)} + \cdots + \delta_d^{(j)}, \tag{2.2.7}$$

i.e., that the $\{\delta^{(j)}\}_{j=1}^{\infty}$ occur in increasing order.

Now define

$$\eta_j := \xi_{\delta^{(j)}} = \xi_{\delta_1^{(j)}} \otimes \cdots \otimes \xi_{\delta_d^{(j)}}; \quad j = 1, 2, \ldots. \tag{2.2.8}$$

We will need to consider multi-indices of arbitrary length. To simplify the notation, we regard multi-indices as elements of the space $(\mathbb{N}_0^{\mathbb{N}})_c$ of all sequences $\alpha = (\alpha_1, \alpha_2, \ldots)$ with elements $\alpha_i \in \mathbb{N}_0$ and with compact support, i.e., with only finitely many $\alpha_i \neq 0$. We write

$$\mathcal{J} = (\mathbb{N}_0^{\mathbb{N}})_c. \tag{2.2.9}$$

Definition 2.2.1 ($m = 1$). Let $\alpha = (\alpha_1, \alpha_2, \ldots) \in \mathcal{J}$. Then we define

$$H_\alpha(\omega) = H_\alpha^{(1)}(\omega) := \prod_{i=1}^{\infty} h_{\alpha_i}(\langle \omega, \eta_i \rangle); \quad \omega \in \mathcal{S}'(\mathbb{R}^d). \tag{2.2.10}$$

Case 2 ($m > 1$). In this case we have to proceed one step further from (2.2.8) to obtain an orthonormal basis for $\mathcal{K} = \bigoplus_{k=1}^{m} L^2(\mathbb{R}^d)$. Define the following elements of $\mathcal{K}$:

$$e^{(1)} = (\eta_1, 0, \dots, 0)$$
$$e^{(2)} = (0, \eta_1, \dots, 0)$$
$$\vdots$$
$$e^{(m)} = (0, 0, \dots, \eta_1).$$

Then repeat with η_1 replaced by η_2:

$$e^{(m+1)} = (\eta_2, 0, \dots, 0)$$
$$\vdots$$
$$e^{(2m)} = (0, 0, \dots, \eta_2),$$

and so on.

In short, for every $k \in \mathbb{N}$ there are unique numbers $i \in \{1, \dots, m\}$ and $j \in \mathbb{N}$ such that $k = i + (j-1)m$. Then we have

$$e^{(k)} = e^{(i+(j-1)m)} = (0, 0, \dots, \eta_j, \dots, 0) = \eta_j \epsilon^{(i)} \in \mathcal{K}, \qquad (2.2.11)$$

where $\epsilon^{(i)}$ is the multi-index with 1 on entry number i and 0 otherwise.

Definition 2.2.2 ($m > 1$). For $\alpha \in \mathcal{J}$ define

$$H_\alpha(\omega) = H_\alpha^{(m)}(\omega) = \prod_{k=1}^{\infty} h_{\alpha_k}(\langle \omega, e^{(k)} \rangle). \qquad (2.2.12)$$

Here $\omega = (\omega_1, \dots, \omega_m) \in \mathcal{S}'$ and

$$\langle \omega, e^{(k)} \rangle = \langle \omega_1, e_1^{(k)} \rangle + \cdots + \langle \omega_m, e_m^{(k)} \rangle = \langle \omega_i, \eta_j \rangle \text{ if } k = i + (j-1)m. \quad (2.2.13)$$

Therefore we can also write

$$H_\alpha^{(m)}(\omega) = \prod_{\substack{k=1 \\ k = i+(j-1)m}}^{\infty} h_{\alpha_k}(\langle \omega_i, \eta_j \rangle); \quad \omega \in \mathcal{S}'. \qquad (2.2.14)$$

For example, if $\alpha = \epsilon^{(k)}$ with $k = i + (j-1)m$, we get

$$H_{\epsilon^{(k)}} = \langle \omega, e^{(k)} \rangle = \langle \omega_i, \eta_j \rangle. \qquad (2.2.15)$$

There is an alternative description of the family $\{H_\alpha^{(m)}\}$ that is natural from a tensor product point of view:

For $\Gamma = (\gamma^{(1)}, \ldots, \gamma^{(m)}) \in \mathcal{J}^m = \mathcal{J} \times \cdots \times \mathcal{J}$, $\omega = (\omega_1, \ldots, \omega_m) \in \mathcal{S}'$ define

$$\mathbf{H}_\Gamma^{(m)}(\omega) = \prod_{i=1}^{m} H_{\gamma^{(i)}}^{(1)}(\omega_i), \qquad (2.2.16)$$

where each $H_{\gamma^{(i)}}^{(1)}(\omega_i)$ is as in (2.2.10). Then we see that

$$\mathbf{H}_\Gamma^{(m)}(\omega) = \prod_{i=1}^{m} \prod_{j=1}^{\infty} h_{\gamma_j^{(i)}}(\langle \omega_i, \eta_j \rangle) = H_\alpha^{(m)}(\omega), \qquad (2.2.17)$$

where $\alpha = (\alpha_1, \alpha_2, \ldots) \in \mathcal{J}$ is related to $\Gamma = (\gamma^{(i)}) \in \mathcal{J}^m$ by

$$\alpha_k = \gamma_j^{(i)} \quad \text{if} \quad i + (j-1)m = k.$$

Theorem 2.2.3. *For any $m \geq 1$ the family $\{H_\alpha\}_{\alpha \in \mathcal{J}} = \{\mathbf{H}_\Gamma\}_{\Gamma \in \mathcal{J}^m}$ constitutes an orthogonal basis for $L^2(\mu_m)$. Moreover, if $\alpha = (\alpha_1, \alpha_2, \ldots) \in \mathcal{J}$, we have the norm expression*

$$\|H_\alpha\|_{L^2(\mu_m)}^2 = \alpha! := \alpha_1! \alpha_2! \cdots . \qquad (2.2.18)$$

Proof First consider the case where we have $m = 1$. Let $\alpha = (\alpha_1, \ldots, \alpha_n)$ and $\beta = (\beta_1, \ldots, \beta_n)$ be two multi-indices. Then using Lemma 2.1.2, where we have $E = E_{\mu_1}$, we get the expression

$$E[H_\alpha H_\beta] = E\left[\prod_{i=1}^{n} h_{\alpha_i}(\langle \omega, \eta_i \rangle) h_{\beta_i}(\langle \omega, \eta_i \rangle) \right]$$

$$= \int_{\mathbb{R}^n} \prod_{i=1}^{n} h_{\alpha_i}(x_i) h_{\beta_i}(x_i) d\lambda_n(x_1, \ldots, x_n)$$

$$= \prod_{i=1}^{n} \int_{\mathbb{R}} h_{\alpha_i}(x_i) h_{\beta_i}(x_i) d\lambda_1(x_i).$$

From the well-known orthogonality relations for Hermite polynomials (see, e.g., (C.10) in Appendix C), we have that

$$\int_{\mathbb{R}} h_j(x) h_k(x) e^{-\frac{1}{2}x^2} dx = \delta_{j,k} \sqrt{2\pi} k!. \qquad (2.2.19)$$

We therefore obtain (2.2.18) and that H_α and H_β are orthogonal if $\alpha \neq \beta$. To prove completeness of the family $\{H_\alpha\}$, we note that by Theorem 2.1.3

any $g \in L^2(\mu_1)$ can be approximated in $L^2(\mu_1)$ by stochastic polynomials of the form
$$p_n(\omega) = f_n(\langle \omega, \eta_1 \rangle, \dots, \langle \omega, \eta_n \rangle).$$

Now the polynomial $f_n(x_1, \dots, x_n)$ can be written as a linear combination of products of Hermite polynomials $h_{\alpha_1}(x_1) h_{\alpha_2}(x_2) \cdots h_{\alpha_n}(x_n)$. Then, of course, $p_n(\omega)$ is the corresponding linear combination of functions $H_\alpha(\omega)$ where $\alpha = (\alpha_1, \dots, \alpha_n)$.

The general case $m \geq 1$ follows from the above case using the tensor product structure. For completeness, we give the details.

With $H_\alpha = H_\alpha^{(m)}$ defined as in Definition 2.2.2, we get, with $\mu = \mu_m$, $E = E_\mu$, $\alpha = (\alpha_1, \dots, \alpha_n), \beta = (\beta_1, \dots, \beta_n)$:

$$E[H_\alpha H_\beta] = E\left[\prod_{k=1}^{n} h_{\alpha_k}(\langle \omega, e^{(k)} \rangle) h_{\beta_k}(\langle \omega, e^{(k)} \rangle) \right]$$

$$= E\left[\prod_{k=1}^{n} h_{\alpha_k}(\langle \omega_{i(k)}, \eta_{j(k)} \rangle) h_{\beta_k}(\langle \omega_{i(k)}, \eta_{j(k)} \rangle) \right]$$

$$= E\left[\prod_{\substack{k=1 \\ i(k)=1}}^{n} h_{\alpha_k}(\langle \omega_1, \eta_{j(k)} \rangle) \cdot h_{\beta_k}(\langle \omega_1, \eta_{j(k)} \rangle) \right.$$

$$\left. \cdots \prod_{\substack{k=1 \\ i(k)=m}}^{n} h_{\alpha_k}(\langle \omega_m, \eta_{j(k)} \rangle) h_{\beta_k}(\langle \omega_m, \eta_{j(k)} \rangle) \right]$$

$$= \prod_{u=1}^{m} E_{\mu_u}\left[\prod_{\substack{k=1 \\ i(k)=u}}^{n} h_{\alpha_k}(\langle \omega_u, \eta_{j(k)} \rangle) h_{\beta_k}(\langle \omega_u, \eta_{j(k)} \rangle) \right]$$

$$= \prod_{u=1}^{m} E_{\mu_u}\left[\prod_{v=1}^{n} \{ h_{\alpha_k}(\langle \omega_u, \eta_v \rangle) \cdot h_{\beta_k}(\langle \omega_u, \eta_v \rangle) \}_{k=u+(v-1)m} \right]$$

$$= \prod_{u=1}^{m} \prod_{v=1}^{n} \int_{\mathbb{R}} \{ h_{\alpha_k}(x_u) h_{\beta_k}(x_u) \}_{k=u+(v-1)m} d\lambda_1(x_u)$$

$$= \prod_{u=1}^{m} \prod_{v=1}^{n} \{ \delta_{\alpha_k, \beta_k} \alpha_k! \}_{k=u+(v-1)m}$$

$$= \prod_{u=1}^{m} \prod_{\substack{k=1 \\ i(k)=u}}^{n} \delta_{\alpha_k, \beta_k} \alpha_k!$$

$$= \prod_{k=1}^{n} \delta_{\alpha_k, \beta_k} \alpha_k! = \begin{cases} \alpha! & \text{if } \alpha = \beta \\ 0 & \text{if } \alpha \neq \beta. \end{cases}$$

We conclude that $\{H_\alpha\}_\alpha$ is an orthogonal family in $L^2(\mu)$ and that (2.2.18) holds.

Finally, since the span of $\{H_\alpha^{(1)}\}_{\alpha\in\mathcal{J}}$ is dense in $L^2(\mu_1)$, it follows by (2.2.17) that the span of $\{H_\alpha^{(m)}\}_{\alpha\in\mathcal{J}}$ is dense in $L^2(\mu_m)$. This completes the proof. □

From now on we fix the parameter dimension $d \geq 1$, the white noise dimension $m \geq 1$ and we fix a state space dimension $N \geq 1$. Let

$$\mathbf{L}^2(\mu_m) = \bigoplus_{k=1}^{N} L^2(\mu_m). \tag{2.2.20}$$

Applying Theorem 2.2.3 to each component of $\mathbf{L}^2(\mu_m)$, we get

Theorem 2.2.4 (Wiener–Itô chaos expansion theorem). *Every* $f \in \mathbf{L}^2(\mu_m)$ *has a unique representation*

$$f(\omega) = \sum_{\alpha\in\mathcal{J}} c_\alpha H_\alpha(\omega) \tag{2.2.21}$$

where $c_\alpha \in \mathbb{R}^N$ *for all* α.

Moreover, we have the isometry

$$\|f\|_{\mathbf{L}^2(\mu_m)}^2 = \sum_{\alpha\in\mathcal{J}} \alpha! c_\alpha^2, \tag{2.2.22}$$

where $c_\alpha^2 = |c_\alpha|^2 = (c_\alpha, c_\alpha)$ denotes the inner product in $\mathbb{R}^N$.

Remark   The major part of this book will be based on this construction. It must be admitted that the definitions behind (2.2.21) are rather complicated. Nevertheless the expression is notationally simple and quite easy to apply as long as we can avoid the underlying structure.

Exercise 2.2.5 ($N = 1, m = 1$)
i) The 1-dimensional smoothed white noise $w(\phi, \omega)$ defined in (2.1.8) has the expansion

$$w(\phi, \omega) = \langle \omega, \phi \rangle = \left\langle \omega, \sum_{j=1}^{\infty} (\phi, \eta_j)\eta_j \right\rangle$$

$$= \sum_{j=1}^{\infty} (\phi, \eta_j)\langle \omega, \eta_j \rangle = \sum_{j=1}^{\infty} (\phi, \eta_j) H_{\epsilon^{(j)}}(\omega),$$

where $\epsilon^{(j)} = (0, 0, \ldots, 1, \ldots)$ with 1 on entry number j, 0 otherwise. The convergence is in $L^2(\mu)$.

In other words,

$$w(\phi, \omega) = \sum_{\alpha} c_\alpha H_\alpha(\omega)$$

with

$$c_\alpha = \begin{cases} (\phi, \eta_j) & \text{if } \alpha = \epsilon^{(j)} \\ 0 & \text{otherwise.} \end{cases}$$ (2.2.23)

ii) The 1-dimensional, d-parameter Brownian motion $B(x,\omega)$ is defined by (2.1.10):

$$B(x,\omega) = \langle \omega, \psi \rangle,$$

where

$$\psi(y) = \chi_{[0,x_1] \times \cdots \times [0,x_d]}(y).$$

Proceeding as above, we write

$$\psi(y) = \sum_{j=1}^{\infty} (\psi, \eta_j) \eta_j(y)$$

$$= \sum_{j=1}^{\infty} \int_0^x \eta_j(u) du \, \eta_j(y),$$

where we have used the multi-index notation

$$\int_0^x \eta_j(u) du = \int_0^{x_d} \cdots \int_0^{x_1} \eta_j(u_1, \ldots, u_d) du_1 \cdots du_d = \prod_{k=1}^d \int_0^{x_k} \xi_{\beta_k^{(j)}}(t_k) dt_k$$

when $x = (x_1, \ldots, x_d)$ (see (2.2.8)). Therefore,

$$B(x,\omega) = \left\langle \omega, \sum_{j=1}^{\infty} \int_0^x \eta_j(u) du \, \eta_j \right\rangle = \sum_{j=1}^{\infty} \int_0^x \eta_j(u) du \langle \omega, \eta_j \rangle,$$

so $B(x,\omega)$ has the expansion

$$B(x,\omega) = \sum_{j=1}^{\infty} \int_0^x \eta_j(u) du \, H_{\epsilon^{(j)}}(\omega).$$ (2.2.24)

Example 2.2.6 ($N = m > 1$).
i) Next consider m-dimensional smoothed white noise defined by (2.1.27):

$$\mathbf{w}(\phi, \omega) = (\langle \omega_1, \phi_1 \rangle, \ldots, \langle \omega_m, \phi_m \rangle),$$

where $\omega = (\omega_1, \ldots, \omega_m) \in \mathcal{S}', \phi = (\phi_1, \ldots, \phi_m) \in \mathcal{S}$.

Using the same procedure as in the previous example, we get

$$\mathbf{w}(\phi,\omega) = \left(\left\langle \omega_1, \sum_{j=1}^{\infty} (\phi_1, \eta_j)\eta_j \right\rangle, \ldots, \left\langle \omega_m, \sum_{j=1}^{\infty} (\phi_m, \eta_j)\eta_j \right\rangle \right)$$

$$= \left(\sum_{j=1}^{\infty} (\phi_1, \eta_j)\langle \omega_1, \eta_j \rangle, \ldots, \sum_{j=1}^{\infty} (\phi_m, \eta_j)\langle \omega_m, \eta_j \rangle \right).$$

Since by (2.2.15)

$$\langle \omega_i, \eta_j \rangle = H_{\epsilon(i+(j-1)m)}(\omega),$$

we conclude that the ith component, $1 \leq i \leq m$, of $\mathbf{w}(\phi,\omega)$, $w_i(\phi,\omega)$, can be written

$$w_i(\phi,\omega) = \sum_{j=1}^{\infty} (\phi_i, \eta_j)\langle \omega_i, \eta_j \rangle$$

$$= \sum_{j=1}^{\infty} (\phi_i, \eta_j) H_{\epsilon(i+(j-1)m)}(\omega). \tag{2.2.25}$$

Thus

$$w_i(\phi,\omega) = (\phi_i, \eta_1) H_{\epsilon(i)} + (\phi_i, \eta_2) H_{\epsilon(i+m)} + \cdots.$$

Note that the expansions of $\{w_i\}_{i=1}^{m}$ involve disjoint families of $\{H_{\epsilon(k)}\}$.

ii) A similar expansion can be found for m-dimensional d-parameter Brownian motion

$$\mathbf{B}(x) = \mathbf{B}(x,\omega) = (B_1(x,\omega), \ldots, B_m(x,\omega))$$

defined by (see (2.1.28))

$$\mathbf{B}(x,\omega) = (\langle \omega_1, \psi \rangle, \ldots, \langle \omega_m, \psi \rangle); \ (\omega_1, \ldots, \omega_m) \in \mathcal{S}',$$

where

$$\psi(y) = \chi_{[0,x_1] \times \cdots \times [0,x_d]}(y); \ y \in \mathbb{R}^d.$$

So from (2.2.24) and (2.2.25) we get

$$\mathbf{B}(x,\omega) = \left(\sum_{j=1}^{\infty} \int_0^x \eta_j(u)du \langle \omega_1, \eta_j \rangle, \ldots, \sum_{j=1}^{\infty} \int_0^x \eta_j(u)du \langle \omega_m, \eta_j \rangle \right)$$

$$= \left(\sum_{j=1}^{\infty} \int_0^x \eta_j(u)du H_{\epsilon(1+(j-1)m)}(\omega), \ldots, \sum_{j=1}^{\infty} \int_0^x \eta_j(u)du H_{\epsilon(jm)}(\omega) \right).$$

Hence the ith component, $B_i(x)$, has expansion

$$B_i(x) = \sum_{j=1}^{\infty} \int_0^x \eta_j(u)du H_{\epsilon^{(i+(j-1)m)}}$$

$$= \int_0^x \eta_1(u)du H_{\epsilon^{(i)}} + \int_0^x \eta_2(u)du H_{\epsilon^{(i+m)}} + \int_0^x \eta_3(u)du H_{\epsilon^{(i+2m)}} + \cdots .$$

$$(2.2.26)$$

Again we note that the expansions of $\{B_i(x)\}_{i=1}^m$ involve disjoint families of $\{H_{\epsilon^{(k)}}\}$.

Note that for white noise and Brownian motion it is natural to have $N = m$. In general, however, one considers functions of white noise or Brownian motion, and in this case N and m need not be related. See Exercise 2.8.

2.2.2 Chaos Expansion in Terms of Multiple Itô Integrals

The chaos expansion (2.2.21)–(2.2.22) has an alternative formulation in terms of iterated Itô integrals. Although this formulation will not play a central role in our presentation, we give a brief review of it here, because it makes it easier for the reader to relate the material of the previous sections of this chapter to other literature of related content. Moreover, we will need this version in Section 2.5.

For convenience of notation we set $N = m = d = 1$ for the rest of this section. For the definition and basic properties of (1-parameter) Itô integrals, the reader is referred to Appendix B. For more information, see, e.g., Chung and Williams (1990), Karatzas and Shreve (1991), or Øksendal (2003). If $\psi(t_1, \ldots, t_n)$ is a symmetric function in its n variables $t_1, \ldots, t_n$, then we define its n-tuple Itô integral by the formula $(n \geq 1)$

$$\int_{\mathbb{R}^n} \psi dB^{\otimes n} := n! \int_{-\infty}^{\infty} \int_{-\infty}^{t_n} \int_{-\infty}^{t_{n-1}} \cdots \int_{-\infty}^{t_2} \psi(t_1, t_2, \ldots, t_n) dB(t_1) dB(t_2) \cdots dB(t_n),$$

$$(2.2.27)$$

where the integral on the right consists of n iterated Itô integrals (note that in each step the corresponding integrand is adapted because of the limits of the preceding integrals). Applying the Itô isometry n times we see that this iterated integral exists iff $\psi \in L^2(\mathbb{R}^n)$, then we have

$$E\left[\left(\int_{\mathbb{R}^n} \psi dB^{\otimes n}\right)^2\right] = n! \int_{\mathbb{R}^n} \psi(t_1, \ldots, t_n)^2 dt_1 \cdots dt_n = n!\|\psi\|^2. \quad (2.2.28)$$

For $n = 0$ we adopt the convention that $\int_{\mathbb{R}^0} \psi dB^{\otimes 0} = \psi = \|\psi\|_{L^2(\mathbb{R}^0)}$ when ψ is constant. Let $\alpha = (\alpha_1, \ldots, \alpha_k)$ be a multi-index, let $n = |\alpha|$ and let $\xi_1, \xi_2, \ldots$ be the Hermite functions defined in (2.2.2). Then by a fundamental result in Itô (1951), we have

$$\int_{\mathbb{R}^n} \xi_1^{\hat{\otimes}\alpha_1} \hat{\otimes} \cdots \hat{\otimes} \xi_k^{\hat{\otimes}\alpha_k} dB^{\otimes n} = \prod_{j=1}^{k} h_{\alpha_j}(\langle \omega, \xi_j \rangle). \qquad (2.2.29)$$

Here $\hat{\otimes}$ denotes the *symmetrized tensor product*. So, for example, if $f, g : \mathbb{R} \to \mathbb{R}$, then

$$(f \otimes g)(x_1, x_2) = f(x_1)g(x_2); \quad (x_1, x_2) \in \mathbb{R}^2$$

and

$$(f \hat{\otimes} g)(x_1, x_2) = \frac{1}{2}[f \otimes g + g \otimes f](x_1, x_2); \quad (x_1, x_2) \in \mathbb{R}^2,$$

and similarly for higher dimensions and for symmetric tensor powers.

Therefore, by comparison with Definition 2.2.1 we can reformulate (2.2.29) as

$$\int_{\mathbb{R}^n} \xi^{\hat{\otimes}\alpha} dB^{\otimes n} = H_\alpha(\omega), \qquad (2.2.30)$$

where we have used multi-index notation

$$\xi^{\hat{\otimes}\alpha} = \xi_1^{\hat{\otimes}\alpha_1} \hat{\otimes} \cdots \hat{\otimes} \xi_k^{\hat{\otimes}\alpha_k}. \qquad (2.2.31)$$

Now assume $f \in L^2(\mu)$ has the Wiener–Itô chaos expansion (2.2.21)

$$f(\omega) = \sum_\alpha c_\alpha H_\alpha(\omega).$$

We rewrite this as

$$f(\omega) = \sum_{n=0}^{\infty} \sum_{|\alpha|=n} c_\alpha H_\alpha(\omega) = \sum_{n=0}^{\infty} \sum_{|\alpha|=n} c_\alpha \int_{\mathbb{R}^n} \xi^{\hat{\otimes}\alpha} dB^{\otimes n}$$

$$= \sum_{n=0}^{\infty} \int_{\mathbb{R}^n} \sum_{|\alpha|=n} c_\alpha \xi^{\hat{\otimes}\alpha} dB^{\otimes n}.$$

Hence

$$f(\omega) = \sum_{n=0}^{\infty} \int_{\mathbb{R}^n} f_n dB^{\otimes n}, \qquad (2.2.32)$$

with

$$f_n = \sum_{|\alpha|=n} c_\alpha \xi^{\hat{\otimes}\alpha} \in \hat{L}^2(\mathbb{R}^n), \qquad (2.2.33)$$

where $\hat{L}^2(\mathbb{R}^n)$ denotes the symmetric functions in $L^2(\mathbb{R}^n)$.

Moreover, from (2.2.22) and (2.2.28) we have

$$\|f\|_{L^2(\mu)}^2 = \sum_{n=0}^{\infty} n! \|f_n\|_{L^2(\mathbb{R}^n)}^2. \tag{2.2.34}$$

We summarize this as follows.

Theorem 2.2.7 (The Wiener–Itô chaos expansion theorem II). *If* $f \in L^2(\mu)$, *then there exists a unique sequence of (deterministic) functions* $f_n \in \hat{L}^2(\mathbb{R}^n)$ *such that*

$$f(\omega) = \sum_{n=0}^{\infty} \int_{\mathbb{R}^n} f_n dB^{\otimes n}. \tag{2.2.35}$$

Moreover, we have the isometry

$$\|f\|_{L^2(\mu)}^2 = \sum_{n=0}^{\infty} n! \|f_n\|_{L^2(\mathbb{R}^n)}^2. \tag{2.2.36}$$

This result extends to arbitrary parameter dimension d. See Itô (1951).

2.3 The Hida Stochastic Test Functions and Stochastic Distributions. The Kondratiev Spaces $(\mathcal{S})_{\rho}^{m;N}, (\mathcal{S})_{-\rho}^{m;N}$

As we saw in the previous section, the growth condition

$$\sum_{\alpha} \alpha! c_{\alpha}^2 < \infty \tag{2.3.1}$$

assures that

$$f(\omega) := \sum_{\alpha} c_{\alpha} H_{\alpha}(\omega) \in L^2(\mu).$$

In the following we will replace condition (2.3.1) by various other conditions. We thus obtain a family of (generalized) function spaces that relates to $L^2(\mu)$ in a natural way. At the same time these spaces form an environment of stochastic test function spaces and stochastic distribution spaces, in a way that is analogous to the spaces $\mathcal{S}(\mathbb{R}^d) \subset L^2(\mathbb{R}^d) \subset \mathcal{S}'(\mathbb{R}^d)$. These spaces provide a favorable setting for the study of stochastic (ordinary and partial) differential equations. They were originally constructed on spaces of sequences by Kondratiev (1978), and later extended by him and others. See Kondratiev et al. (1994) and the references therein.

Let us first recall the characterizations of $\mathcal{S}(\mathbb{R}^d)$ and $\mathcal{S}'(\mathbb{R}^d)$ in terms of Fourier coefficients: As in (2.2.7) we let $\{\delta^{(j)}\}_{j=1}^{\infty} = \{(\delta_1^{(j)}, \ldots, \delta_d^{(j)})\}_{j=1}^{\infty}$ be a fixed ordering of all d-dimensional multi-indices $\delta = (\delta_1, \ldots, \delta_d) \in \mathbb{N}^d$ satisfying (2.2.7). In general, if $\alpha = (\alpha_1, \ldots, \alpha_j, \ldots) \in \mathcal{J}$, $\beta = (\beta_1, \ldots, \beta_j, \ldots) \in (\mathbb{R}^{\mathbb{N}})_c$ are two finite sequences, we will use the notation

$$\alpha^{\beta} = \alpha_1^{\beta_1} \alpha_2^{\beta_2} \cdots \alpha_j^{\beta_j} \cdots \quad \text{where} \quad \alpha_j^0 = 1. \tag{2.3.2}$$

Theorem 2.3.1 Reed and Simon (1980), Theorem V. 13–14.

a) Let $\phi \in L^2(\mathbb{R}^d)$, so that

$$\phi = \sum_{j=1}^{\infty} a_j \eta_j, \tag{2.3.3}$$

where $a_j = (\phi, \eta_j)$; $j = 1, 2, \ldots$, are the Fourier coefficients of ϕ with respect to $\{\eta_j\}_{j=1}^{\infty}$, with η_j as in (2.2.8). Then $\phi \in \mathcal{S}(\mathbb{R}^d)$ if and only if

$$\sum_{j=1}^{\infty} a_j^2 (\delta^{(j)})^{\gamma} < \infty \tag{2.3.4}$$

for all d-dimensional multi-indices $\gamma = (\gamma_1, \ldots, \gamma_d)$.

b) The space $\mathcal{S}'(\mathbb{R}^d)$ can be identified with the space of all formal expansions

$$T = \sum_{j=1}^{\infty} b_j \eta_j \tag{2.3.5}$$

such that

$$\sum_{j=1}^{\infty} b_j^2 (\delta^{(j)})^{-\theta} < \infty \tag{2.3.6}$$

for some d-dimensional multi-index $\theta = (\theta_1, \ldots, \theta_d)$.

If (2.3.6) holds, then the action of $T \in \mathcal{S}'(\mathbb{R}^d)$ given by (2.3.5) on $\phi \in \mathcal{S}(\mathbb{R}^d)$ given by (2.3.2) reads

$$\langle T, \phi \rangle = \sum_{j=1}^{\infty} a_j b_j. \tag{2.3.7}$$

We now formulate a stochastic analogue of Theorem 2.3.1. The following quantity is crucial:

If $\gamma = (\gamma_1, \ldots, \gamma_j, \ldots) \in (\mathbb{R}^{\mathbb{N}})_c$ (i.e., only finitely many of the real numbers γ_j are nonzero), we write

$$(2\mathbb{N})^{\gamma} := \prod_j (2j)^{\gamma_j}. \tag{2.3.8}$$

As before, d is the parameter dimension, m is the dimension of the white noise vector, $\mu_m = \mu_1 \times \cdots \times \mu_1$ as in (2.1.24), and N is the state space dimension.

Definition 2.3.2. The Kondratiev spaces of stochastic test function and stochastic distributions.

a) *The stochastic test function spaces*
Let N be a natural number. For $0 \le \rho \le 1$, let

$$(\mathcal{S})_\rho^N = (\mathcal{S})_\rho^{m;N}$$

consist of those

$$f = \sum_\alpha c_\alpha H_\alpha \in \mathbf{L}^2(\mu_m) = \bigoplus_{k=1}^N L^2(\mu_m) \quad \text{with } c_\alpha \in \mathbb{R}^N$$

such that

$$\|f\|_{\rho,k}^2 := \sum_\alpha c_\alpha^2 (\alpha!)^{1+\rho} (2\mathbb{N})^{k\alpha} < \infty \quad \text{for all } k \in \mathbb{N} \qquad (2.3.9)$$

where

$$c_\alpha^2 = |c_\alpha|^2 = \sum_{k=1}^N (c_\alpha^{(k)})^2 \text{ if } c_\alpha = (c_\alpha^{(1)}, \dots, c_\alpha^{(N)}) \in \mathbb{R}^N.$$

b) *The stochastic distribution spaces*
For $0 \le \rho \le 1$, let

$$(\mathcal{S})_{-\rho}^N = (\mathcal{S})_{-\rho}^{m;N}$$

consist of all formal expansions

$$F = \sum_\alpha b_\alpha H_\alpha \quad \text{with} \quad b_\alpha \in \mathbb{R}^N$$

such that

$$\|F\|_{-\rho,-q}^2 := \sum_\alpha b_\alpha^2 (\alpha!)^{1-\rho} (2\mathbb{N})^{-q\alpha} < \infty \quad \text{for some } q \in \mathbb{N}. \qquad (2.3.10)$$

The family of seminorms $\|f\|_{\rho,k}$; $k \in \mathbb{N}$ gives rise to a topology on $(\mathcal{S})_\rho^N$, and we can regard $(\mathcal{S})_{-\rho}^N$ as the dual of $(\mathcal{S})_\rho^N$ by the action

$$\langle F, f \rangle = \sum_\alpha (b_\alpha, c_\alpha) \alpha! \qquad (2.3.11)$$

if

$$F = \sum_\alpha b_\alpha H_\alpha \in (\mathcal{S})_{-\rho}^N; \; f = \sum_\alpha c_\alpha H_\alpha \in (\mathcal{S})_\rho^N$$

and (b_α, c_α) is the usual inner product in $\mathbb{R}^N$. Note that this action is well defined since

$$
\sum_\alpha |(b_\alpha, c_\alpha)| \alpha! = \sum_\alpha |(b_\alpha, c_\alpha)| (\alpha!)^{\frac{(1-\rho)}{2}} (\alpha!)^{\frac{(1+\rho)}{2}} (2\mathbb{N})^{-\frac{q\alpha}{2}} (2\mathbb{N})^{\frac{q\alpha}{2}}
$$

$$
\leq \left(\sum_\alpha b_\alpha^2 (\alpha!)^{1-\rho} (2\mathbb{N})^{-q\alpha} \right)^{\frac{1}{2}} \left(\sum_\alpha c_\alpha^2 (\alpha!)^{1+\rho} (2\mathbb{N})^{q\alpha} \right)^{\frac{1}{2}} < \infty
$$

for q large enough. When the value of m is clear from the context we simply write $(\mathcal{S})_\rho^N$, $(\mathcal{S})_{-\rho}^N$ instead of $(\mathcal{S})_\rho^{m;N}$, $(\mathcal{S})_{-\rho}^{m;N}$, respectively. If $N = 1$, we write $(\mathcal{S})_\rho$, $(\mathcal{S})_{-\rho}$ instead of $(\mathcal{S})_\rho^1$, $(\mathcal{S})_{-\rho}^1$, respectively.

Remarks

a) Note that for general $\rho \in [0, 1]$ we have

$$
(\mathcal{S})_1^N \subset (\mathcal{S})_\rho^N \subset (\mathcal{S})_0^N \subset \mathbf{L}^2(\mu_m) \subset (\mathcal{S})_{-0}^N \subset (\mathcal{S})_{-\rho}^N \subset (\mathcal{S})_{-1}^N. \quad (2.3.12)
$$

From (2.3.11) we see that if $F = (F_1, \ldots, F_N)$ and $G = (G_1, \ldots, G_N)$ both belong to $\mathbf{L}^2(\mu_m)$, then the action of F on G is given by

$$
\langle F, G \rangle = E \left[\sum_{i=1}^N F_i G_i \right]. \quad (2.3.13)
$$

b) In some cases it is useful to consider various generalizations of the spaces $(\mathcal{S})_\rho^{m;N}$, $(\mathcal{S})_{-\rho}^{m;N}$. For example, the coefficients c_α, b_α may not be constants, but may depend on some random parameter $\hat{\omega}$ that is independent of our white noise. In these cases we assume that $b_\alpha(\hat{\omega}), c_\alpha(\hat{\omega}) \in L^2(\nu)$, where ν is the probability measure for $\hat{\omega}$. Then the definitions above apply, with the modification that in (2.3.9) we replace c_α^2 by $\|c_\alpha\|_{L^2(\nu)}^2$ and in (2.3.11) we interpret (b_α, c_α) as

$$
E_\nu[b_\alpha c_\alpha] = \int_{\hat{\Omega}} b_\alpha(\hat{\omega}) c_\alpha(\hat{\omega}) d\nu(\hat{\omega}) = (b_\alpha, c_\alpha)_{L^2(\nu)}. \quad (2.3.14)
$$

Another useful generalization (where the c_α are elements of Sobolev spaces) is discussed in Våge (1996a).

c) The quantity $(2\mathbb{N})^\alpha$ in (2.3.8) will be applied frequently, so it is useful to have some estimates of it.

First note that if $\alpha = \epsilon^{(k)}$, we get

$$
(2\mathbb{N})^{\epsilon^{(k)}} = 2k. \quad (2.3.15)
$$

Next we state the following result from Zhang (1992).

Proposition 2.3.3 Zhang (1992). *We have that*

$$\sum_{\alpha \in \mathcal{J}} (2\mathbb{N})^{-q\alpha} < \infty \tag{2.3.16}$$

if and only if $q > 1$.

Proof   First assume $q > 1$. If $\alpha = (\alpha_1, \alpha_2, \ldots) \in \mathcal{J}$, define

$$\text{Index } \alpha = \max\{j; \alpha_j \neq 0\}.$$

Consider

$$a_n := \sum_{\substack{\alpha \\ \text{Index } \alpha = n}} (2\mathbb{N})^{-q\alpha} = \sum_{\substack{\alpha_1, \ldots, \alpha_{n-1} \geq 0 \\ \alpha_n \geq 1}} \prod_{j=1}^{n} (2j)^{-q\alpha_j}$$

$$= \left[\prod_{j=1}^{n-1} \left(\sum_{\alpha_j=0}^{\infty} (2j)^{-q\alpha_j} \right) \right] \left(\sum_{\alpha_n=1}^{\infty} (2n)^{-q\alpha_n} \right)$$

$$= \frac{1}{((2n)^q - 1)} \prod_{j=1}^{n-1} \frac{(2j)^q}{((2j)^q - 1)} = \frac{1}{(2n)^q} \prod_{j=1}^{n} \frac{(2j)^q}{((2j)^q - 1)}.$$

This gives

$$\frac{a_n}{a_{n+1}} - 1 = \frac{(2n+2)^q - 1}{(2n)^q} - 1 = \left(1 + \frac{1}{n} \right)^q - (2n)^{-q} - 1. \tag{2.3.17}$$

In particular,

$$\frac{a_n}{a_{n+1}} - 1 \geq \frac{q}{n} - (2n)^{-q}.$$

Hence

$$\liminf_{n \to \infty} n \left(\frac{a_n}{a_{n+1}} - 1 \right) \geq q > 1$$

and, therefore, by Abel's criterion for convergence,

$$\sum_{\alpha} (2\mathbb{N})^{-q\alpha} = \sum_{n=0}^{\infty} a_n < \infty,$$

as claimed. Conversely, if $q = 1$, then, by (2.3.17) above,

$$\frac{a_n}{a_{n+1}} = 1 + \frac{1}{2n}, \quad \text{so}$$

$$\lim_{n \to \infty} n \left(\frac{a_n}{a_{n+1}} - 1 \right) = \lim_{n \to \infty} n \cdot \frac{1}{2n} = \frac{1}{2} < 1.$$

Hence $\sum_{n=0}^{\infty} a_n = \infty$ by Abel's criterion. $\qquad \square$

The following useful result relates the sum in (2.3.16) to sums of the type appearing in Theorem 2.3.1. It was pointed out to us by Y. Hu (private communication).

Lemma 2.3.4.   *Let* $\delta^{(j)} = (\delta_1^{(j)}, \delta_2^{(j)}, \ldots, \delta_d^{(j)})$ *be as in (2.2.7). For all* $j \in \mathbb{N}$ *and all* $d \in \mathbb{N}$ *we have*

$$j^{\frac{1}{d}} \leq \delta_1^{(j)} \cdot \delta_2^{(j)} \cdots \delta_d^{(j)} \leq j^d.$$

Proof   The case $d = 1$ is trivial, so we fix $d \geq 2$. Since $\delta_k^{(j)} \leq j$ (by (2.2.7)), the second inequality is immediate. To prove the first inequality, we fix j and set

$$M = \delta_1^{(j)} + \delta_2^{(j)} + \cdots + \delta_d^{(j)}.$$

Note that $M \geq d$. Consider the minimization problem

$$\inf \left\{ f(x_1, \ldots, x_d); \ x_i \in [1, \infty) \text{ for all } i; \sum_{i=1}^{d} x_i = M \right\},$$

where $f(x) = x_1 x_2 \cdots x_d$. Clearly a minimum exists. Using the Lagrange multiplier method we see that the only candidate for a minimum point when $x_i > 1$ for all i is $(x_1, x_2, \ldots, x_d) = (M/d, \ldots, M/d)$, which gives the value

$$f\left(\frac{M}{d}, \ldots, \frac{M}{d}\right) = \left(\frac{M}{d}\right)^d.$$

If one or several x_i's have the value 1, then the minimization problem can be reduced to the case when d and M are replaced by $d - 1$ and $M - 1$, respectively. Since

$$\left(\frac{M-1}{d-1}\right)^{d-1} \leq \left(\frac{M}{d}\right)^d,$$

we conclude by induction on d that

$$x_1 \cdots x_d \geq M - d + 1 \text{ for all } (x_1, \ldots, x_d) \in [1, \infty)^d \text{ with } \sum_{i=1}^{d} x_i = M.$$

$$(2.3.18)$$

To finish the proof of the lemma we now compare M and j:

Since $\delta_1^{(j)} + \cdots + \delta_d^{(j)} = M$ and the sequence $\{(\delta_1^{(i)}, \ldots, \delta_d^{(i)})\}_{i=1}^{\infty}$ is increasing (see (2.2.7)), we know that

$$\delta_1^{(i)} + \cdots + \delta_d^{(i)} \leq M \quad \text{for all} \quad i < j.$$

Now (by a known result in combinatorics) the total number of multi-indices $(\delta_1, \ldots, \delta_d) \in \mathbb{N}^d$ such that $\delta_1 + \delta_2 + \cdots + \delta_d \leq M$ is equal to

$$\sum_{n=d}^{M} \binom{n-1}{d-1} = \binom{M}{d}.$$

Therefore

$$j \le \binom{M}{d} = \frac{M(M-1)\cdots(M-d+1)}{d!} \le (M-d+1)^d$$

or

$$M - d + 1 \ge j^{\frac{1}{d}}.$$

Combined with (2.3.18) this gives

$$\delta_1^{(j)}\delta_2^{(j)}\cdots\delta_d^{(j)} \ge j^{\frac{1}{d}}. \qquad \square$$

As a consequence of this, we obtain the following alternative characterization of the spaces $(\mathcal{S})_\rho^N, (\mathcal{S})_{-\rho}^N$. This characterization has often been used as a definition of the Kondratiev spaces (see, e.g., Holden et al. (1995a), and the references therein). As usual we let $(\delta_1^{(j)}, \ldots, \delta_d^{(j)})$ be as in (2.2.7).

In this connection the following notation is convenient:

With $(\delta_1^{(j)}, \ldots, \delta_d^{(j)})$ as in (2.2.7), let $\Delta = (\Delta_1, \ldots, \Delta_k, \ldots) \in \mathbb{N}^\mathbb{N}$ be the sequence defined by

$$\Delta_j = 2^d \delta_1^{(j)} \delta_2^{(j)} \cdots \delta_d^{(j)}; \; j = 1, 2, \ldots. \qquad (2.3.19)$$

Then if $\alpha = (\alpha_1, \ldots, \alpha_j, \ldots) \in (\mathbb{R}^\mathbb{N})_c$, we define

$$\Delta^\alpha = \Delta_1^{\alpha_1} \Delta_2^{\alpha_2} \cdots \Delta_j^{\alpha_j} \cdots = \prod_{j=1}^{\infty} (2^d \delta_1^{(j)} \cdots \delta_d^{(j)})^{\alpha_j}, \qquad (2.3.20)$$

in accordance with the general multi-index notation (2.3.2).

Corollary 2.3.5. *Let $0 \le \rho \le 1$. Then we have*

a) $f = \sum_\alpha c_\alpha H_\alpha$ *(with $c_\alpha \in \mathbb{R}^N$ for all α) belongs to $(\mathcal{S})_\rho^N$ if and only if*

$$\sum_\alpha c_\alpha^2 (\alpha!)^{1+\rho} \Delta^{k\alpha} < \infty \quad \text{for all} \;\; k \in \mathbb{N}. \qquad (2.3.21)$$

b) *The formal expansion $F = \sum_\alpha b_\alpha H_\alpha$ (with $b_\alpha \in \mathbb{R}^N$ for all α) belongs to $(\mathcal{S})_{-\rho}^N$ if and only if*

$$\sum_\alpha b_\alpha^2 (\alpha!)^{1-\rho} \Delta^{-q\alpha} < \infty \quad \text{for some} \;\; q \in \mathbb{N}. \qquad (2.3.22)$$

Proof By the second inequality of Lemma 2.3.4 we see that

$$\Delta^{k\alpha} = \prod_{j=1}^{\infty}(2^d\delta_d^{(j)}\cdots\delta_d^{(j)})^{k\alpha_j} \le \prod_{j=1}^{\infty}(2^dj^d)^{k\alpha_j} = \prod_{j=1}^{\infty}(2j)^{dk\alpha_j} \qquad (2.3.23)$$

for all $\alpha \in \mathcal{J}$ and all $k \in \mathbb{N}$.

From the first inequality of Lemma 2.3.4 we get

$$\prod_{j=1}^{\infty}(2^d\delta_1^{(j)}\cdots\delta_d^{(j)})^{k\alpha_j} \ge \prod_{j=1}^{\infty}(2j^{\frac{d-1}{d}})^{k\alpha_j} \ge \prod_{j=1}^{\infty}(2j)^{\frac{d-1}{d}\cdot k\alpha_j} \qquad (2.3.24)$$

for all $\alpha \in \mathcal{J}$ and all $k \in \mathbb{N}$. These two inequalities show the equivalence of the criterion in a) with (2.3.9) and the equivalence of the criterion in b) with (2.3.10). □

Example 2.3.6.   If $\phi \in \mathcal{S}(\mathbb{R}^d)$, then

$$w(\phi,\omega) \in (\mathcal{S})_1.$$

Indeed, by (2.2.23) we have

$$w(\phi,\omega) = \sum_{j=1}^{\infty}(\phi,\eta_j)H_{\epsilon^{(j)}}(\omega) = \sum_{\alpha}c_\alpha H_\alpha,$$

so

$$\sum_{\alpha}c_\alpha^2(\alpha!)^2\Delta^{k\alpha} = \sum_{j=1}^{\infty}(\phi,\eta_j)^2(2^d\delta_1^{(j)}\cdots\delta_d^{(j)})^k$$

$$= 2^{dk}\sum_{j=1}^{\infty}(\phi,\eta_j)^2(\delta_1^{(j)}\cdots\delta_d^{(j)})^k < \infty$$

by Theorem 2.3.1a. Hence $w(\phi,\omega) \in (\mathcal{S})_1$ by Corollary 2.3.5. □

Note that with our notation $(\mathcal{S})_0$ and $(\mathcal{S})_{-0}$ are different spaces. In fact, they coincide with the Hida spaces $(\mathcal{S})$ and $(\mathcal{S})^*$, respectively, which we describe below.

Remark The definition of stochastic test function and distribution spaces used in Kondratiev et al. (1994), and Kondratiev et al. (1995a), does not coincide with ours. However, the two definitions are equivalent, as we will now show.

Let us first recall Kondratiev's definition. Let $\phi \in L^2(\mu_1)$ be given by

$$\phi = \sum_{\alpha}c_\alpha H_\alpha(\omega).$$

For $p \in \mathbb{R}$ define

$$\mathcal{K}_p := \{\phi \in L^2(\mu_1) \; ; \; \|\phi\|_p^2 < +\infty\}$$

where

$$\|\phi\|_p^2 = \sum_{n=0}^{\infty} (n!)^2 \sum_{\substack{\alpha \\ |\alpha|=n}} c_\alpha^2 (2\mathbb{N})^{\alpha p} < +\infty.$$

The *Kondratiev test function space* $(\mathcal{K})_1$ is defined as

$$(\mathcal{K})_1 = \bigcap_p \mathcal{K}_p, \quad \text{the projective limit of } \mathcal{K}_p.$$

The *Kondratiev distribution space* $(\mathcal{K})_{-1}$ is the inductive limit of $\mathcal{K}_{-p}$, the dual of $\mathcal{K}_p$.

According to our definition,

$$\|\phi\|_{1,p}^2 = \sum_\alpha c_\alpha^2 (\alpha!)^2 (2\mathbb{N})^{\alpha p} = \sum_{n=0}^{\infty} \sum_{\substack{\alpha \\ |\alpha|=n}} (\alpha!)^2 c_\alpha^2 (2\mathbb{N})^{\alpha p}.$$

Obviously $\alpha! \leq n!$ if $|\alpha| = n$. Therefore

$$\|\phi\|_{1,p} \leq \|\phi\|_p.$$

Hence

$$(\mathcal{K})_1 \subset (\mathcal{S})_1.$$

On the other hand, if $\alpha_1 + \alpha_2 + \cdots + \alpha_m = n, \alpha_i \geq 1$ we have

$$
\begin{aligned}
n! &= \alpha_1!(\alpha_1 + 1)(\alpha_1 + 2)\cdots(\alpha_1 + \alpha_2)(\alpha_1 + \alpha_2 + 1)\cdots(\alpha_1 + \cdots + \alpha_m) \\
&\leq \alpha_1! \alpha_1 (\alpha_2 + 1)! \alpha_2 (\alpha_3 + 1)! \cdots \alpha_m(\alpha_m + 1)! \\
&\leq \prod_{j=1}^{m} (\alpha_j(\alpha_j + 1))\alpha_j! = \left(\prod_{j=1}^{m} \alpha_j!\right) \prod_{j=1}^{m} 2\alpha_j^2 \\
&\leq \alpha! \prod_{0=1}^{m} (2j)^{2\alpha_j} \quad (\text{since } 2\alpha_j^2 \leq (2j)^{2\alpha_j}) = \alpha!(2\mathbb{N})^{2\alpha}.
\end{aligned}
$$

Hence

$$\|\phi\|_p^2 = \sum_{n=0}^{\infty} (n!)^2 \sum_{|\alpha|=n} c_\alpha^2 (2\mathbb{N})^{\alpha p} \leq \sum_{n=0}^{\infty} \sum_{|\alpha|=n} (\alpha!)^2 (2\mathbb{N})^{4\alpha} c_\alpha^2 (2\mathbb{N})^{\alpha p}$$

$$= \sum_\alpha (\alpha!)^2 c_\alpha^2 (2\mathbb{N})^{\alpha(p+4)} = \|\phi\|_{1,p+4}^2.$$

This shows $(\mathcal{S})_1 \subset (\mathcal{K})_1$, and hence $(\mathcal{S})_1 = (\mathcal{K})_1$.

2.3.1 The Hida Test Function Space $(\mathcal{S})$ and the Hida Distribution Space $(\mathcal{S})^*$

There is an extensive literature on these spaces. See Hida et al. (1993), and the references therein. According to the characterization in Zhang (1992), we can describe these spaces, generalized to arbitrary dimension m, as follows:

Proposition 2.3.7 Zhang (1992). a) *The Hida test function space* $(\mathcal{S})^N$ *consists of those*

$$f = \sum_\alpha c_\alpha H_\alpha \in L^2(\mu_m) \quad \text{with} \quad c_\alpha \in \mathbb{R}^N$$

such that

$$\sup_\alpha \{c_\alpha^2 \alpha!(2\mathbb{N})^{k\alpha}\} < \infty \quad \text{for all} \quad k < \infty. \tag{2.3.25}$$

b) *The Hida distribution space* $(\mathcal{S})^{*,N}$ *consists of all formal expansions*

$$F = \sum_\alpha b_\alpha H_\alpha \quad \text{with} \quad b_\alpha \in \mathbb{R}^N$$

such that

$$\sup_\alpha \{b_\alpha^2 \alpha!(2\mathbb{N})^{-q\alpha}\} < \infty \quad \text{for some } q < \infty. \tag{2.3.26}$$

Hence, after comparison with Definition 2.3.2, we see that

$$(\mathcal{S})^N = (\mathcal{S})_0^{m;N} \quad \text{and} \quad (\mathcal{S})^{*,N} = (\mathcal{S})_{-0}^{m;N}. \tag{2.3.27}$$

If $N = 1$, we write

$$(\mathcal{S})^1 = (\mathcal{S}) \quad \text{and} \quad (\mathcal{S})^{*,1} = (\mathcal{S})^*.$$

Corollary 2.3.8. *For $N = 1$ and $p \in (1, \infty)$ we have*

$$(\mathcal{S}) \subset L^p(\mu_m) \subset (\mathcal{S})^*. \tag{2.3.28}$$

Proof We give a proof in the case $m = d = 1$. Since $L^p(\mu) \supset L^{p'}(\mu)$ for $p' > p$ and the dual of $L^p(\mu)$ is $L^q(\mu)$ with $1/p + 1/q = 1$ if $1 < p < \infty$, it suffices to prove that

$$(\mathcal{S}) \subset L^p(\mu) \quad \text{for all } p \in (1, \infty).$$

To this end choose

$$f = \sum_\alpha c_\alpha H_\alpha \in (\mathcal{S}).$$

Then

$$\|f\|_{L^p(\mu)} \le \sum_\alpha |c_\alpha| \|H_\alpha\|_{L^p(\mu)}.$$

If $\alpha = (\alpha_1, \ldots, \alpha_k)$, we have $H_\alpha(\omega) = h_{\alpha_1}(\langle \omega, \xi_1 \rangle) h_{\alpha_2}(\langle \omega, \xi_2 \rangle) \cdots h_{\alpha_k}$ $(\langle \omega, \xi_k \rangle)$ for $\omega \in \mathcal{S}'(\mathbb{R})$. Hence, by independence,

$$\|H\|^p_{L^p(\mu)} = E[|H_\alpha|^p]$$

$$= \prod_{j=1}^k E[h^p_{\alpha_j}(\langle \omega, \xi_j \rangle)]. \tag{2.3.29}$$

Note that by (2.2.29), we have

$$h_{\alpha_j}(\langle \omega, \xi_j \rangle) = \int_{\mathbb{R}^{\alpha_j}} \xi_j^{\hat{\otimes} \alpha_j}(x) dB^{\otimes \alpha_j}(x). \tag{2.3.30}$$

By the Carlen–Kree estimates in Carlen and Kree (1991), we have, in general, for $p \ge 1$, $n \in \mathbb{N}$, $\phi \in L^2(\mathbb{R})$,

$$\left\| \int_{\mathbb{R}^n} \phi^{\hat{\otimes} n}(x) dB^{\otimes n}(x) \right\|_{L^p(\mu)} \le \sqrt{n!}(\theta_p \sqrt{ep})^n (2\pi n)^{-\frac{1}{4}} \|\phi\|^n, \tag{2.3.31}$$

where

$$\theta_p = 1 + \sqrt{1 + \frac{1}{p}}. \tag{2.3.32}$$

Applied to (2.3.29)–(2.3.30), this gives

$$\|H\|_{L^p(\mu)} \le \prod_{j=1}^k \sqrt{\alpha_j!}(\theta_p \sqrt{ep})^{\alpha_j} (2\pi\alpha_j)^{-\frac{1}{4}}$$

$$\le \sqrt{\alpha!}(\theta_p \sqrt{ep})^{|\alpha|}.$$

Hence, by (2.3.25),

$$\|f\|_{L^p(\mu)} \le \sum_\alpha |c_\alpha| \sqrt{\alpha!}(\theta_p \sqrt{ep})^{|\alpha|}$$

$$\le \sum_\alpha |c_\alpha| \sqrt{\alpha!}(2\mathbb{N})^{k\alpha} (\theta_p \sqrt{ep})^{|\alpha|} (2\mathbb{N})^{-k\alpha}$$

$$\le \sup_\alpha \{|c_\alpha| \sqrt{\alpha!}(2\mathbb{N})^{k\alpha}\} \sum_\alpha (\theta_p \sqrt{ep})^{|\alpha|} (2\mathbb{N})^{-k\alpha}$$

$$< \infty$$

for k large enough. $\square$

2.3.2 Singular White Noise

One of the many useful properties of $(\mathcal{S})^*$ is that it contains the *singular* or *pointwise white noise.*

Definition 2.3.9. **a)** The *1-dimensional (d-parameter) singular white noise process* is defined by the formal expansion

$$W(x) = W(x, \omega) = \sum_{k=1}^{\infty} \eta_k(x) H_{\epsilon^{(k)}}(\omega); \ x \in \mathbb{R}^d, \qquad (2.3.33)$$

where $\{\eta_k\}_{k=1}^{\infty}$ is the basis of $L^2(\mathbb{R}^d)$ defined in (2.2.8) while $H_\alpha = H_\alpha^{(1)}$ is defined by (2.2.10).

b) The *m-dimensional (d-parameter) singular white noise process* is defined by

$$\mathbf{W}(x) = \mathbf{W}(x, \omega) = (W_1(x, \omega), \dots, W_m(x, \omega)),$$

where the *i*th component $W_i(x)$, of $\mathbf{W}(x)$, has expansion

$$W_i(x) = \sum_{j=1}^{\infty} \eta_j(x) H_{\epsilon_{i+(j-1)m}}$$

$$= \eta_1(x) H_{\epsilon^{(i)}} + \eta_2(x) H_{\epsilon_{i+m}} + \eta_3(x) H_{\epsilon_{i+2m}} + \cdots . \qquad (2.3.34)$$

(Compare with the expansion (2.2.25) we have for smoothed *m*-dimensional white noise.)

Proposition 2.3.10.

$$\mathbf{W}(x, \omega) \in (\mathcal{S})^{*, m} \quad \text{for each} \quad x \in \mathbb{R}^d.$$

Proof   (i) $m = 1$. We must show that the expansion (2.3.34) satisfies condition (2.3.10) for $\rho = 0$, i.e.,

$$\sum_{k=1}^{\infty} \eta_k^2(x)(2k)^{-q} < \infty \qquad (2.3.35)$$

for some $q \in \mathbb{N}$. By (2.2.5) and (2.2.8) we have $|\eta_k^2(x)| \leq C$ for all $k = 1, 2, \dots, x \in \mathbb{R}^d$ for a constant C.

Therefore, by Proposition 2.3.3, the series in (2.3.35) converges for all $q > 1$.

(ii) $m > 1$. The proof in this case is similar to the above, replacing η_k by $e^{(k)}$. $\square$

Remark Using (2.2.11) we may rewrite (2.3.34) as

$$\mathbf{W}(x,\omega) = \sum_{\substack{i=1,\dots,m \\ j=1,2,\cdots}} e^{(i+(j-1)m)}(x) H^{(m)}_{\epsilon_{i+(j-1)m}}(\omega)$$

$$= \left(\sum_{j=1}^{\infty} \eta_j(x) H^{(1)}_{\epsilon_{1+(j-1)m}}(\omega), \dots, \sum_{j=1}^{\infty} \eta_j(x) H^{(1)}_{\epsilon_{m+(j-1)m}}(\omega) \right).$$

$$(2.3.36)$$

By comparing the expansion (2.3.33) for singular white noise $W(x)$ with the expansion (2.2.24) for Brownian motion $B(x)$, we see that

$$W_i(x) = \frac{\partial^d}{\partial x_1 \cdots \partial x_d} B_i(x) \quad \text{in } (\mathcal{S})^*; \quad \text{for } 1 \le i \le m = N, d \ge 1 \quad (2.3.37)$$

In particular,

$$W(t) = \frac{d}{dt} B(t) \text{ in } (\mathcal{S})^* \ (d = m = N = 1) \tag{2.3.38}$$

See Exercise 2.30. See also (2.5.27).

Thus we may say that m-dimensional singular white noise $\mathbf{W}(x,\omega)$ consists of m independent copies of 1-dimensional singular white noise. Here "independence" is interpreted in the sense that if we truncate the summations over j to a finite number of terms, then the components are independent when they are regarded as random variables in $L^2(\mu_m) = L^2(\mu_1 \times \cdots \times \mu_1)$.

In spite of Proposition 2.3.10 and the fact that also many other important Brownian functionals belong to $(\mathcal{S})^*$ (see Hida et al. (1993)), the space $(\mathcal{S})^*$ turns out to be too small for the purpose of solving stochastic ordinary and partial differential equations. We will return to this in Chapters 3 and 4, where we will give examples of such equations with no solution in $(\mathcal{S})^*$ but a unique solution in $(\mathcal{S})_{-1}$.

2.4 The Wick Product

The Wick product was introduced in Wick (1950) as a tool to renormalize certain infinite quantities in quantum field theory. In stochastic analysis the Wick product was first introduced by Hida and Ikeda (1965). A systematic, general account of the traditions of both mathematical physics and probability theory regarding this subject was given in Dobrushin and Minlos (1977). In Meyer and Yan (1989), this kind of construction was extended to cover Wick products of Hida distributions. We should point out that this (stochastic) Wick product does not in general coincide with the Wick product in physics, as defined, e.g., in Simon (1974). See also the survey in Gjessing et al. (1993).

Today the Wick product is also important in the study of stochastic (ordinary and partial) differential equations. In general, one can say that the use of this product corresponds to – and extends naturally – the use of Itô integrals. We now explain this in more detail.

The (stochastic) Wick product can be defined in the following way:

Definition 2.4.1. The *Wick product* $F \diamond G$ of two elements

$$F = \sum_\alpha a_\alpha H_\alpha, \ G = \sum_\alpha b_\alpha H_\alpha \in (\mathcal{S})_{-1}^{m;N} \quad \text{with} \quad a_\alpha, b_\alpha \in \mathbb{R}^N \qquad (2.4.1)$$

is defined by

$$F \diamond G = \sum_{\alpha,\beta} (a_\alpha, b_\beta) H_{\alpha+\beta}. \qquad (2.4.2)$$

With this definition the Wick product can be described in a very simple manner. What is not obvious from the construction, however, is that $F \diamond G$ in fact does not depend on our particular choice of base elements for $L^2(\mu)$. It is possible to give a direct proof of this, but the details are tedious. A sketch of a proof is given in Appendix D.

In the $L^2(\mu)$ case the basis independence of the Wick product can also be seen from the following formulation of Wick multiplication in terms of multiple Itô integrals (see Theorem 2.2.7).

Proposition 2.4.2. *Let $N = m = d = 1$. Assume that $f, g \in L^2(\mu)$ have the following representation in terms of multiple Itô integrals:*

$$f(\omega) = \sum_{i=0}^\infty \int_{\mathbb{R}^i} f_i dB^{\otimes i}, \qquad g(\omega) = \sum_{j=0}^\infty \int_{\mathbb{R}^j} g_j dB^{\otimes j}.$$

Suppose $f \diamond g \in L^2(\mu)$. Then

$$(f \diamond g)(\omega) = \sum_{n=0}^\infty \int_{\mathbb{R}^n} \sum_{i+j=n} f_i \hat\otimes g_j dB^{\otimes n}. \qquad (2.4.3)$$

Proof   By (2.4.2) we have

$$(f \diamond g)(\omega) = \sum_{\alpha,\beta} a_\alpha b_\beta H_{\alpha+\beta}(\omega)$$

and by (2.2.30) we have

$$H_{\alpha+\beta}(\omega) = \int_{\mathbb{R}^{|\alpha+\beta|}} \xi_\alpha^{\hat\otimes\alpha} \hat\otimes \xi_\beta^{\hat\otimes\beta} dB^{\otimes|\alpha+\beta|}.$$

Combining this with (2.2.33), we get

$$(f \diamond g)(\omega) = \sum_{n=0}^{\infty} \sum_{|\alpha+\beta|=n} a_\alpha b_\beta \left(\int_{\mathbb{R}^n} \xi_\alpha^{\hat\otimes\alpha} \hat\otimes \xi_\beta^{\hat\otimes\beta} dB^{\otimes n} \right)$$

$$= \sum_{n=0}^{\infty} \int_{\mathbb{R}^n} \left(\sum_{i+j=n} \sum_{|\alpha|=i} a_\alpha \xi_\alpha^{\hat\otimes\alpha} \hat\otimes \sum_{|\beta|=j} b_\beta \xi_\beta^{\hat\otimes\beta} \right) dB^{\otimes n}$$

$$= \sum_{n=0}^{\infty} \int_{\mathbb{R}^n} \left(\sum_{i+j=n} f_i \hat\otimes g_j \right) dB^{\otimes n},$$

as claimed.   □

Example 2.4.3. Let $0 \le t_0 < t_1 < \infty$ and assume that $h(\omega) \in L^2(\mu_1)$ is $\mathcal{F}_{t_0}$-measurable. Then

$$h \diamond (B(t_1) - B(t_0)) = h \cdot (B(t_1) - B(t_0)). \qquad (2.4.4)$$

Proof If

$$h(\omega) = \sum_{i=0}^{\infty} \int_{\mathbb{R}^i} h_i(x) dB^{\otimes i}(x),$$

then each of the functions $h_i(x)$ must satisfy

$h_i(x) = 0$ almost surely outside $\{x; x_j \le t_0 \text{ for } j = 1, 2, \ldots, n\}$.

Therefore the symmetrized tensor product of $\chi_{[t_0,t_1]}(s)$ and $h_i(x_1, \ldots, x_n)$ is given by (with $x_{n+1} = s$)

$$(\chi_{[t_0,t_1]} \hat\otimes h_i)(x_1, \ldots, x_{n+1}) = \frac{h(y)}{n+1} \chi_{[t_0,t_1]}(\max_j \{x_j\}),$$

where $y = (y_1, y_2, \ldots, y_n)$ is an arbitrary permutation of the remaining x_i when $\tilde{y} := x_{\hat{j}} := \max_{1 \le i \le n+1} \{x_i\}$ is removed. (For almost all $(x_1, \ldots, x_{n+1})$ there is a unique such $\hat{j}$.)

Since

$$B(t_1) - B(t_0) = \int_{\mathbb{R}} \chi_{[t_0,t_1]}(s) dB(s),$$

we get, by (2.4.3),

$$h \diamond (B(t_1) - B(t_0))$$

$$= \sum_{n=0}^{\infty} \int_{\mathbb{R}^{n+1}} (\chi_{[t_0,t_1]} \hat\otimes h_i)(x_1, \ldots, x_{n+1}) dB^{\otimes(n+1)}(x_1, \ldots, x_{n+1})$$

$$= \sum_{n=0}^{\infty} (n+1)! \int_0^{t_1} \int_0^{x_{n-1}} \cdots \int_0^{x_1} \frac{1}{n+1} \chi_{[t_0,t_1]}(\widetilde{y}) h(y) dB(x_1) \cdots dB(x_{n+1})$$

$$= \sum_{n=0}^{\infty} n! \int_{t_0}^{t_1} \left[\int_0^{t_0} \int_0^{x_2} h(x_1,\ldots,x_n) dB(x_1) dB(x_2) \cdots \right] dB(x_{n+1})$$

$$= \sum_{n=0}^{\infty} n! \left(\int \int_{0 \leq x_1 \leq x_2 \leq \cdots \leq x_n} h(x_1,\ldots,x_n) dB(x_1) \cdots dB(x_n) \right) \left(\int_{t_0}^{t_1} dB(x_{n+1}) \right)$$

$$= \sum_{n=0}^{\infty} \int_{\mathbb{R}^n} h(x) dB^{\otimes n}(x) \cdot (B(t_1) - B(t_0)) = h \cdot (B(t_1) - B(t_0)).$$

$\square$

An important property of the spaces $(\mathcal{S})_{-1}, (\mathcal{S})_1$ and $(\mathcal{S})^*, (\mathcal{S})$ is that they are closed under Wick products.

Lemma 2.4.4.

a) $F, G \in (\mathcal{S})_{-1}^{m;N} \Rightarrow F \diamond G \in (\mathcal{S})_{-1}^{m;1}$;

b) $f, g \in (\mathcal{S})_1^{m;N} \Rightarrow f \diamond g \in (\mathcal{S})_1^{m;1}$;

c) $F, G \in (\mathcal{S})^{*,N} \Rightarrow F \diamond G \in (\mathcal{S})^{*,1}$;

d) $f, g \in (\mathcal{S})^N \Rightarrow f \diamond g \in (\mathcal{S})$.

Proof We may assume $N = 1$.

a) Take $F = \sum_\alpha a_\alpha H_\alpha, G = \sum_\beta b_\beta H_\beta \in (\mathcal{S})_{-1}$. This means that there exist q_1 such that

$$\sum_\alpha a_\alpha^2 (2\mathbb{N})^{-q_1 \alpha} < \infty \quad \text{and} \quad \sum_\beta b_\beta^2 (2\mathbb{N})^{-q_1 \beta} < \infty. \tag{2.4.5}$$

We note that $F \diamond G = \sum_{\alpha,\beta} a_\alpha b_\beta H_{\alpha+\beta} = \sum_\gamma (\sum_{\alpha+\beta=\gamma} a_\alpha b_\beta) H_\gamma$ and then set $c_\gamma = \sum_{\alpha+\beta=\gamma} a_\alpha b_\beta$. With $q = q_1 + k$ we have

$$\sum_\gamma (2\mathbb{N})^{-q\gamma} c_\gamma^2 = \sum_\gamma (2\mathbb{N})^{-k\gamma} (2\mathbb{N})^{-q_1\gamma} \left(\sum_{\alpha+\beta=\gamma} a_\alpha b_\beta \right)^2$$

$$\leq \sum_\gamma (2\mathbb{N})^{-k\gamma} (2\mathbb{N})^{-q_1\gamma} \left(\sum_{\alpha+\beta=\gamma} a_\alpha^2 \right) \left(\sum_{\alpha+\beta=\gamma} b_\beta^2 \right)$$

$$= \sum_\gamma (2\mathbb{N})^{-k\gamma} \left(\sum_{\alpha+\beta=\gamma} a_\alpha^2 (2\mathbb{N})^{-q_1\alpha} \right) \left(\sum_{\alpha+\beta=\gamma} b_\beta^2 (2\mathbb{N})^{-q_1\beta} \right)$$

$$\leq \left(\sum_\gamma (2\mathbb{N})^{-k\gamma} \right) \left(\sum_\alpha a_\alpha^2 (2\mathbb{N})^{-q_1\alpha} \right) \left(\sum_\beta b_\beta^2 (2\mathbb{N})^{-q_1\beta} \right) < \infty$$

$$\tag{2.4.6}$$

for $k > 1$, by Proposition 2.3.3. The proofs of **b)**, **c)** and **d)** are similar. $\square$

The following basic algebraic properties of the Wick product follow directly from the definition.

Lemma 2.4.5.

a) (*Commutative law*) $F, G \in (\mathcal{S})_{-1}^{m;N} \Rightarrow F \diamond G = G \diamond F.$
b) (*Associative law*)

$$F, G, H \in (\mathcal{S})_{-1}^{m;1} \Rightarrow F \diamond (G \diamond H) = (F \diamond G) \diamond H.$$

c) (*Distributive law*)

$$F, A, B \in (\mathcal{S})_{-1}^{m;N} \Rightarrow F \diamond (A + B) = F \diamond A + F \diamond B.$$

The *Wick powers* $F^{\diamond k}$; $k = 0, 1, 2, \ldots$ of $F \in (\mathcal{S})_{-1}$ are defined inductively as follows:

$$\begin{cases} F^{\diamond 0} = 1 \\ F^{\diamond k} = F \diamond F^{\diamond(k-1)} \quad \text{for } k = 1, 2, \ldots. \end{cases} \tag{2.4.7}$$

More generally, if $p(x) = \sum_{n=0}^{N} a_n x^n$; $a_n \in \mathbb{R}$, $x \in \mathbb{R}$, is a polynomial, then we define its *Wick version*

$$p^{\diamond} : (\mathcal{S})_{-1} \to (\mathcal{S})_{-1}$$

by

$$p^{\diamond}(F) = \sum_{n=0}^{N} a_n F^{\diamond n} \quad \text{for} \quad F \in (\mathcal{S})_{-1}. \tag{2.4.8}$$

Later we will extend this construction to more general functions than polynomials (see Section 2.6).

2.4.1 Some Examples and Counterexamples

For simplicity, we will assume that we have $N = m = d = 1$ in this paragraph. If $F, G \in L^p(\mu)$ for $p > 1$, then it also makes sense to consider the ordinary (pointwise) product

$$(F \cdot G)(\omega) = F(\omega) \cdot G(\omega).$$

How does this product compare to the Wick product $(F \diamond G)(\omega)$? This is a difficult question in general. Let us first consider some illustrating examples:

Example 2.4.6. Suppose at least one of F and G is deterministic, e.g., that $F = a_0 \in \mathbb{R}$. Then

$$F \diamond G = F \cdot G.$$

Hence the Wick product coincides with the ordinary product in the deterministic case. In particular, if $F = 0$, then $F \diamond G = 0$.

Example 2.4.7.   Suppose $F, G \in L^2(\mu)$ are both Gaussian, i.e.,

$$F(\omega) = a_0 + \sum_{k=1}^{\infty} a_k H_{\epsilon^{(k)}}(\omega), \; G(\omega) = b_0 + \sum_{l=1}^{\infty} b_l H_{\epsilon_l}(\omega), \qquad (2.4.9)$$

where

$$\sum_{k=1}^{\infty} a_k^2 < \infty, \; \sum_{l=1}^{\infty} b_l^2 < \infty.$$

Then we have

$$(F \diamond G)(\omega) = a_0 b_0 + \sum_{k,l=1}^{\infty} a_k b_l H_{\epsilon^{(k)} + \epsilon_l}(\omega).$$

Now

$$h_{\epsilon^{(k)} + \epsilon_l} = \begin{cases} h_{\epsilon^{(k)}} h_{\epsilon_l} & \text{for } k \neq l \\ h_{\epsilon^{(k)}}^2 - 1 & \text{for } k = l. \end{cases}$$

Hence

$$(F \diamond G)(\omega) = F(\omega) \cdot G(\omega) - \sum_{k=1}^{\infty} a_k b_k. \qquad (2.4.10)$$

This result can be restated in terms of Itô integrals, as follows: We may write

$$F(\omega) = a_0 + \int_{\mathbb{R}} f(t) dB(t), \qquad (2.4.11)$$

where $f(t) = \sum_{k=1}^{\infty} a_k \xi_k(t) \in L^2(\mathbb{R})$, and, similarly,

$$G(\omega) = b_0 + \int_{\mathbb{R}} g(t) dB(t), \qquad (2.4.12)$$

with $g(t) = \sum_{k=1}^{\infty} b_k \xi_k(t) \in L^2(\mathbb{R})$. Then (2.4.10) states that

$$\left(\int_{\mathbb{R}} f(t) dB(t) \right) \diamond \left(\int_{\mathbb{R}} g(t) dB(t) \right)$$

$$= \left(\int_{\mathbb{R}} f(t) dB(t) \right) \cdot \left(\int_{\mathbb{R}} g(t) dB(t) \right) - \int_{\mathbb{R}} f(t) g(t) dt. \qquad (2.4.13)$$

In particular, choosing $f = g = \chi_{[0,t]}$ we obtain

$$B(t)^{\diamond 2} = B(t)^2 - t. \qquad (2.4.14)$$

Note, in particular, that $B(t)^{\diamond 2}$ is *not* positive (but see Example 2.6.15). Similarly, for the smoothed white noise we obtain

$$w(\phi) \diamond w(\psi) = w(\phi) \cdot w(\psi) - (\phi, \psi) \qquad (2.4.15)$$

for $\phi, \psi \in L^2(\mathbb{R}^d)$ with $(\phi, \psi) = \int_{\mathbb{R}^d} \phi(x)\psi(x)dx$. (See Exercise 2.9.)

Note that if $\psi = \phi$ and $\|\phi\| = 1$, this can be written

$$w(\phi)^{\diamond 2} = h_2(w(\phi)); \quad \|\phi\| = 1. \qquad (2.4.16)$$

This suggests the general formula

$$w(\phi)^{\diamond n} = h_n(w(\phi)); \quad \|\phi\| = 1. \qquad (2.4.17)$$

To prove (2.4.17) we use that the Wick product is independent of the choice of basis elements of $L^2(\mathbb{R}^d)$. (See Appendix D.) In this case, where $w(\phi)$ and its Wick powers all belong to $L^2(\mu)$, the basis independence follows from Proposition 2.4.2. Therefore, we may assume that $\phi = \eta_1$, and then

$$w(\phi)^{\diamond n} = h_1(\langle \omega, \eta_1 \rangle)^{\diamond n} = H_{\epsilon_1}^{\diamond n}(\omega) = H_{n\epsilon_1}(\omega)$$
$$= h_n(\langle \omega, \eta_1 \rangle) = h_n(w(\phi)).$$

Example 2.4.8 Gjessing (1993).   The $L^p(\mu)$ spaces are not closed under Wick products.

For example, choose $\phi \in \mathcal{S}(\mathbb{R}^d)$ with $\|\phi\|_{L^2} = 1$, put $\theta(\omega) = \langle \omega, \phi \rangle$ and define

$$X(\omega) = \begin{cases} 1 & \text{if } \langle \omega, \phi \rangle \geq 0 \\ 0 & \text{if } \langle \omega, \phi \rangle < 0. \end{cases}$$

Then

$$X \in L^\infty(\mu) \quad \text{but} \quad X^{\diamond 2} \notin L^2(\mu).$$

Example 2.4.9 Gjessing (1993). Independence of X and Y is not enough to ensure that

$$X \diamond Y = X \cdot Y.$$

To see this, let X, θ be as in the previous example. Then $Y = \theta^{\diamond 2} = \theta^2 - 1$ is independent of X, but $X \diamond Y$ and $X \cdot Y$ are not equal. In fact, they do not even have the same second moments. See, however, Propositions 8.2 and 8.3 in Benth and Potthoff (1996).

Example 2.4.10 Gjessing (1993). The Wick product is not local, i.e., the value of $(X \diamond Y)(\omega_0)$ is not (in general) determined by the values of X and Y in a neighborhood V of ω_0 in $\mathcal{S}'(\mathbb{R}^d)$.

2.5 Wick Multiplication and Hitsuda/Skorohod Integration

In this section we put $N = m = d = 1$ for simplicity. One of the most striking features of the Wick product is its relation to Hitsuda/Skorohod integration. In short, this relation can be expressed as

$$\int_{\mathbb{R}} Y(t)\delta B(t) = \int_{\mathbb{R}} Y(t) \diamond W(t)dt. \qquad (2.5.1)$$

Here the left-hand side denotes the Hitsuda/Skorohod integral of the stochastic process $Y(t) = Y(t, \omega)$ (which coincides with the Itô integral if $Y(t)$ is adapted; see Appendix B), while the right-hand side is to be interpreted as an $(\mathcal{S})^*$-valued (Pettis) integral. Strictly speaking the right-hand side of (2.5.1) represents a generalization of the Hitsuda/Skorohod integral. For simplicity we will call this generalization the Skorohod integral.

The relation (2.5.1) explains why the Wick product is so natural and important in stochastic calculus. It is also the key to the fact that Itô calculus (with Itô's formula, etc.) with ordinary multiplication is equivalent to ordinary calculus with Wick multiplication. To illustrate the content of this statement, consider the example with $Y(t) = B(t) \cdot \chi_{[0,T]}(t)$ in (2.5.1): Then the left hand side becomes, by Itô's formula,

$$\int_0^T B(t)dB(t) = \frac{1}{2}B^2(T) - \frac{1}{2}T \quad \text{(assuming } B(0) = 0\text{)}, \qquad (2.5.2)$$

while (formal) Wick calculation makes the right hand side equal to

$$\int_0^T B(t) \diamond W(t)dt = \int_0^T B(t) \diamond B'(t)dt = \frac{1}{2}B(T)^{\diamond 2}, \qquad (2.5.3)$$

which is equal to (2.5.2) by virtue of (2.4.14).

This computation will be made rigorous later (Example 2.5.11), and we will illustrate applications of this principle in Chapters 3 and 4.

Various versions of (2.5.1) have been proved by several authors. A version involving the operator ∂_t^* is proved in Hida et al. (1993), see Theorem 8.7 and subsequent sections. In Lindstrøm et al. (1992), a formula of the type (2.5.1) is proved, but under stronger conditions than necessary. In Benth (1993), the result was extended to be valid under the sole condition that the left hand side exists. The proof we present here is based on Benth (1993). First we recall the definition of the Skorohod integral:

Let $Y(t) = Y(t, \omega)$ be a stochastic process such that

$$E[Y(t)^2] < \infty \quad \text{for all } t. \tag{2.5.4}$$

Then, by Theorem 2.2.7, $Y(t)$ has a Wiener–Itô chaos expansion

$$Y(t) = \sum_{n=0}^{\infty} \int_{\mathbb{R}^n} f_n(s_1, \ldots, s_n, t) dB^{\otimes n}(s_1, \ldots, s_n), \tag{2.5.5}$$

where $f_n(\cdot, t) \in \hat{L}^2(\mathbb{R}^n)$ for $n = 0, 1, 2, \ldots$ and for each t. Let

$$\hat{f}_n(s_1, \ldots, s_n, s_{n+1})$$

be the symmetrization of $f_n(s_1, \ldots, s_{n+1})$ wrt the $n+1$ variables $s_1, \ldots, s_n$, s_{n+1}.

Definition 2.5.1. Assume that

$$\sum_{n=0}^{\infty} (n+1)! \|\hat{f}_n\|^2_{L^2(\mathbb{R}^{n+1})} < \infty. \tag{2.5.6}$$

Then we define *the Skorohod integral* of $Y(t)$, denoted by

$$\int_{\mathbb{R}} Y(t) \delta B(t),$$

by

$$\int_{\mathbb{R}} Y(t) \delta B(t) = \sum_{n=0}^{\infty} \int_{\mathbb{R}^{n+1}} \hat{f}_n(s_1, \ldots, s_{n+1}) dB^{\otimes(n+1)}(s_1, \ldots, s_{n+1}). \tag{2.5.7}$$

By (2.5.6) and (2.5.7) the Skorohod integral belongs to $L^2(\mu)$ and

$$\left\| \int_{\mathbb{R}} Y(t) \delta B(t) \right\|^2_{L^2(\mu)} = \sum_{n=0}^{\infty} (n+1)! \|\hat{f}_n\|^2_{L^2(\mathbb{R}^{n+1})}. \tag{2.5.8}$$

Note that we do *not* require that the process be adapted. In fact, the Skorohod integral may be regarded as an extension of the Itô integral to non-adapted (anticipating) integrands. This was proved in Nualart and Zakai (1986). See also Theorem 8.5 in Hida et al. (1993), and the references there. For completeness we include a proof here.

First we need a result (of independent interest) about how to characterize adaptedness of a process in terms of the coefficients of its chaos expansion.

Lemma 2.5.2. *Suppose $Y(t)$ is a stochastic process with $E[Y^2(t)] < \infty$ for all t and with the multiple Itô integral expansion*

$$Y(t) = \sum_{n=0}^{\infty} \int_{\mathbb{R}^n} f_n(x,t) dB^{\otimes n}(x) \quad \text{with} \quad f_n(\cdot,t) \in \hat{L}^2(\mathbb{R}^n) \quad \text{for all} \quad n.$$

(2.5.9)

Then $Y(t)$ is $\mathcal{F}_t$-adapted if and only if

$$\text{supp} f_n(\cdot,t) \subset \{x \in \mathbb{R}_+^n; \ x_i \le t \quad \text{for} \quad i = 1,2,\ldots,n\}, \tag{2.5.10}$$

for all n.

Here support is interpreted as essential support with respect to Lebesgue measure:

$$\text{supp } f_n(\cdot,t) = \bigcap \{F; \ F \text{ closed}, \ f_n(x,t) = 0 \quad \text{for a.e.} \quad x \notin F\}.$$

Proof We first observe that for all n and all $f \in \hat{L}^2(\mathbb{R}^n)$, we have

$$E\left[\int_{\mathbb{R}^n} f(x) dB^{\otimes n}(x) \Big| \mathcal{F}_t\right]$$

$$= E\left[n! \int_{-\infty}^{\infty} \int_{-\infty}^{t_n} \int_{-\infty}^{t_2} f(t_1,\ldots,t_n) dB(t_1) dB(t_n) \Big| \mathcal{F}_t\right]$$

$$= n! \int_0^t \int_0^{t_n} \int_0^{t_2} f(t_1,\ldots,t_n) dB(t_1) dB(t_n) = \int_{\mathbb{R}^n} f(x) \chi_{[0,t]^n}(x) dB^{\otimes n}(x).$$

Therefore we get

$Y(t)$ is $\mathcal{F}_t$-adapted

$$\Leftrightarrow E\left[Y(t) | \mathcal{F}_t\right] = Y_t \quad \text{for all } t$$

$$\Leftrightarrow \sum_{n=0}^{\infty} E\left[\int_{\mathbb{R}^n} f_n(x,t) dB^{\otimes n}(x) | \mathcal{F}_t\right] = \sum_{n=0}^{\infty} \int_{\mathbb{R}^n} f_n(x,t) dB^{\otimes n}(x)$$

$$\Leftrightarrow \sum_{n=0}^{\infty} \int_{\mathbb{R}^n} f_n(x,t) \chi_{[0,t]^n}(x) dB^{\otimes n}(x) = \sum_{n=0}^{\infty} \int_{\mathbb{R}^n} f_n(x,t) dB^{\otimes n}(x)$$

$$\Leftrightarrow f_n(x,t) \chi_{[0,t]^n}(x) = f_n(x,t) \quad \text{for all } t \text{ and almost all } x,$$

by the uniqueness of the expansion. □

The corresponding characterization for Hermite chaos expansions is

Lemma 2.5.3. *Suppose $Y(t)$ is a stochastic process with $E[Y^2(t)] < \infty$ for all t and with the Hermite chaos expansion*

$$Y(t) = \sum_{\alpha} c_{\alpha}(t) H_{\alpha}(\omega). \qquad (2.5.11)$$

Then $Y(t)$ is $\mathcal{F}_t$-adapted if and only if

$$supp\left\{ \sum_{|\alpha|=n} c_{\alpha}(t)\xi^{\hat{\otimes}\alpha}(x) \right\} \subset \{x \in \mathbb{R}^n; \ x_i \le t \quad for \quad i = 1,\dots,n\} \qquad (2.5.12)$$

for all n.

Proof   This follows from Lemma 2.5.2 and (2.2.33). $\square$

Proposition 2.5.4. *Suppose $Y(t)$ is an $\mathcal{F}_t$-adapted stochastic process such that*

$$\int_{\mathbb{R}} E[Y^2(t)]dt < \infty.$$

Then $Y(t)$ is both Skorohod-integrable and Itô integrable, and the two integrals coincide:

$$\int_{\mathbb{R}} Y(t)\delta B(t) = \int_{\mathbb{R}} Y(t)dB(t). \qquad (2.5.13)$$

Proof   Suppose $Y(t)$ has the expansion

$$Y(t) = \sum_{n=0}^{\infty} \int_{\mathbb{R}^n} f_n(x,t)dB^{\otimes n}(x); \ f_n(\cdot,t) \in \hat{L}^2(\mathbb{R}^n) \quad \text{for all} \quad n.$$

Since $Y(t)$ is adapted we know that $f_n(x_1, x_2, \dots, x_n, t) = 0$ if $\max_{1 \le i \le n}\{x_i\} > t$, a.e.

Therefore, the symmetrization $\hat{f}_n(x_1, \dots, x_n, t)$ of $f_n(x_1, \dots, x_n, t)$ satisfies (with $x_{n+1} = t$)

$$\hat{f}_n(x_1, \dots, x_n, x_{n+1}) = \frac{1}{n+1} f(y_1, \dots, y_n, \max_{1 \le i \le n+1}\{x_i\}),$$

where $(y_1, \dots, y_n)$ is an arbitrary permutation of the remaining x_j when the maximum value $x_{\hat{j}} := \max_{1 \le i \le n+1}\{x_i\}$ is removed. This maximum is obtained for a unique $\hat{j}$, for almost all $x \in \mathbb{R}^{n+1}$ with respect to Lebesgue measure.

Hence the Itô integral of $Y(t)$ is

$$\int_{\mathbb{R}} Y(t)dB(t)$$

$$= \sum_{n=0}^{\infty} \int_{\mathbb{R}} \left(\int_{\mathbb{R}^n} f_n(x_1, \ldots, x_n, t)dB^{\otimes n}(x) \right) dB(t)$$

$$= \sum_{n=0}^{\infty} n! \int_{\mathbb{R}} \left(\int_{-\infty}^{t} \int_{-\infty}^{x_n} \cdots \int_{-\infty}^{x_2} f_n(x_1, \ldots, x_n, t)dB(x_1) \cdots dB(x_n) \right) dB(t)$$

$$= \sum_{n=0}^{\infty} n!(n+1) \int_{-\infty}^{\infty} \int_{-\infty}^{x_{n+1}} \int_{-\infty}^{x_n} \cdots \int_{-\infty}^{x_2} \hat{f}_n(x_1, \ldots, x_n, x_{n+1})dB(x_1) \cdot dB(x_n)dB(x_{n+1})$$

$$= \sum_{n=0}^{\infty} \int_{\mathbb{R}^{n+1}} \hat{f}_n(x_1, \ldots, x_n, x_{n+1})dB^{\otimes(n+1)}(x_1, \ldots, x_n, x_{n+1})$$

$$= \int_{\mathbb{R}} Y(t)\delta B(t),$$

as claimed.   □

We now proceed to consider integrals with values in $(\mathcal{S})^*$.

Definition 2.5.5. A function $Z(t) : \mathbb{R} \to (\mathcal{S})^*$ (also called an $(\mathcal{S})^*$-valued process) is called $(\mathcal{S})^*$-*integrable* if

$$\langle Z(t), f \rangle \in L^1(\mathbb{R}, dt) \quad \text{for all} \quad f \in (\mathcal{S}). \tag{2.5.14}$$

Then the $(\mathcal{S})^*$-integral of $Z(t)$, denoted by $\int_{\mathbb{R}} Z(t)dt$, is the (unique) $(\mathcal{S})^*$-element such that

$$\left\langle \int_{\mathbb{R}} Z(t)dt, f \right\rangle = \int_{\mathbb{R}} \langle Z(t), f \rangle dt; \ f \in (\mathcal{S}). \tag{2.5.15}$$

Remark   It is a consequence of Proposition 8.1 in Hida et al. (1993) that (2.5.15) defines $\int_{\mathbb{R}} Z(t)dt$ as an element in $(\mathcal{S})^*$.

Lemma 2.5.6.   *Assume that $Z(t) \in (\mathcal{S})^*$ has the chaos expansion*

$$Z(t) = \sum_{\alpha} c_\alpha(t) H_\alpha, \tag{2.5.16}$$

where

$$\sum_{\alpha} \alpha! \|c_\alpha\|_{L^1(\mathbb{R})}^2 (2\mathbb{N})^{-p\alpha} < \infty \quad \text{for some} \quad p < \infty. \tag{2.5.17}$$

Then $Z(t)$ is $(\mathcal{S})^*$-integrable and

$$\int_{\mathbb{R}} Z(t)dt = \sum_\alpha \int_{\mathbb{R}} c_\alpha(t)dt H_\alpha. \qquad (2.5.18)$$

Proof   Let $f = \sum_\alpha a_\alpha H_\alpha \in (\mathcal{S})$. Then by (2.5.17)

$$\int_{\mathbb{R}} |\langle Z(t), f\rangle| dt = \int_{\mathbb{R}} \left| \sum_\alpha \alpha! a_\alpha c_\alpha(t) \right| dt \leq \sum_\alpha \alpha! |a_\alpha| \|c_\alpha\|_{L^1(\mathbb{R})}$$

$$= \sum_\alpha \sqrt{\alpha!} |a_\alpha| (2\mathbb{N})^{\frac{\alpha p}{2}} \sqrt{\alpha!} \|c_\alpha\|_{L^1(\mathbb{R})} (2\mathbb{N})^{-\frac{\alpha p}{2}}$$

$$\leq \left(\sum_\alpha \alpha! a_\alpha^2 (2\mathbb{N})^{\alpha p} \right)^{\frac{1}{2}} \left(\sum_\alpha \alpha! \|c_\alpha\|_{L^1(\mathbb{R})}^2 (2\mathbb{N})^{-\alpha p} \right)^{\frac{1}{2}} < \infty.$$

Hence $Z(t)$ is $(\mathcal{S})^*$-integrable and

$$\left\langle \int_{\mathbb{R}} Z(t)dt, f \right\rangle = \int_{\mathbb{R}} \langle Z(t), f\rangle dt = \int_{\mathbb{R}} \sum_\alpha \alpha! a_\alpha c_\alpha(t) dt$$

$$= \sum_\alpha \alpha! a_\alpha \int_{\mathbb{R}} c_\alpha(t) dt = \left\langle \sum_\alpha \int_{\mathbb{R}} c_\alpha(t)dt H_\alpha, f \right\rangle,$$

which proves (2.5.18). $\square$

Lemma 2.5.7.   *Suppose*

$$Y(t) = \sum_\alpha c_\alpha(t) H_\alpha \in (\mathcal{S})^* \quad \text{for all } t \in \mathbb{R},$$

and that there exists $q < \infty$ such that

$$K := \sup_\alpha \{\alpha! \|c_\alpha\|_{L^1(\mathbb{R})}^2 (2\mathbb{N})^{-q\alpha}\} < \infty. \qquad (2.5.19)$$

Choose $\phi \in \mathcal{S}(\mathbb{R})$. Then $Y(t) \diamond W(t)$ and $Y(t) \diamond W_\phi(t)$ (with $W_\phi(t)$ as in (2.1.15)) are both $(\mathcal{S})^*$-integrable and

$$\int_{\mathbb{R}} Y(t) \diamond W(t)dt = \sum_{\alpha,k} \int_{\mathbb{R}} c_\alpha(t)\xi_k(t)dt H_{\alpha+\epsilon^{(k)}} \qquad (2.5.20)$$

and

$$\int_{\mathbb{R}} Y(t) \diamond W_\phi(t)dt = \sum_{\alpha,k} \int_{\mathbb{R}} c_\alpha(t)(\phi_t(\cdot), \xi_k)dt H_{\alpha+\epsilon^{(k)}}. \qquad (2.5.21)$$

Proof We prove (2.5.20), the proof of (2.5.21) being similar. Since

$$Y(t) \diamond W(t) = \sum_{\alpha,k} c_\alpha(t)\xi_k(t)H_{\alpha+\epsilon^{(k)}} = \sum_\beta \sum_{\substack{\alpha,k \\ \alpha+\epsilon^{(k)}=\beta}} c_\alpha(t)\xi_k(t)H_\beta,$$

the result follows from Lemma 2.5.6 if we can verify that

$$M(p) := \sum_\beta \beta! \left\| \sum_{\substack{\alpha,k \\ \alpha+\epsilon^{(k)}=\beta}} c_\alpha(t)\xi_k(t) \right\|_{L^1(\mathbb{R})}^2 (2\mathbb{N})^{-p\beta} < \infty$$

for some $p < \infty$.

By (2.2.5) we have, for some constant $C < \infty$,

$$\int_\mathbb{R} |c_\alpha(t)||\xi_k(t)|dt \le C \int_\mathbb{R} |c_\alpha(t)|dt = C\|c_\alpha\|_{L^1(\mathbb{R})}.$$

Note that for each β, α there is at most one k such that $\alpha + \epsilon^{(k)} = \beta$. Therefore

$$\left\| \sum_{\substack{\alpha,k \\ \alpha+\epsilon^{(k)}=\beta}} c_\alpha(t)\xi_k(t) \right\|_{L^1(\mathbb{R})}^2 \le \left[\sum_{\substack{\alpha,k \\ \alpha+\epsilon^{(k)}=\beta}} \|c_\alpha\xi_k\|_{L^1(\mathbb{R})} \right]^2$$

$$\le C^2 \left[\sum_{\substack{\alpha,k \\ \alpha+\epsilon^{(k)}=\beta}} \|c_\alpha\|_{L^1(\mathbb{R})} \right]^2$$

$$\le C^2 (l(\beta))^2 \sum_{\substack{\alpha \\ \exists k, \alpha+\epsilon^{(k)}=\beta}} \|c_\alpha\|_{L^1(\mathbb{R})}^2,$$

where $l(\beta)$ is the number of nonzero elements of β, i.e., $l(\beta)$ is the length of β. We conclude that

$$M(2q) \le C^2 \sum_{\alpha,k} (\alpha+\epsilon^{(k)})!(l(\alpha+\epsilon^{(k)}))^2 \|c_\alpha\|_{L^1(\mathbb{R})}^2 (2\mathbb{N})^{-2q(\alpha+\epsilon^{(k)})}$$

$$\le C^2 K \sum_{\alpha,k} \frac{(\alpha+\epsilon^{(k)})!}{\alpha!}(l(\alpha+\epsilon^{(k)}))^2 (2\mathbb{N})^{-q\alpha}(2\mathbb{N})^{-2q\epsilon^{(k)}}$$

$$\le C^2 K \sum_{\alpha,k} (|\alpha|+1)^3 2^{-|\alpha|q} k^{-2q} < \infty \quad \text{for} \quad q > \frac{1}{2}. \qquad \square$$

Corollary 2.5.8.   *Let $Y(t) = \sum_\alpha c_\alpha(t)H_\alpha$ be a stochastic process such that $\int_a^b E[Y_t^2]dt < \infty$ for some $a, b \in \mathbb{R}, a < b$. Then $Y(t) \diamond W(t)$ is $(\mathcal{S})^*$-integrable over $[a, b]$ and*

$$\int_a^b Y(t) \diamond W(t) dt = \sum_{\alpha,k} \int_a^b c_\alpha(t) \xi_k(t) dt H_{\alpha+\epsilon^{(k)}}. \qquad (2.5.22)$$

Proof We have

$$\sum_\alpha \alpha! \int_a^b c_\alpha^2(t) dt = \int_a^b E[Y(t)^2] dt < \infty,$$

hence (2.5.19) holds, so by Lemma 2.5.7 the corollary follows.

We are now ready to prove the main result of this section.   □

Theorem 2.5.9. *Assume that* $Y(t) = \sum_\alpha c_\alpha(t) H_\alpha$ *is a Skorohod-integrable stochastic process. Let* $a, b \in \mathbb{R}$, $a < b$. *Then* $Y(t) \diamond W(t)$ *is* $(\mathcal{S})^*$*-integrable over* $[a, b]$ *and we have*

$$\int_a^b Y(t) \delta B(t) = \int_a^b Y(t) \diamond W(t) dt. \qquad (2.5.23)$$

Proof By the preceding corollary and by replacing $c_\alpha(t)$ by $c_\alpha(t)\chi_{[a,b]}(t)$, we see that it suffices to verify that

$$\int_\mathbb{R} Y(t) \delta B(t) = \sum_{\alpha,k} (c_\alpha, \xi_k) H_{\alpha+\epsilon^{(k)}}, \qquad (2.5.24)$$

where $(c_\alpha, \xi_k) = \int_\mathbb{R} c_\alpha(t)\xi_k(t) dt$. This will be done by computing the left-hand side explicitly: Let $Y(t) = \sum_{n=0}^\infty \int_{\mathbb{R}^n} f_n(u_1, \ldots, u_n, t) dB^{\otimes n}(u_1, \ldots, u_n)$.
Then by (2.2.33) we have

$$Y(t) = \sum_{n=0}^\infty \int_{\mathbb{R}^n} \sum_{|\alpha|=n} c_\alpha(t) \xi^{\hat{\otimes}\alpha}(u_1, \ldots, u_n) dB^{\otimes n}(u_1, \ldots, u_n)$$

$$= \sum_{n=0}^\infty \int_{\mathbb{R}^n} \sum_{|\alpha|=n} \sum_{k=1}^\infty (c_\alpha, \xi_k) \xi_k(t) \xi^{\hat{\otimes}\alpha}(u_1, \ldots, u_n) dB^{\otimes n}(u_1, \ldots, u_n).$$

Now the symmetrization of

$$\xi_k(u_0) \xi^{\hat{\otimes}\alpha}(u_1, \ldots, u_n) = \xi_k(u_0)(\xi_1^{\otimes\alpha_1} \hat{\otimes} \cdots \hat{\otimes} \xi_j^{\otimes\alpha_j})(u_1, \ldots, u_n),$$

where $\alpha = (\alpha_1, \ldots, \alpha_j)$, as a function of $u_0, \ldots, u_n$ is simply

$$\xi^{\hat{\otimes}(\alpha+\epsilon^{(k)})} = \xi_1^{\otimes\alpha_1} \hat{\otimes} \cdots \hat{\otimes} \xi_k^{\otimes(\alpha_k+1)} \hat{\otimes} \cdots \hat{\otimes} \xi_j^{\otimes\alpha_j}. \qquad (2.5.25)$$

Therefore the Skorohod integral of $Y(s)$ becomes, by (2.5.7) and (2.2.30),

$$\int_{\mathbb{R}} Y(t)\delta B(t) = \sum_{n=0}^{\infty} \int_{\mathbb{R}^{n+1}} \sum_{|\alpha|=n} \sum_{k=1}^{\infty} (c_\alpha, \xi_k) \xi^{\hat{\otimes}(\alpha+\epsilon^{(k)})} dB^{\otimes(n+1)}$$

$$= \sum_{n=0}^{\infty} \sum_{|\alpha|=n} \sum_{k=1}^{\infty} (c_\alpha, \xi_k) H_{\alpha+\epsilon^{(k)}} = \sum_{\alpha,k} (c_\alpha, \xi_k) H_{\alpha+\epsilon^{(k)}},$$

as claimed. $\qquad \square$

To illustrate the contents of these results, we consider some simple examples.

Example 2.5.10.   It is immediate from the definition that

$$\int_0^t 1\delta B(s) = B(t)$$

(assuming, as before, $B(0) = 0$), so from Theorem 2.5.9 we have

$$B(t) = \int_0^t W(s)ds. \tag{2.5.26}$$

In other words, we have proved that as elements of $(\mathcal{S})^*$ we have

$$\frac{dB(t)}{dt} = W(t), \tag{2.5.27}$$

where differentiation is in $(\mathcal{S})^*$ (compare with (2.1.17)).

More generally, if we choose $Y(t)$ to be deterministic, $Y(t,\omega) = \psi(t) \in L^2(\mathbb{R})$, then by Theorem 2.5.9 and Proposition 2.5.4

$$\int_{\mathbb{R}} \psi(t)dB(t) = \int_{\mathbb{R}} \psi(t)\delta B(t) = \int_{\mathbb{R}} \psi(t) \diamond W(t)dt. \tag{2.5.28}$$

Example 2.5.11.   Let us apply Theorem 2.5.9 and Corollary 2.5.8 to compute the Skorohod integral

$$\int_0^t B(s)\delta B(s) = \int_0^t B(s) \diamond W(s)ds.$$

From Example 2.2.5 we know that

$$B(s) = \sum_{j=1}^{\infty} \int_0^s \xi_j(r)dr H_{\epsilon^{(j)}}(\omega),$$

which substituted in (2.5.22) gives

$$\int_0^t B(s)\delta B(s) = \sum_{j,k} \int_0^t \int_0^s \xi_j(r)dr\xi_k(s)ds H_{\epsilon^{(j)}+\epsilon^{(k)}}.$$

Integration by parts gives

$$\int_0^t \int_0^s \xi_j(r)dr\xi_k(s)ds = \int_0^t \xi_j(r)dr \int_0^t \xi_k(s)ds - \int_0^t \int_0^s \xi_k(s)ds\xi_j(r)dr.$$

Hence, by the symmetry of j and k,

$$\int_0^t B(s)\delta B(s) = \frac{1}{2}\sum_{j,k} \int_0^t \xi_j(r)dr \int_0^t \xi_k(s)ds H_{\epsilon^{(j)}+\epsilon^{(k)}}.$$

By the Wick product definition (2.4.2) this is equal to $\frac{1}{2}B(t) \diamond B(t)$. Hence we obtain, using (2.4.14), the familiar formula

$$\int_0^t B(s)dB(s) = \int_0^t B(s)\delta B(s) = \frac{1}{2}B(t)^{\diamond 2} = \frac{1}{2}B^2(t) - \frac{1}{2}t.$$

We can more easily obtain this formula if we use (2.5.27) and work in $(\mathcal{S})^*$. Then

$$\int_0^t B(s)\delta B(s) = \int_0^t B(s) \diamond W(s)ds = \int_0^t B(s) \diamond B'(s)ds$$

$$= \left| \frac{1}{2}B(s)^{\diamond 2} \right|_0^t = \frac{1}{2}B(t)^{\diamond 2} = \frac{1}{2}B^2(t) - \frac{1}{2}t.$$

Corollary 2.5.12. *Suppose that $Y(t) = Y(t,\omega)$ is Skorohod-integrable, that $h(\omega) \in (\mathcal{S})^*$ does not depend on t and that $h \diamond Y(t)$ is Skorohod-integrable. Then for $a < b$ we have*

$$\int_a^b h \diamond Y(t)\delta B(t) = h \diamond \int_a^b Y(t)\delta B(t). \qquad (2.5.29)$$

Proof By Theorem 2.5.9 we have

$$\int_a^b h \diamond Y(t)\delta B(t) = \int_a^b h \diamond Y(t) \diamond W(t)dt = h \diamond \int_a^b Y(t)\delta B(t).$$

$\square$

Example 2.5.13.   Choose $Y(t) = \chi_{[a,b]}(t)h(\omega)$, with $h \in L^2(\mu_1), a < b$. Then the Skorohod integral becomes

$$\int_a^b h(\omega)\delta B_s(\omega) = \int_a^b h(\omega) \diamond W(s)ds = h(\omega) \diamond (B(b) - B(a)). \qquad (2.5.30)$$

Finally, we state and prove a smoothed version of Theorem 2.5.9.

Theorem 2.5.14.   *Let $Y(t) = Y(t, \omega)$ be a stochastic process such that*

$$\int_{\mathbb{R}} E[Y^2(t)]dt < \infty. \qquad (2.5.31)$$

Choose $\phi \in \mathcal{S}(\mathbb{R})$ and let

$$(\phi * Y)(t, \omega) = \int_{\mathbb{R}} \phi(t - s)Y(s, \omega)ds \qquad (2.5.32)$$

be the convolution of ϕ and $Y(\cdot, \omega)$, for almost all ω. Suppose $(\phi * Y)(t)$ is Skorohod-integrable. Then $Y(t) \diamond W_\phi(t)$ is $(\mathcal{S})^*$-integrable, and we have

$$\int_{\mathbb{R}} (\phi * Y)(t)\delta B(t) = \int_{\mathbb{R}} Y(t) \diamond W_\phi(t)dt. \qquad (2.5.33)$$

Proof Suppose $Y(s)$ has the expansion

$$Y(s) = \sum_\alpha c_\alpha(s)H_\alpha.$$

Then by (2.5.31) and Lemma 2.5.7 $Y(t) \diamond W_\phi(t)$ is $(\mathcal{S})^*$-integrable. Applying Theorem 2.5.9 with $Y(t)$ replaced by $(\phi * Y)(t)$, we get, by (2.1.18),

$$\int_{\mathbb{R}} (\phi * Y)(t)\delta B(t) = \int_{\mathbb{R}} (\phi * Y)(t) \diamond W(t)dt$$

$$= \int_{\mathbb{R}} \left(\int_{\mathbb{R}} \phi(t - s)Y(s)ds \right) \diamond W(t)dt$$

$$= \int_{\mathbb{R}} Y(s) \diamond \int_{\mathbb{R}} \phi(t - s)W(t)dtds = \int_{\mathbb{R}} Y(s) \diamond W_\phi(s)ds.$$

$\square$

2.6 The Hermite Transform

Since the Wick product satisfies all the ordinary algebraic rules for multiplication, one can carry out calculations in much the same way as with usual products. Problems arise, however, when limit operations are involved. To handle these situations it is convenient to apply a transformation, called the Hermite transform or the $\mathcal{H}$-transform, which converts Wick products into ordinary (complex) products and convergence in $(\mathcal{S})_{-1}$ into bounded, pointwise convergence in a certain neighborhood of 0 in $\mathbb{C}^N$. This transform, which first appeared in Lindstrøm et al. (1991), has been applied by the authors in many different connections. We will see several of these applications later. We first give the definition and some of its basic properties.

Definition 2.6.1. Let $F = \sum_\alpha b_\alpha H_\alpha \in (\mathcal{S})_{-1}^N$ with $b_\alpha \in \mathbb{R}^N$ as in Definition 2.3.2. Then the *Hermite transform of F*, denoted by $\mathcal{H}F$ or $\tilde{F}$, is defined by

$$\mathcal{H}F(z) = \tilde{F}(z) = \sum_\alpha b_\alpha z^\alpha \in \mathbb{C}^N \quad \text{(when convergent)}, \qquad (2.6.1)$$

where $z = (z_1, z_2, \ldots) \in \mathbb{C}^{\mathbb{N}}$ (the set of all sequences of complex numbers) and

$$z^\alpha = z_1^{\alpha_1} z_2^{\alpha_2} \cdots z_n^{\alpha_n} \cdots \qquad (2.6.2)$$

if $\alpha = (\alpha_1, \alpha_2, \ldots) \in \mathcal{J}$, where $z_j^0 = 1$.

Example 2.6.2 ($N = m = 1$).

i) The 1-dimensional smoothed white noise $w(\phi)$ has chaos expansion (see (2.2.23))

$$w(\phi, \omega) = \sum_{j=1}^n (\phi, \eta_j) H_{\epsilon^{(j)}}(\omega), \qquad (2.6.3)$$

and therefore the Hermite transform $\tilde{w}(\phi)$ of $w(\phi)$ is

$$\tilde{w}(\phi)(z) = \sum_{j=1}^\infty (\phi, \eta_j) z_j, \qquad (2.6.4)$$

which is convergent for all $z = (z_1, z_2, \ldots) \in (\mathbb{C}^{\mathbb{N}})_c$.

ii) The 1-dimensional (d-parameter) Brownian motion $B(x)$ has chaos expansion (see (2.2.24))

$$B(x, \omega) = \sum_{j=1}^\infty \int_0^x \eta_j(u) du H_{\epsilon^{(j)}}(\omega), \qquad (2.6.5)$$

and therefore

$$\widetilde{B}(x)(z) = \sum_{j=1}^{\infty} \int_0^x \eta_j(u) du z_j; \ z = (z_1, z_2, \ldots) \in (\mathbb{C}^N)_c, \tag{2.6.6}$$

where $(\mathbb{C}^N)_c$ is the set of all finite sequences in $\mathbb{C}^N$.

iii) The 1-dimensional singular white noise $W(x, \omega)$ has the expansion (see (2.2.23))

$$W(x, \omega) = \sum_{j=1}^{\infty} \eta_j(x) H_{\epsilon^{(j)}}(\omega), \tag{2.6.7}$$

and therefore

$$\widetilde{W}(x)(z) = \sum_{j=1}^{\infty} \eta_j(x) z_j; \ z = (z_1, z_2, \ldots) \in (\mathbb{C}^N)_c. \tag{2.6.8}$$

Example 2.6.3 ($N = m > 1$).

i) The m-dimensional smoothed white noise $\mathbf{w}(\phi)$ has chaos expansion (see (2.2.25))

$$\mathbf{w}(\phi, \omega) = (w_1(\phi, \omega), \ldots, w_m(\phi, \omega)),$$

with

$$w_i(\phi, \omega) = w(\phi_i, \omega_i)$$
$$= \sum_{j=1}^{\infty} (\phi_i, \eta_j) H_{\epsilon_{i+(j-1)m}}(\omega); \ 1 \leq i \leq m. \tag{2.6.9}$$

Hence the Hermite transform of coordinate $w_i(\phi, \omega)$ of $\mathbf{w}(\phi, \omega)$ is, for $z \in (\mathbb{C}^N)_c$,

$$\widetilde{w}_i(\phi)(z) = \sum_{j=1}^{\infty} (\phi_i, \eta_j) z_{(j-1)i+m}$$
$$= (\phi_i, \eta_1) z_i + (\phi_i, \eta_2) z_{i+m} + (\phi_i, \eta_3) z_{i+2m} + \cdots ; 1 \leq i \leq m. \tag{2.6.10}$$

Note that different components of $\mathbf{w}$ involve disjoint families of z_k-variables when we take the $\mathcal{H}$-transform.

ii) For the m-dimensional d-parameter Brownian motion

$$\mathbf{B}(x, \omega) = (B_1(x, \omega), \ldots, B_m(x, \omega))$$

we have (see (2.2.26))

$$B_i(x, \omega) = \sum_{j=1}^{\infty} \int_0^x \eta_j(u)du \, H_{\epsilon_{i+(j-1)m}} \qquad (2.6.11)$$

and hence

$$\widetilde{B}_i(x)(z) = \sum_{j=1}^{\infty} \int_0^x \eta_j(u)du \, z_{i+(j-1)m}; \ 1 \le i \le m. \qquad (2.6.12)$$

iii) The m-dimensional singular white noise

$$\mathbf{W}(x, \omega) = (W_1(x, \omega), \ldots, W_m(x, \omega))$$

has expansion (see (2.3.34))

$$W_i(x) = \sum_{j=1}^{\infty} \eta_j(x) H_{\epsilon_{i+(j-1)m}}; \ 1 \le i \le m, \qquad (2.6.13)$$

and therefore

$$\widetilde{W}_i(x)(z) = \sum_{j=1}^{\infty} \eta_j(x) z_{i+(j-1)m}; \ 1 \le i \le m, \ z \in (\mathbb{C}^{\mathbb{N}})_c. \qquad (2.6.14)$$

Note that if $F = \sum_\alpha b_\alpha H_\alpha \in (\mathcal{S})_{-\rho}^N$ for $\rho < 1$, then $(\mathcal{H}F)(z_1, z_2, \ldots)$ converges for all finite sequences $(z_1, z_2, \ldots)$ of complex numbers. To see this we write

$$\sum_\alpha |b_\alpha||z^\alpha| = \sum_\alpha |b_\alpha|(\alpha!)^{\frac{(1-\rho)}{2}}(\alpha!)^{\frac{(\rho-1)}{2}}|z^\alpha|(2\mathbb{N})^{-\frac{\alpha q}{2}}(2\mathbb{N})^{\frac{\alpha q}{2}}$$

$$\le \left(\sum_\alpha |b_\alpha|^2(\alpha!)^{1-\rho}(2\mathbb{N})^{-\alpha q} \right)^{\frac{1}{2}} \left(\sum_\alpha |z^\alpha|^2(\alpha!)^{\rho-1}(2\mathbb{N})^{\alpha q} \right)^{\frac{1}{2}}.$$
$$(2.6.15)$$

Now if $z = (z_1, \ldots, z_n)$ with $|z_j| \le M$, then

$$\sum_\alpha |z^\alpha|^2(\alpha!)^{\rho-1}(2\mathbb{N})^{\alpha q} \le \sum_\alpha M^{2|\alpha|}(\alpha!)^{\rho-1}2^{q|\alpha|}n^{q|\alpha|} < \infty \qquad (2.6.16)$$

for all $q < \infty$. If q is large enough, then by Proposition 2.3.3, the expression (2.6.15) is finite.

If $F \in (\mathcal{S})_{-1}^N$, however, we can only obtain convergence of $\mathcal{H}F(z_1, z_2, \ldots)$ in a neighborhood of the origin. We have

$$\sum_\alpha |b_\alpha \| z^\alpha| \le \left(\sum_\alpha b_\alpha^2 (2\mathbb{N})^{-\alpha q} \right)^{\frac{1}{2}} \left(\sum_\alpha |z^\alpha|^2 (2\mathbb{N})^{\alpha q} \right)^{\frac{1}{2}} \qquad (2.6.17)$$

where the first factor on the right hand side converges for q large enough. For such a value of q we have convergence of the second factor if $z \in \mathbb{C}^\mathbb{N}$ with

$$|z_j| < (2j)^{-q} \quad \text{for all } j.$$

Definition 2.6.4.   For $0 < R,\ q < \infty$, define the infinite-dimensional neighborhoods $\mathbb{K}_q(R)$ of 0 in $\mathbb{C}^\mathbb{N}$ by

$$\mathbb{K}_q(R) = \left\{ (\zeta_1, \zeta_2, \dots) \in \mathbb{C}^\mathbb{N}; \sum_{\alpha \neq 0} |\zeta^\alpha|^2 (2\mathbb{N})^{q\alpha} < R^2 \right\}. \qquad (2.6.18)$$

Note that

$$q \le Q,\ r \le R \Rightarrow \mathbb{K}_Q(r) \subset \mathbb{K}_q(R). \qquad (2.6.19)$$

For any $q < \infty, \delta > 0$ and natural number k, there exists $\epsilon > 0$ such that

$$z = (z_1, \dots, z_k) \in \mathbb{C}^k \quad \text{and} \quad |z_i| < \epsilon;\ 1 \le i \le k \Rightarrow z \in \mathbb{K}_q(\delta). \qquad (2.6.20)$$

The conclusions above can be stated as follows:

Proposition 2.6.5.   **a)** *If $F \in (\mathcal{S})_{-\rho}^N$ for some $\rho \in [-1, 1)$, then the Hermite transform $(\mathcal{H}F)(z)$ converges for all $z \in (\mathbb{C}^\mathbb{N})_c$.*

b) *If $F \in (\mathcal{S})_{-1}^N$, then there exists $q < \infty$ such that $(\mathcal{H}F)(z)$ converges for all $z \in \mathbb{K}_q(R)$ for all $R < \infty$.*

One of the reasons why the Hermite transform is so useful, is the following result, which is an immediate consequence of Definition 2.4.1 and Definition 2.6.1.

Proposition 2.6.6.   *If $F, G \in (\mathcal{S})_{-1}^N$, then*

$$\mathcal{H}(F \diamond G)(z) = \mathcal{H}F(z) \cdot \mathcal{H}G(z) \qquad (2.6.21)$$

for all z such that $\mathcal{H}F(z)$ and $\mathcal{H}G(z)$ exist. The product on the right hand side of (2.6.21) is the complex bilinear product between two elements of $\mathbb{C}^N$ defined by

$$(\zeta_1, \dots, \zeta_N) \cdot (w_1, \dots, w_N) = \sum_{i=1}^N \zeta_i w_i;\ \zeta_i, w_i \in \mathbb{C}.$$

Note that there is no complex conjugation in this definition.

Example 2.6.7.   Referring to Examples 2.6.2 and 2.6.3, we get the following Hermite transforms:

(i) $\mathcal{H}(w^{\diamond 2}(\phi))(z) = \sum_{j,k=1}^{\infty} (\phi, \eta_j)(\phi, \eta_k) z_j z_k$

(ii) $\mathcal{H}(B^{\diamond 2}(x))(z) = \sum_{j,k=1}^{\infty} \left(\int_0^x \eta_j(u) du \right) \left(\int_0^x \eta_k(u) du \right) z_j z_k$

(iii) $\mathcal{H}(W^{\diamond 3}(x))(z) = \sum_{i,j,k=1}^{\infty} \eta_i(x) \eta_j(x) \eta_k(x) z_i z_j z_k$

(iv) $\mathcal{H}(W_1(x) \diamond W_2(x))(z) = \left(\sum_{j=1}^{\infty} \eta_j(x) z_{2j-1} \right) \cdot \left(\sum_{k=1}^{\infty} \eta_k(x) z_{2k} \right)$

$$= \sum_{j,k=1}^{\infty} \eta_j(x) \eta_k(x) z_{2j-1} z_{2k}; \ z = (z_1, z_2, \ldots).$$

The Characterization Theorem for $(\mathcal{S})_{-1}^N$

Proposition 2.6.5 states that the $\mathcal{H}$-transform of any $F \in (\mathcal{S})_{-1}^N$ is a $\mathbb{C}^N$-valued analytic function on $\mathbb{K}_q(R)$ for all $R < \infty$, if $q < \infty$ is large enough. It is natural to ask if the converse is true: Is every $\mathbb{C}^N$-valued analytic function g on $\mathbb{K}_q(R)$ (for some $R < \infty, q < \infty$) the $\mathcal{H}$-transform of some element in $(\mathcal{S})_{-1}^N$? The answer is yes, if we add the condition that g be bounded on some $\mathbb{K}_q(R)$ (see Theorem 2.6.11).

To prove this, we first establish some auxiliary results. We say that a formal power series in infinitely many complex variables $z_1, z_2, \ldots$

$$g(z) = \sum_{\alpha} a_{\alpha} z^{\alpha}; \ a_{\alpha} \in \mathbb{C}^N, z = (z_1, z_2, \ldots)$$

is *convergent* at z if

$$\sum_{\alpha} |a_{\alpha}||z^{\alpha}| < \infty. \tag{2.6.22}$$

If this holds, the series has a well-defined sum that we denote by $g(z)$.

Proposition 2.6.8.   Let $g(z) = \sum_{\alpha} a_{\alpha} z^{\alpha}$, $a_{\alpha} \in \mathbb{C}^N$, $z = (z_1, z_2, \ldots)$ be a formal power series in infinitely many variables. Suppose there exist $q < \infty, M < \infty$ and $\delta > 0$ such that $g(z)$ is convergent for $z \in \mathbb{K}_q(\delta)$ and $|g(z)| \leq M$ for all $z \in \mathbb{K}_q(\delta)$.

Then

$$\sum_\alpha |a_\alpha z^\alpha| \le M \, A(q) \quad \text{for all} \quad z \in \mathbb{K}_{3q}(\delta),$$

where, by Proposition 2.3.3,

$$A(q) := \sum_\alpha (2\mathbb{N})^{-q\alpha} < \infty \quad \text{for} \quad q > 1.$$

To prove this proposition we need the two lemmas below.

Lemma 2.6.9. *Suppose $f(z) = \sum_{k=0}^\infty a_k z^k$ is an analytic function in one complex variable z such that*

$$\sup_{|z| \le r} |f(z)| \le M. \tag{2.6.23}$$

Then $|a_k z^k| \le M$ for all k and all z with $|z| \le r$.

Proof By the Cauchy formula

$$f^{(k)}(z) = \frac{k!}{2\pi i} \int_{|\zeta|=r} \frac{f(\zeta)}{(\zeta - z)^{k+1}} d\zeta, \quad \text{for} \quad |z| < r, \tag{2.6.24}$$

and we have

$$|f^{(k)}(0)| \le k! r^{-k} M. \tag{2.6.25}$$

Hence

$$|a_k z^k| = \left| \frac{f^{(k)}(0)}{k!} z^k \right| \le M \left| \frac{z}{r} \right|^k \le M, \quad \text{for} \quad |z| \le r. \tag{2.6.26}$$

$\square$

Lemma 2.6.10. *Let $g(z) = \sum_\alpha a_\alpha z^\alpha$ be an analytic function in n complex variables such that there exists $M < \infty$ and $c_1, \ldots, c_n > 0, \delta > 0$ such that*

$$|g(z)| \le M, \tag{2.6.27}$$

when $z \in \mathbb{K} := \{z = (z_1, \ldots, z_n) \in \mathbb{C}^n;\ c_1 |z_1|^2 + \cdots + c_n |z_n|^2 \le \delta^2\}$. Then $|a_\alpha z^\alpha| \le M$ for $z \in \mathbb{K}$, for all α.

Proof Use the previous lemma and induction. For example, for $n = 2$ the proof is the following: Write $g(z_1, z_2) = \sum_{k=0}^\infty A_k(z_2) z_1^k$. Fix z_2 such that $c_2 z_2^2 \le \delta^2$, and let

$$f(z_1) = \sum_{k=0}^\infty A_k(z_2) z_1^k, \quad \text{for} \quad (z_1, z_2) \in \mathbb{K}. \tag{2.6.28}$$

By the previous lemma we have $|A_k(z_2)z_1^k| \leq M$ for $(z_1, z_2) \in \mathbb{K}$. Applying the same lemma to

$$f(z_2) = A_k(z_2) = \sum_{l=0}^{\infty} a_{kl} z_2^l, \tag{2.6.29}$$

we get that $|a_{kl} z_2^l| \leq \frac{M}{|z_1^k|}$ or $|a_{kl} z_1^k z_2^l| \leq M$, as claimed.   □

Proof of Proposition 2.6.8 We may assume $N = 1$. Without loss of generality we can assume that q is so large that $\sum_{\alpha}(2\mathbb{N})^{-q\alpha} < \infty$, and we put $Q = 3q$. Then by Lemma 2.6.10 we have

$$|a_\alpha w^\alpha| \leq M \quad \text{for all} \quad w \in \mathbb{K}_q(\delta). \tag{2.6.30}$$

Choose $z \in \mathbb{K}_{3q}(\delta)$. Then if

$$w_j = (2j)^q z_j, \tag{2.6.31}$$

we have

$$\sum_\alpha |w^\alpha|^2 (2\mathbb{N})^{q\alpha} = \sum_\alpha (2\mathbb{N})^{3q\alpha} |z^\alpha|^2 < \delta, \tag{2.6.32}$$

so $w \in \mathbb{K}_q(\delta)$. Therefore

$$\sum_\alpha |a_\alpha\| z^\alpha| \leq \left(\sum_\alpha |a_\alpha|^2 |z^\alpha|^2 (2\mathbb{N})^{q\alpha} \right)^{\frac{1}{2}} \left(\sum_\alpha (2\mathbb{N})^{-q\alpha} \right)^{\frac{1}{2}}$$

$$= \left(\sum_\alpha |a_\alpha|^2 |w^\alpha|^2 (2\mathbb{N})^{-q\alpha} \right) \left(\sum_\alpha (2\mathbb{N})^{q\alpha} \right)^{\frac{1}{2}}$$

$$\leq M \sum_\alpha (2\mathbb{N})^{-q\alpha}. \tag{2.6.33}$$

□

Theorem 2.6.11 (Characterization theorem for $(\mathcal{S})_{-1}^N$). **a)** If $F(\omega) = \sum_\alpha a_\alpha H_\alpha(\omega) \in (\mathcal{S})_{-1}^N$, where $a_\alpha \in \mathbb{R}^n$, then there is $q < \infty, M_q < \infty$ such that

$$|\widetilde{F}(z)| \leq \sum_\alpha |a_\alpha\| z^\alpha| \leq M_q \left(\sum_\alpha (2\mathbb{N})^{q\alpha} |z^\alpha|^2 \right)^{\frac{1}{2}} \quad \text{for all} \quad z \in (\mathbb{C}^\mathbb{N})_c. \tag{2.6.34}$$

In particular, $\widetilde{F}$ is a bounded analytic function on $\mathbb{K}_q(R)$ for all $R < \infty$.
b) Conversely, suppose $g(z) = \sum_\alpha b_\alpha z^\alpha$ is a given power series of $z \in (\mathbb{C}^\mathbb{N})_c$ with $b_\alpha \in \mathbb{R}^N$ such that there exists $q < \infty$ and $\delta > 0$, such that $g(z)$ is absolutely convergent when $z \in \mathbb{K}_q(\delta)$ and

$$\sup_{z \in \mathbb{K}_q(\delta)} |g(z)| < \infty. \tag{2.6.35}$$

Then there exists a unique $G \in (\mathcal{S})_{-1}^N$ such that $\widetilde{G} = g$, namely

$$G(\omega) = \sum_{\alpha} b_{\alpha} H_{\alpha}(\omega). \tag{2.6.36}$$

c) *Let $F = \sum_{\alpha} c_{\alpha} H_{\alpha}(\omega) \in (\mathcal{S})_{-1,-q}$. Then we have*

$$\sup_{z \in \mathbb{K}_q(R)} |\mathcal{H}F(z)| \leq R\|F\|_{-1,-q} \text{ for all } R > 0$$

d) *Suppose there exist $q > 1, \delta > 0$ such that*

$$M_q(\delta) := \sup_{z \in \mathbb{K}_q(\delta)} \left| \sum_{\alpha} c_{\alpha} z^{\alpha} \right| < \infty$$

Then there exists $r \geq q$ such that

$$S_r := \sup_{\alpha} |c_{\alpha}| (2\mathbb{N})^{-r\alpha} \leq M_q(\delta) A(q),$$

where

$$A(q) := \sum_{\alpha} (2\mathbb{N})^{-q\alpha} \text{ (see Proposition 2.6.8)},$$

and such that $F := \sum_{\alpha} c_{\alpha} H_{\alpha}$ satisfies

$$\|F\|_{-1,-2r} \leq A(q) \sup_{z \in \mathbb{K}_q(\delta)} |\mathcal{H}F(z)|$$

e) *For all $R > 0, q > 1$ there exist $r \geq q$ such that*

$$\sup_{z \in \mathbb{K}_q(R)} |\mathcal{H}F(z)| \leq R\|F\|_{-1,-q} \leq R\|F\|_{-1,-2r} \leq RA(q) \sup_{z \in \mathbb{K}_q(R)} |\mathcal{H}F(z)|$$

Proof   a) We have

$$|\widetilde{F}(z)| \leq \sum_{\alpha} |a_{\alpha}||z^{\alpha}| \leq \left(\sum_{\alpha} |a_{\alpha}|^2 (2\mathbb{N})^{-q\alpha} \right)^{\frac{1}{2}} \left(\sum_{\alpha} |z^{\alpha}|^2 (2\mathbb{N})^{q\alpha} \right)^{\frac{1}{2}}.$$

$$\tag{2.6.37}$$

Since $F \in (\mathcal{S})_{-1}^N$, we see that $M_q^2 := \sum_{\alpha} |a_{\alpha}|^2 (2\mathbb{N})^{-q\alpha} < \infty$ if q is large enough.

b) Conversely, assume that (2.6.35) holds. For $r < \infty$ and k a natural number, choose $\zeta = \zeta^{(r,k)} = (\zeta_1, \zeta_2, \ldots, \zeta_k)$ with

$$\zeta_j = (2j)^{-r}; \ 1 \leq j \leq k. \tag{2.6.38}$$

Then

$$\sum_\alpha |\zeta^\alpha|^2 (2\mathbb{N})^{r\alpha} \le \sum_\alpha (2\mathbb{N})^{-r\alpha} < \delta^2 \qquad (2.6.39)$$

if r is large enough, say $r \ge q_1$. Hence

$$\zeta \in \mathbb{K}_r(\delta) \quad \text{for} \quad r \ge q_1.$$

By Proposition 2.6.8 we have

$$\sum_\alpha |b_\alpha||z^\alpha| \le M\, A(q) \quad \text{for} \quad z \in \mathbb{K}_{3q}(\delta),$$

where $M = \sup\{|g(z)|; z \in \mathbb{K}_q(\delta)\}$. Hence, if $r \ge \max(3q, q_1)$, we get

$$\sum_{\substack{\alpha \\ \text{Index } \alpha \le k}} |b_\alpha|(2\mathbb{N})^{-r\alpha} = \sum_{\substack{\alpha \\ \text{Index } \alpha \le k}} |b_\alpha| \zeta^\alpha$$

$$= \sum_{\substack{\alpha \\ \text{Index } \alpha \le k}} |b_\alpha||\zeta^\alpha| \le \sum_\alpha |b_\alpha||z^\alpha|$$

$$\le M\, A(q), \text{ for } z \in \mathbb{K}_{3q}(\delta), \qquad (2.6.40)$$

where Index α is the position of the last nonzero element in the sequence $(\alpha_1, \alpha_2, \dots)$.
Now let $k \to \infty$ to deduce that

$$K := \sup_\alpha |b_\alpha|(2\mathbb{N})^{-r\alpha} < \infty. \qquad (2.6.41)$$

This gives

$$\sum_\alpha |b_\alpha|^2 (2\mathbb{N})^{-2r\alpha} \le K \sum_\alpha |b_\alpha|(2\mathbb{N})^{-r\alpha} < \infty, \qquad (2.6.42)$$

and hence $G := \sum_\alpha b_\alpha H_\alpha \in (\mathcal{S})_{-1}$ as claimed.

c) If $z \in \mathbb{K}_q(R)$ we have

$$|\mathcal{H}F(z)| = \left| \sum_\alpha c_\alpha z^\alpha \right|$$

$$\le \sum_\alpha |c_\alpha|(2\mathbb{N})^{-\frac{q}{2}\alpha} |z^\alpha|(2\mathbb{N})^{\frac{q}{2}\alpha}$$

$$\le \left(\sum_\alpha |z^\alpha|^2 (2\mathbb{N})^{q\alpha} \right)^{\frac{1}{2}} \left(\sum_\alpha |c^\alpha|^2 (2\mathbb{N})^{-q\alpha} \right)^{\frac{1}{2}}$$

$$\le R\|F\|_{-1,-q}$$

d) Suppose $M_q(\delta) < \infty$. Then it follows as in (2.6.41) that there exists $r \geq q$ such that $S_r \leq M_q(\delta)A(q)$ and

$$\left\| \sum_\alpha c_\alpha H_\alpha \right\|_{-1,-2r}^2 = \sum_\alpha |c_\alpha|^2 (2\mathbb{N})^{-2r\alpha}$$

$$\leq S_r \sum_\alpha |c_\alpha|(2\mathbb{N})^{-r\alpha}$$

$$\leq S_r M_q(\delta)A(q) \leq A^2(q)M_q^2(\delta)$$

e) This is a synthesis of c) and d).

$\square$

From this we deduce the following useful result:

Theorem 2.6.12. Kondratiev et al. (1994), Theorem 12 (Analytic functions operate on $\mathcal{H}$-transforms). *Suppose* $g = \mathcal{H}X$ *for some* $X \in (\mathcal{S})_{-1}^N$, *and let* $M \in \mathbb{N}$. *Let* f *be a* $\mathbb{C}^M$-*valued analytic function on a neighborhood* U *of* $\zeta_0 := g(0)$ *in* $\mathbb{C}^N$ *such that the Taylor expansion of* f *around* ζ_0 *has real coefficients. Then there exists a unique* $Y \in (\mathcal{S})_{-1}^M$ *such that*

$$\mathcal{H}Y = f \circ g. \tag{2.6.43}$$

Proof Let $r > 0$ be such that

$$\{\zeta \in \mathbb{C}^N; |\zeta - \zeta_0| < r\} \subset U.$$

Then choose $q < \infty$ such that $g(z)$ is a bounded analytic function on $\mathbb{K}_q(1)$ and such that

$$|g(z) - \zeta_0| < \frac{r}{2} \quad \text{for} \quad z \in \mathbb{K}_q(1).$$

(This is possible by the estimate (2.6.37)). Then $f \circ g$ is a bounded analytic function on $\mathbb{K}_q(1)$, so the result follows from Theorem 2.6.11. $\square$

Definition 2.6.13 (Generalized expectation). Let $X = \sum_\alpha c_\alpha H_\alpha \in (\mathcal{S})_{-1}^N$. Then the vector $c_0 = \widetilde{X}(0) \in \mathbb{R}^N$ is called the *generalized expectation* of X and is denoted by $E[X]$. In the case when $X = F \in L^p(\mu)$ for some $p > 1$ then the generalized expectation of F coincides with the usual expectation

$$E[F] = \int_{\mathcal{S}'} F(\omega)d\mu(\omega).$$

To see this we note that for $N = 1$ the action of $F \in L^p(\mu)$ on $f \in L^p(\mu)^* = L^q(\mu)$ (where $1/p + 1/q = 1$) is given by

$$\langle F, f \rangle = E[Ff],$$

so that, in particular,

$$E[F] = \langle F, 1 \rangle.$$

On the other hand, if $F = \sum_\alpha c_\alpha H_\alpha$, then by (2.3.11)

$$\langle F, 1 \rangle = c_0 = \widetilde{F}(0).$$

So

$$E[F] = c_0 = \widetilde{F}(0) \tag{2.6.44}$$

for $F \in L^p(\mu), p > 1$, as claimed.

In fact, (2.6.43) also holds if $F \in L^1(\mu) \cap (\mathcal{S})_{-1}$. (See Exercise 2.10.)
Note that from this definition we have

$$E[X \diamond Y] = (E[X], E[Y]) \quad \text{for all} \quad X, Y \in (\mathcal{S})_{-1}^N, \tag{2.6.45}$$

where $(\cdot, \cdot)$ denotes the inner product in $\mathbb{R}^N$, and, in particular,

$$E[X \diamond Y] = E[X]E[Y]; \ X, Y \in (\mathcal{S})_{-1}. \tag{2.6.46}$$

Thanks to Theorem 2.6.12 we can construct the Wick versions $f^\diamond$ of analytic functions f as follows:

Definition 2.6.14 (Wick versions of analytic functions). Let $X \in (\mathcal{S})_{-1}^N$ and let $f : U \to \mathbb{C}^M$ be an analytic function, where U is a neighborhood of $\zeta_0 := E[X]$. Assume that the Taylor series of f around ζ_0 has coefficients in $\mathbb{R}^M$. Then the *Wick version* $f^\diamond(X)$ of f applied to X is defined by

$$f^\diamond(X) = \mathcal{H}^{-1}(f \circ \widetilde{X}) \in (\mathcal{S})_{-1}^M. \tag{2.6.47}$$

In other words, if f has the power series expansion

$$f(z) = \sum a_\alpha (z - \zeta_0)^\alpha \quad \text{with} \quad a_\alpha \in \mathbb{R}^M,$$

then

$$f^\diamond(X) = \sum a_\alpha (X - \zeta_0)^{\diamond\alpha} \in (\mathcal{S})_{-1}^M. \tag{2.6.48}$$

Example 2.6.15. If the function $f : \mathbb{C}^N \to \mathbb{C}^M$ is *entire*, i.e., analytic in the whole space $\mathbb{C}^N$, then $f^\diamond(X)$ is defined for all $X \in (\mathcal{S})_{-1}^N$. For example,

i) *The Wick exponential* of $X \in (\mathcal{S})_{-1}$ is defined by

$$\exp^\diamond X = \sum_{n=0}^\infty \frac{1}{n!} X^{\diamond n}. \tag{2.6.49}$$

Using the Hermite transform we see that the Wick exponential has the same algebraic properties as the usual exponential. For example,

$$\exp^\diamond[X + Y] = \exp^\diamond[X] \diamond \exp^\diamond[Y] \; ; \; X, Y \in (\mathcal{S})_{-1}. \tag{2.6.50}$$

ii) The analytic logarithm, $f(z) = \log z$, is well-defined in any simply connected domain $U \subset \mathbb{C}$ not containing the origin. If we require that $1 \in U$, then we can choose the branch of $f(z) = \log z$ with $f(1) = 0$. For any $X \in (\mathcal{S})_{-1}$ with $E[X] \neq 0$, choose a simply connected $U \subset \mathbb{C} \setminus \{0\}$ such that $\{1, E[X]\} \subset U$ and define the *Wick-logarithm* of X, $\log^\diamond X$, by

$$\log^\diamond X = \mathcal{H}^{-1}(\log(\widetilde{X}(z)) \in (\mathcal{S})_{-1}. \tag{2.6.51}$$

If $E[X] \neq 0$, we have

$$\exp^\diamond(\log^\diamond(X)) = X. \tag{2.6.52}$$

For all $X \in (\mathcal{S})_{-1}$, we have

$$\log^\diamond(\exp^\diamond X) = X. \tag{2.6.53}$$

Moreover, if $E[X] \neq 0$ and $E[Y] \neq 0$, then

$$\log^\diamond(X \diamond Y) = \log^\diamond X + \log^\diamond Y. \tag{2.6.54}$$

iii) Similarly, if $E[X] \neq 0$, we can define the *Wick inverse* $X^{\diamond(-1)} \in (\mathcal{S})_{-1}$, having the property that

$$X \diamond X^{\diamond(-1)} = 1.$$

More generally, if $E[X] \neq 0$, we can define the *Wick powers* $X^{\diamond r} \in (\mathcal{S})_{-1}$ for all real numbers r.

Remark Note that, with the generalized expectation $E[Y]$ defined for $Y \in (\mathcal{S})_{-1}$ as in Definition 2.6.13, we have

$$E[\exp^\diamond[X]] = \exp[E[X]]; \; X \in (\mathcal{S})_{-1}, \tag{2.6.55}$$

simply because

$$E[\exp^\diamond[X]] = \mathcal{H}(\exp^\diamond[X])(0) = \exp[\mathcal{H}(X)(0)] = \exp[E[X]].$$

Positive Noise

An important special case of the Wick exponential is obtained by choosing X to be smoothed white noise $w(\phi)$. Since $w(\phi, \cdot) \in L^2(\mu)$, the usual exponential function exp can also be applied to $w(\phi, \omega)$ for almost all ω, and the relation between these two quantities is given by the following result.

Lemma 2.6.16.

$$\exp^\diamond[w(\phi,\omega)] = \exp\left[w(\phi,\omega) - \frac{1}{2}\|\phi\|^2\right]; \phi \in L^2(\mathbb{R}^d) \qquad (2.6.56)$$

where $\|\phi\| = \|\phi\|_{L^2(\mathbb{R}^d)}$.

Proof By basis independence, which in this $L^2(\mu)$-case follows from Proposition 2.4.2 (see Appendix D for the general case), we may assume that $\phi = c\eta_1$, in which case we get

$$\exp^\diamond[w(\phi)] = \sum_{n=0}^\infty \frac{1}{n!}w(\phi)^{\diamond n} = \sum_{n=0}^\infty \frac{1}{n!}c^n\langle\omega,\eta_1\rangle^{\diamond n}$$

$$= \sum_{n=0}^\infty \frac{c^n}{n!}H_{\epsilon_1}^{\diamond n}(\omega) = \sum_{n=0}^\infty \frac{c^n}{n!}H_{n\epsilon_1}(\omega)$$

$$= \sum_{n=0}^\infty \frac{c^n}{n!}h_n(\langle\omega,\eta_1\rangle) = \exp\left[c\langle\omega,\eta_1\rangle - \frac{1}{2}c^2\right]$$

$$= \exp\left[w(\phi) - \frac{1}{2}\|\phi\|^2\right],$$

where we have used the generating property of the Hermite polynomials (see Appendix C). $\qquad\qquad\qquad\Box$

In particular, (2.6.55) shows that $\exp^\diamond w(\phi)$ is positive for all $\phi \in L^2(\mu)$ and all ω. Moreover, if

$$W_\phi(x,\omega) := w(\phi_x,\omega); \ x \in \mathbb{R}^d$$

is the smoothed white noise process defined in (2.1.15), then the process

$$K_\phi(x,\omega) := \exp^\diamond[W_\phi(x,\omega)] = \exp\left[W_\phi(x,\omega) - \frac{1}{2}\|\phi\|^2\right] \qquad (2.6.57)$$

has the following three properties (compare with (2.1.20)–(2.1.22)):

If supp $\phi_{x_1} \cap$ supp $\phi_{x_2} = \emptyset$, then $K_\phi(x_1,\cdot)$ and $K_\phi(x_2,\cdot)$ are independent.
$$\qquad\qquad\qquad\qquad\qquad\qquad\qquad\qquad\qquad\qquad\qquad\qquad (2.6.58)$$

$\{K_\phi(x,\cdot)\}_{x\in\mathbb{R}^d}$ is a stationary process. $\qquad\qquad\qquad\qquad\qquad (2.6.59)$

For each $x \in \mathbb{R}^d$ the random variable $K_\phi(x,\cdot) > 0$ has a *lognormal* distribution (i.e., $\log K_\phi(x,\cdot)$ has a normal distribution) and $E[K_\phi(x,\cdot)] = 1$,
Var$[K_\phi(x,\cdot)] = \exp[\|\phi\|^2] - 1$. $\qquad\qquad\qquad\qquad\qquad\qquad (2.6.60)$

Properties (2.6.57) and (2.6.58) follow directly from the corresponding properties (2.1.20) and (2.1.21) for $W_\phi(x,\cdot)$. The first parts of (2.6.59)

follow from (2.6.56) and the fact that $E[W_\phi(x,\cdot)^{\diamond k}] = 0$ for all $k \geq 1$.
The last part of (2.6.59) is left as an exercise for the reader (Exercise 2.11).
These three properties make $K_\phi(x,\omega)$ a good mathematical model for many
cases of "positive noise" occurring in various applications. In particular, the
function $K_\phi(x,\omega)$ is suitable as a model for the *stochastic permeability* of
a heterogeneous, isotropic rock. See (1.1.5) and Section 4.6. We shall call
$K_\phi(x,\cdot)$ the *smoothed positive noise process*. Similarly, we call

$$K(x,\cdot) = \exp^\diamond[W(x,\cdot)] \in (\mathcal{S})^* \tag{2.6.61}$$

the *singular positive noise process*. Computer simulations of the 1-parameter
(i.e., $d = 1$) positive noise process $K_\phi(x,\omega)$ for a given ϕ are shown in
Figure 2.2.

Computer simulations of the 2-parameter (i.e., $d = 2$) positive noise
process $K_\phi(x,\omega)$ where $\phi(y) = \epsilon \chi_{[0,h] \times [0,h]}(y)$; $y \in \mathbb{R}^2$ are shown on
Figure 2.3.

The Positive Noise Matrix

When the (deterministic) medium is anisotropic, the non-negative permeability function $k(x)$ in Darcy's law (1.1.5) must be replaced by a permeability
matrix $K(x) = [K_{ij}(x)] \in \mathbb{R}^{d \times d}$. The interpretation of the (i,j)th element,
K_{ij}, is that
$K_{ij}(x)$ = velocity of fluid at x in direction i induced by a pressure gradient
of unit size in direction j.

Physical arguments lead to the conclusion that $K(x) = [K_{ij}(x)]$ should be
a symmetric, non-negative definite matrix for each x.

For a stochastic anisotropic medium it is natural to represent the stochastic
permeability matrix as follows (Gjerde (1995a), Øksendal (1994b)):

Let $\mathbf{W}(x) \in (\mathcal{S})_{-0}^{N;N}$ be N-dimensional, d-parameter white noise with the
value $N = 1/2d(d+1)$. Define

$$\mathbf{K}(x) := \exp^\diamond[\mathcal{W}(x)]; \tag{2.6.62}$$

where

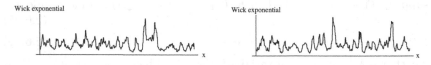

Wick exponential   Wick exponential

Fig. 2.2 Two sample paths of the Wick exponential of the 1-parameter white
noise process.

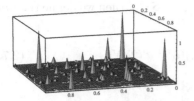 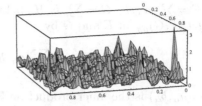

Fig. 2.3 Two sample paths of positive noise $K_\phi(x,\omega)$, $(h=1/50,\ \epsilon=0.05)$ and $(h=1/20,\ \epsilon=0.1)$.

$$
\mathcal{W}(x) = \begin{bmatrix}
W_{1,1}(x) & W_{1,2}(x) & \cdots & W_{1,d}(x) \\
W_{1,2}(x) & W_{2,2}(x) & \cdots & W_{2,d}(x) \\
\vdots & & \ddots & \vdots \\
W_{1,d}(x) & \cdots & & W_{d,d}(x)
\end{bmatrix}
\tag{2.6.63}
$$

and $W_{ij}(x)$; $1 \le i \le j \le d$ are the $1/2d(d+1)$ independent components of $\mathbf{W}(x)$, in some (arbitrary) order.

Here the Wick exponential is to be interpreted in the Wick matrix sense, i.e.,

$$
\exp^\diamond[\mathbf{M}] = \sum_{n=0}^{\infty} \frac{1}{n!}\mathbf{M}^{\diamond n}
\tag{2.6.64}
$$

when $\mathbf{M} \in (\mathcal{S})_{-1}^{m;k \times k}$ is a stochastic distribution matrix. It follows from Theorem 2.6.12 that $\exp^\diamond \mathbf{M}$ exists as an element of $(\mathcal{S})_{-1}^{m;k \times k}$.

We call $\mathbf{K}(x)$ the (singular) *positive noise matrix*. It will be used in Section 4.7.

Similarly, one can define the *smoothed positive noise matrix*

$$
\mathbf{K}_\phi(x) = \exp^\diamond[\mathcal{W}_\phi(x)],
\tag{2.6.65}
$$

where the entries of the matrix $\mathcal{W}_\phi(x)$ are the components of the $1/2d(d+1)$-dimensional smoothed white noise process $\mathbf{W}_\phi(x)$.

2.7 The $(\mathcal{S})_{\rho,r}^N$ Spaces and the $\mathcal{S}$-Transform

Sometimes the following spaces, which are intermediate spaces to the spaces $(\mathcal{S})_\rho^N, (\mathcal{S})_{-\rho}^N$, are convenient to work in (see Våge (1996a)).

Definition 2.7.1. For $\rho \in [-1, 1]$ and $r \in \mathbb{R}$, let $(\mathcal{S})_{\rho,r}^N$ consist of those $F = \sum_\alpha a_\alpha H_\alpha \in (\mathcal{S})_\rho^N$ (with $a_\alpha \in \mathbb{R}^N$ for all α) such that

$$
\|F\|_{\rho,r}^2 := \sum_\alpha a_\alpha^2 (\alpha!)^{1+\rho} (2\mathbb{N})^{r\alpha} < \infty.
\tag{2.7.1}
$$

If $F = \sum_\alpha a_\alpha H_\alpha, G = \sum_\alpha b_\alpha H_\alpha$ belong to $(\mathcal{S})^N_{\rho,r}$, then we define the *inner product* $(F, G)_{\rho,r}$ of F and G by

$$(F, G)_{\rho,r} = \sum_\alpha (a_\alpha, b_\alpha)(\alpha!)^{1+\rho}(2\mathbb{N})^{r\alpha}, \qquad (2.7.2)$$

where (a_α, b_α) is the inner product on $\mathbb{R}^N$.

Note that if $\rho \in [0, 1]$, then $(\mathcal{S})^N_\rho$ is the projective limit (intersection) of the spaces $\{(\mathcal{S})^N_{\rho,r}\}_{r\geq 0}$, while $(\mathcal{S})_{-\rho}$ is the inductive limit (union) of $\{(\mathcal{S})_{-\rho,-r}\}_{r\geq 0}$.

Lemma 2.7.2 Våge (1996a).   *For every pair $(\rho, r) \in [-1, 1] \times \mathbb{R}$ the space $(\mathcal{S})^N_{\rho,r}$ equipped with the inner product (2.7.2) is a separable Hilbert space.*

Proof   We first prove completeness: Fix ρ, r and suppose $F_k = \sum_\alpha a^{(k)}_\alpha H_\alpha$ is a Cauchy sequence in $(\mathcal{S})^N_{\rho,r}, k = 1, 2, \ldots$. Then $\{a^{(k)}_\alpha\}^\infty_{k=1}$ is a Cauchy sequence in $\mathbb{R}^N$ (with the usual norm), so $a^{(k)}_\alpha \to a_\alpha$, say, as $k \to \infty$. Define

$$F = \sum_\alpha a_\alpha H_\alpha.$$

We must prove that $f \in (\mathcal{S})^N_{\rho,r}$ and that $F_k \to F$ in $(\mathcal{S})^N_{\rho,r}$. To this end let $\epsilon > 0$ and $n \in \mathbb{N}$. Then there exists $M \in \mathbb{N}$ such that

$$\sum_{\alpha \in \Gamma_n} (a^{(i)}_\alpha - a^{(j)}_\alpha)^2 (\alpha!)^{1+\rho}(2\mathbb{N})^{r\alpha}$$

$$\leq \sum_\alpha (a^{(i)}_\alpha - a^{(j)}_\alpha)^2 (\alpha!)^{1+\rho}(2\mathbb{N})^{r\alpha} < \epsilon^2 \quad \text{for} \quad i, j \geq M,$$

where

$$\Gamma_n = \{\alpha = (\alpha_1, \ldots, \alpha_n); \alpha_j \in \{0, 1, \ldots, n\}, \; j = 1, \ldots, n\}. \qquad (2.7.3)$$

If we let $i \to \infty$, we see that

$$\sum_{\alpha \in \Gamma_n} (a_\alpha - a^{(j)}_\alpha)^2 (\alpha!)^{1+\rho}(2\mathbb{N})^{r\alpha} < \epsilon^2 \quad \text{for} \quad j \geq M.$$

Letting $n \to \infty$, we obtain that

$$F - F_j \in (\mathcal{S})^N_{\rho,r}$$

and that

$$F_j \to F \quad \text{in} \quad (\mathcal{S})^N_{\rho,r}.$$

Finally, the separability follows from the fact that $\{H_\alpha\}$ is a countable dense subset of $(\mathcal{S})^N_{\rho,r}$.   □

Example 2.7.3. Singular white noise $W(x, \omega)$ belongs to $(\mathcal{S})_{-0,-q}$ for all $q > 1$. This follows from the proof of Proposition 2.3.10.

The $\mathcal{S}$-Transform

The Hermite transform is closely related to the $\mathcal{S}$-transform. See Hida et al. (1993), and the references therein. For completeness, we give a short introduction to the $\mathcal{S}$-transform here.

Earlier we saw that if $\phi \in \mathcal{S}(\mathbb{R}^d)$, then $\langle \omega, \phi \rangle^{\diamond n} = w(\phi, \omega)^{\diamond n} \in (\mathcal{S})_1$, for all natural numbers n (Example 2.3.4). It is natural to ask if we also have $\exp^{\diamond}[w(\phi, \omega)] \in (\mathcal{S})_1$, at least if $\|\phi\|_{L^2(\mathbb{R}^d)}$ is small enough. This is not the case. However, we have the following:

Lemma 2.7.4. **a)** *Let* $\phi \in \mathcal{S}(\mathbb{R}^d)$ *and* $q < \infty$. *Then there exists* $\epsilon > 0$ *such that for* $\lambda \in \mathbb{R}$ *with* $|\lambda| < \epsilon$ *we have*

$$\exp^{\diamond}[\lambda w(\phi, \cdot)] \in (\mathcal{S})_{1,q}. \tag{2.7.4}$$

b) *For all* $\rho < 1$ *we have*

$$\exp^{\diamond}[\lambda w(\phi, \cdot)] \in (\mathcal{S})_{\rho} \tag{2.7.5}$$

for all $\lambda \in \mathbb{R}$.

Proof Choose $\lambda_1, \ldots, \lambda_k \in \mathbb{R}$ and consider

$$\exp^{\diamond}[\langle \omega, \lambda_1 \eta_1 + \cdots + \lambda_k \eta_k \rangle]$$

$$= \exp^{\diamond} \left(\sum_{j=1}^{k} \lambda_j \langle \omega, \eta_j \rangle \right)$$

$$= \exp^{\diamond} \left(\sum_{j=1}^{k} \lambda_j H_{\epsilon^{(j)}}(\omega) \right) = \sum_{n=0}^{\infty} \frac{1}{n!} \left(\sum_{j=1}^{k} \lambda_j H_{\epsilon^{(j)}} \right)^{\diamond n}$$

$$= \sum_{n=0}^{\infty} \frac{1}{n!} \left(\sum_{\substack{\alpha_i = 1 \\ 1 \leq i \leq k}}^{n} \frac{n!}{\alpha_1! \cdots \alpha_k!} \lambda_1^{\alpha_1} \cdots \lambda_k^{\alpha_k} H_{\alpha_1 \epsilon_1 + \cdots + \alpha_k \epsilon^{(k)}} \right)$$

$$= \sum_{n=0}^{\infty} \sum_{\substack{\alpha_i = 1 \\ 1 \leq i \leq k}}^{n} \frac{1}{\alpha_1! \cdots \alpha_k!} \lambda_1^{\alpha_1} \cdots \lambda_k^{\alpha_k} H_{\alpha_1 \epsilon_1 + \cdots + \alpha_k \epsilon^{(k)}}$$

$$= \sum_{n=0}^{\infty} \sum_{\substack{|\alpha| = n \\ \text{Index } \alpha \leq k}} \frac{1}{\alpha!} \lambda^{\alpha} H_{\alpha} = \sum_{\substack{\alpha \\ \text{Index } \alpha \leq k}} \frac{1}{\alpha!} \lambda^{\alpha} H_{\alpha} =: \sum_{\alpha} a_{\alpha}^{(k)} H_{\alpha},$$

$$\tag{2.7.6}$$

where $\lambda^\alpha = \lambda_1^{\alpha_1}\lambda_2^{\alpha_2}\cdots\lambda_k^{\alpha_k}$. Hence by Lemma 2.3.4

$$\sum_\alpha (a_\alpha^{(k)})^2(\alpha!)^2(2\mathbb{N})^{q\alpha}$$

$$= \sum_{n=0}^{\infty}\sum_{\substack{|\alpha|=n\\ \text{Index } \alpha\le k}} \left(\frac{1}{\alpha!}\right)^2 \lambda^{2\alpha}(\alpha!)^2(2\mathbb{N})^{q\alpha}$$

$$= \sum_{\text{Index } \alpha\le k} \lambda^{2\alpha}(2\mathbb{N})^{q\alpha}$$

$$= \sum_{\text{Index } \alpha\le k} \lambda_1^{2\alpha_1}\cdots\lambda_k^{2\alpha_k}2^{q\alpha_1}4^{q\alpha_2}\cdots(2k)^{q\alpha_k}$$

$$\le \left(\sum_{\alpha_1=0}^{\infty}\left(\lambda_1^2(2^d\delta_1^{(1)}\cdots\delta_d^{(1)}\cdots\delta_d^{(1)})^{q'}\right)^{\alpha_1}\right)$$

$$\cdots\left(\sum_{\alpha_k=1}^{\infty}\left(\lambda_k^2(2^d\delta_1^{(k)}\cdots\delta_1^{(k)}\cdots\delta_d^{(k)})^{q'}\right)^{\alpha_k}\right)$$

$$= \prod_{j=1}^{k}\frac{1}{1-\Lambda_j} < \infty, \tag{2.7.7}$$

if

$$\Lambda_j := \lambda_j^2(2^d\delta_1^{(j)}\cdots\delta_d^{(j)})^{q'} < 1, \tag{2.7.8}$$

where $q' = (d/d-1)q$ if $d\ge 2$, $q'=q$ if $d=1$. Now choose $\phi \in \mathcal{S}(\mathbb{R}^d)$. Then by Theorem 2.3.1 there exists $M < \infty$ such that

$$(\phi,\eta_j)^2 \le M^2(2^d\delta_1^{(j)}\cdots\delta_d^{(j)})^{-q'} \quad \text{for all} \quad j.$$

Hence, if $\lambda \in \mathbb{R}$ with $|\lambda|$ small enough, we have

$$\lambda^2(\phi,\eta_j)^2 \le \frac{1}{2}\left(2^d\delta_1^{(j)}\cdots\delta_d^{(j)}\right)^{-q'} \quad \text{for all} \quad j. \tag{2.7.9}$$

Therefore, if we define

$$\lambda_j^2 := \lambda^2(\phi,\eta_j)^2,$$

we see that (2.7.8) holds, and we can apply the above argument.

Then, if we write $(\phi,\eta)^\alpha = (\phi,\eta_1)^{\alpha_1}\cdots(\delta,\eta_k)^{\alpha_k}$ when $\alpha = (\alpha_1,\ldots,\alpha_k)$, we get

$$\exp^\diamond[\lambda\langle\omega,\phi\rangle] = \sum_\alpha \frac{1}{\alpha!}\lambda^{|\alpha|}(\phi,\eta)^\alpha H_\alpha$$

$$=: \sum_\alpha c_\alpha^{(\lambda)} H_\alpha, \tag{2.7.10}$$

and hence, by (2.7.7) and (2.7.9),

$$\sum_\alpha (c_\alpha^{(A)})^2 (\alpha!)(2\mathbb{N})^{k\alpha} = \lim_{k\to\infty} \prod_{j=1}^k \frac{1}{1-\Lambda_j} \leq \prod_{j=1}^\infty (1 + 2\Lambda_j)$$

$$= \exp\left[\sum_{j=1}^\infty \log(1 + 2\Lambda_j)\right] \leq \exp\left[\sum_{j=1}^\infty 2\Lambda_j\right] < \infty,$$

by (2.3.3).   □

If $F \in (S)_{-1}$, then there exists $q < \infty$ such that $F \in (S)_{-1,-q}$. Hence we can make the following definition:

Definition 2.7.5 (The S-transform).   (i) Let $F \in (S)_{-1}$ and let $\phi \in S(\mathbb{R}^d)$. Then the S-transform of F at $\lambda\phi$, $(SF)(\lambda\phi)$, is defined, for all real numbers λ with $|\lambda|$ small enough, by

$$(SF)(\lambda\phi) = \langle F, \exp^\diamond[w(\lambda\phi, \cdot)]\rangle, \qquad\qquad (2.7.11)$$

where $\langle \cdot, \cdot \rangle$ denotes the action of $F \in (S)_{-1,-q}$ on $\exp^\diamond[w(\lambda\phi, \cdot)]$, which belongs to $((S)_{-1,-q})^* = (S)_{1,q}$ for $|\lambda|$ small enough, by Lemma 2.7.4.

(ii) Let $F \in (S)_{-\rho}$ for some $\rho < 1$ and let $\phi \in S(\mathbb{R}^d)$. Then the S-transform of F at $\lambda\phi$ is defined by

$$(SF)(\lambda\phi) = \langle F, \exp^\diamond[w(\lambda\phi, \cdot)]\rangle \qquad\qquad (2.7.12)$$

for all $\lambda \in \mathbb{R}$.

In terms of the chaos expansion we can express the S-transform as follows:

Proposition 2.7.6.   *Suppose $F = \sum_\alpha a_\alpha H_\alpha \in (S)_{-1}$, and let $\phi \in S(\mathbb{R}^d)$. Then, if $\lambda \in \mathbb{R}$ with $|\lambda|$ small enough, we have*

$$(SF)(\lambda\phi) = \sum_\alpha \lambda^{|\alpha|} a_\alpha(\phi, \eta)^\alpha; \quad \phi \in S(\mathbb{R}^d), \qquad (2.7.13)$$

where $(\phi, \eta)^\alpha = (\phi, \eta_1)^{\alpha_1} (\phi, \eta_2)^{\alpha_2} \cdots$.

Proof By (2.7.10) and (2.3.11) we have

$$(SF)(\lambda\phi) = \langle F, \exp^\diamond[w(\lambda\phi, \cdot)]\rangle$$

$$= \sum_\alpha a_\alpha \left(\frac{1}{\alpha!}\lambda^{|\alpha|}(\phi, \eta)^\alpha\right)\alpha! = \sum_\alpha \lambda^{|\alpha|} a_\alpha(\phi, \eta)^\alpha.$$

                                                                                                   □

Corollary 2.7.7. *As a function of λ the expression $(\mathcal{S}F)(\lambda\phi)$ is real analytic and hence has an analytic extension to all $\lambda \in \mathbb{C}$ with $|\lambda|$ small enough. If $F \in (\mathcal{S})_{-\rho}$ for some $\rho < 1$, then $(\mathcal{S}F)(\lambda\phi)$ extends to an entire function of $\lambda \in \mathbb{C}$.*

From now on we will consider the $\mathcal{S}$-transforms $(\mathcal{S}F)(\lambda\phi)$ to be these analytic extensions.

Example 2.7.8.   i) The $\mathcal{S}$-transform of smoothed white noise $F = w(\psi, \cdot)$, where $\psi \in L^2(\mathbb{R}^d)$, is, by (2.2.23) and (2.7.13),

$$(\mathcal{S}w(\psi, \cdot))(\lambda\phi) = \sum_{j=1}^{\infty} \lambda(\psi, \eta_j)(\phi, a_j) = (\lambda\phi, \psi); \quad \phi \in \mathcal{S}(\mathbb{R}^d), \lambda \in \mathbb{C}. \quad (2.7.14)$$

ii) The $\mathcal{S}$-transform of singular white noise $F = W(x, \cdot)$ is, by (2.3.33) and (2.7.13),

$$(\mathcal{S}W(x))(\lambda\phi) = \sum_{j=1}^{\infty} \lambda\eta_j(x)(\phi, \eta_j) = \lambda\phi(x); \quad \phi \in \mathcal{S}(\mathbb{R}^d), \lambda \in \mathbb{C}. \quad (2.7.15)$$

An important property of the $\mathcal{S}$-transform follows: (Compare it with Proposition 2.6.6.)

Proposition 2.7.9.   *Suppose $F, G \in (\mathcal{S})_{-1}$ and $\phi \in \mathcal{S}(\mathbb{R}^d)$. Then, if $|\lambda|$ is small enough,*

$$\mathcal{S}(F \diamond G)(\lambda\phi) = (\mathcal{S}F)(\lambda\phi) \cdot (\mathcal{S}G)(\lambda\phi). \quad (2.7.16)$$

Proof Suppose $F = \sum_\alpha a_\alpha H_\alpha$, $G = \sum_\beta b_\beta H_\beta$. Then by (2.7.13)

$$(\mathcal{S}F)(\lambda\phi) \cdot (\mathcal{S}G)(\lambda\phi) = \left(\sum_\alpha \lambda^{|\alpha|} a_\alpha(\phi, \eta)^\alpha\right)\left(\sum_\beta \lambda^{|\beta|} b_\beta(\phi, \eta)^\beta\right)$$

$$= \sum_{\alpha,\beta} \lambda^{|\alpha+\beta|} a_\alpha b_\beta(\phi, \eta)^{\alpha+\beta}$$

$$= \sum_\gamma \lambda^{|\gamma|} \left(\sum_{\alpha+\beta=\gamma} a_\alpha b_\beta\right)(\phi, \eta)^\gamma$$

$$= \mathcal{S}(F \diamond G)(\lambda\phi).$$

$\square$

The relation between the $\mathcal{S}$-transform and the $\mathcal{H}$-transform is the following:

Theorem 2.7.10.   *Let $F \in (\mathcal{S})_{-1}$. Then*

$$(\mathcal{H}F)(z_1, z_2, \ldots, z_k) = (\mathcal{S}F)(z_1\eta_1 + z_2\eta_2 + \cdots + z_k\eta_k) \quad (2.7.17)$$

for all $(z_1, \ldots, z_k) \in \mathbb{C}^k$ with $|z_j| < (2^d \delta_1^{(j)} \cdots \delta_d^{(j)})^{-q'}$; $1 \leq j \leq k$, where $q < \infty$ is so large that

$$F \in (\mathcal{S})_{-1,-q}$$

with $q' = d/d - 1$ if $d \leq 2$, $q' = q$ if $d = 1$.

Proof By (2.6.18) and (2.7.7), both sides of (2.7.11) are defined for all such $z = (z_1, \ldots, z_k) \in \mathbb{C}^k$. Suppose F has the chaos expansion

$$F = \sum_\alpha b_\alpha H_\alpha.$$

Then by (2.7.13) we have

$$(\mathcal{S}F)(z_1 \eta_1 + \cdots + z_k \eta_k) = \sum_\alpha b_\alpha (z_1 \eta_1 + \cdots + z_k \eta_k, \eta)^\alpha$$

$$= \sum_\alpha b_\alpha z_1^{\alpha_1} z_2^{\alpha_2} \cdots z_k^{\alpha_k} = \sum_\alpha b_\alpha z^\alpha = (\mathcal{H}F)(z),$$

as claimed. □

2.8 The Topology of $(\mathcal{S})_{-1}^N$

The topologies of $(\mathcal{S})_\rho^N$ and $(\mathcal{S})_{-\rho}^N$; $0 \leq \rho \leq 1$ are defined by the corresponding families of seminorms given in (2.3.9) and (2.3.10), respectively. Since we will often be working with the $\mathcal{H}$-transforms of elements of $(\mathcal{S})_{-1}^N$, it is useful to have a description of the topology in terms of the transforms. Such a description is

Theorem 2.8.1. *The following are equivalent:*

a) $X_n \to X$ in $(\mathcal{S})_{-1}^N$;
b) *there exist $\delta > 0, q < \infty$ such that $\tilde{X}_n(z) \to \tilde{X}(z)$ uniformly in $\mathbb{K}_q(\delta)$;*
c) *there exist $\delta > 0, q < \infty$ such that $\tilde{X}_n(z) \to \tilde{X}(z)$ pointwise boundedly in $\mathbb{K}_q(\delta)$.*

It suffices to prove this when $N = 1$. We need the following result (Recall that a bounded linear operator $A : H_1 \to H_2$ where H_i are Hilbert spaces $i = 1, 2$, is called a *Hilbert-Schmidt operator* if the series $\sum_{i,j} |(Ae_i, f_j)|^2$ converges whenever $\{e_i\}$ and $\{f_j\}$ are orthonormal bases for H_1 and H_2, respectively.):

Lemma 2.8.2. $(\mathcal{S})_1$ *is a nuclear space.*

Proof Define

$$(\mathcal{S})_{1,r} = \left\{ f = \sum_\alpha c_\alpha H_\alpha; \|f\|_{1,r}^2 < \infty \right\} \tag{2.8.1}$$

where $\|f\|_{\rho,r}$ is defined by Definition 2.7.1, so that

$$\|f\|_{1,r}^2 = \sum_\alpha c_\alpha^2(\alpha!)^2(2\mathbb{N})^{r\alpha}. \tag{2.8.2}$$

Then $(\mathcal{S})_{1,r}$ is a Hilbert space with inner product

$$\langle f, g\rangle_{1,r} = \sum_\alpha a_\alpha b_\alpha(\alpha!)^2(2\mathbb{N})^{r\alpha} \tag{2.8.3}$$

when $f = \sum_\alpha a_\alpha H_\alpha \in (\mathcal{S})_{1,r}$, $g = \sum_\beta b_\beta H_\beta \in (\mathcal{S})_{1,r}$.

Therefore the family of functions

$$H_{\alpha,r} = \frac{1}{\alpha!}(2\mathbb{N})^{-\frac{r\alpha}{2}} H_\alpha; \alpha \in \mathcal{J}$$

constitutes an orthonormal basis for $(\mathcal{S})_{1,r}$. By definition, $(\mathcal{S})_1$ is the projective limit of $(\mathcal{S})_{1,r}$, i.e.,

$$(\mathcal{S})_1 = \bigcap_{r=1}^\infty (\mathcal{S})_{1,r}.$$

If $r_2 > r_1 + 1$, then

$$\sum_\alpha \|H_{\alpha,r_2}\|_{1,r_1}^2 = \sum_\alpha \frac{1}{(\alpha!)^2}(2\mathbb{N})^{-r_2\alpha}(\alpha!)^2(2\mathbb{N})^{r_1\alpha}$$

$$= \sum_\alpha (2\mathbb{N})^{-(r_2-r_1)\alpha} < \infty,$$

by Proposition 2.3.3. This means that the imbedding $(\mathcal{S})_{1,r_2} \subset (\mathcal{S})_{1,r_1}$ is Hilbert–Schmidt if $r_2 > r_1 + 1$ and hence $(\mathcal{S})_1$ is a nuclear space. □

Proof of Theorem 2.8.1. a) $\Rightarrow$ **b).** First note that the dual $(\mathcal{S})_{-1,-r}$ of the space $(\mathcal{S})_{1,r}$ is defined by

$$(\mathcal{S})_{-1,-r} = \left\{ F = \sum_\alpha c_\alpha H_\alpha; \ \|F\|_{-1,-r}^2 := \sum_\alpha c_\alpha^2 (2\mathbb{N})^{-r\alpha} < \infty \right\}.$$

Assume that $X_n = \sum_\alpha b_\alpha^{(n)} H_\alpha \to X = \sum_\alpha b_\alpha H_\alpha$ in $(\mathcal{S})_{-1}$. Since $(\mathcal{S})_1$ is nuclear (Hida (1980)), this implies that there exists r_0 such that $X_n \to X$ in $(\mathcal{S})_{-1,-r_0}$ as $n \to \infty$. From this we deduce that

$$M^2 := \sup_n \{ \|X_n\|_{-1-r_0}^2 \}$$

$$= \sup_n \left\{ \sum_\alpha |b_\alpha^{(n)}|^2 (2\mathbb{N})^{-r_0\alpha} \right\} < \infty.$$

Hence

$$|\tilde{X}_n(z)| = \left| \sum_\alpha b^{(n)}_\alpha z^\alpha \right| = \left| \sum_\alpha b^{(n)}_\alpha (2\mathbb{N})^{-\frac{r_0\alpha}{2}} (2\mathbb{N})^{\frac{r_0\alpha}{2}} z^\alpha \right|$$

$$\leq \left(\sum_\alpha |b^{(n)}_\alpha|^2 (2\mathbb{N})^{-r_0\alpha} \right)^{\frac{1}{2}} \cdot \left(\sum_\alpha |z^\alpha|^2 (2\mathbb{N})^{r_0\alpha} \right)^{\frac{1}{2}}$$

$$\leq M(1+R)$$

if $z \in \mathbb{K}_{r_0}(R)$, so $\{\tilde{X}_n(z)\}$ is a bounded sequence on $\mathbb{K}_{r_0}(R)$ for all R.

Moreover, since $X_n \to X$ in $(\mathcal{S})_{-1,r_0}$, we have, by the same procedure as above, that

$$|\tilde{X}_n(z) - \tilde{X}(z)| = \left| \sum_\alpha (b^{(n)}_\alpha - b_\alpha) z^\alpha \right|$$

$$\leq \left(\sum_\alpha |b^{(n)}_\alpha - b_\alpha|^2 (2\mathbb{N})^{-r_0\alpha} \right)^{\frac{1}{2}} \cdot \left(\sum_\alpha |z^\alpha|^2 (2\mathbb{N})^{r_0\alpha} \right)^{\frac{1}{2}}$$

$$\leq (1+R)\|X_n - X\|_{-1,-r_0} \to 0 \quad \text{as} \quad n \to \infty,$$

uniformly for $z \in \mathbb{K}_{r_0}(R)$, for each $R < \infty$.

b) $\Rightarrow$ a). Suppose there exist $\delta > 0, q < \infty, M < \infty$ such that $\tilde{X}_n(z) \to \tilde{X}(z)$ for $z \in \mathbb{K}_q(\delta)$ and $|\tilde{X}_n(z)| \leq M$ for all $n = 1, 2, \ldots, z \in \mathbb{K}_q(\delta)$.

For $r < \infty$ and k a natural number, choose $\zeta = \zeta^{(r,k)} = (\zeta_1, \ldots, \zeta_k)$ with

$$\zeta_j = (2j)^{-r} \quad \text{for} \quad j = 1, \ldots, k.$$

Then

$$\sum_\alpha (2\mathbb{N})^{r\alpha} |\zeta^\alpha|^2 \leq \sum_\alpha (2\mathbb{N})^{-r\alpha} < \delta^2$$

for r large enough, say $r \geq q_1$.

Hence $\zeta \in \mathbb{K}_r(\delta)$ for $r \geq q_1$. Write $X_n = \sum_\alpha b^{(n)}_\alpha H_\alpha$. Since $|\tilde{X}_n(z)| \leq M$ for $z \in \mathbb{K}_q(\delta)$, we have by Proposition 2.6.8

$$\sum_\alpha |b^{(n)}_\alpha||z^\alpha| \leq MA(q) \quad \text{for all} \quad z \in \mathbb{K}_{3q}(\delta).$$

Thus if $r \geq \max(3q, q_1)$, we get

$$\sum_{\substack{\alpha \\ \text{Index } \alpha \leq k}} |b^{(n)}_\alpha|(2\mathbb{N})^{-r\alpha} = \sum_{\substack{\alpha \\ \text{Index } \alpha \leq k}} |b^{(n)}_\alpha|\zeta^\alpha$$

$$\sum_{\substack{\alpha \\ \text{Index } \alpha \leq k}} |b^{(n)}_\alpha||\zeta^\alpha| \leq \sum_\alpha |b^{(n)}_\alpha||\zeta^\alpha| \leq MA(q).$$

Letting $k \to \infty$ we deduce that

$$K := \sup_{\alpha} |b_{\alpha}^{(n)}| (2\mathbb{N})^{-r\alpha} < \infty,$$

which implies

$$\sum_{\alpha} |b_{\alpha}^{(n)}|^2 (2\mathbb{N})^{-2r\alpha} \leq K \sum_{\alpha} |b_{\alpha}^{(n)}| (2\mathbb{N})^{-r\alpha} < K M A(q).$$

So

$$\|X_n\|_{-1,-2r} \leq K M A(q) \quad \text{for all} \quad n.$$

A similar argument applied to $X_n - X$ instead of X_n gives the estimate

$$\|X_n - X\|_{-1,-2r} \leq K A(q) \sup_{z \in \mathbb{K}_q(\delta)} |\widetilde{X}_n(z) - \widetilde{X}(z)|.$$

The proof of the equivalence of **b)** and **c)** follows the familiar argument from the finite-dimensional case and is left as an exercise. $\square$

Stochastic Distribution Processes

As mentioned in the introduction, one advantage of working in the general space $(\mathcal{S})^N_{-1}$ of stochastic distributions is that it contains the solutions of many stochastic differential equations, both ordinary and partial and in arbitrary dimension. Moreover, if the objects of such equations are regarded as $(\mathcal{S})^N_{-1}$-valued, then differentiation can be interpreted in the usual strong sense in $(\mathcal{S})^N_{-1}$. This makes the following definition natural.

Definition 2.8.3. A measurable function

$$u : \mathbb{R}^d \to (\mathcal{S})^N_{-1}$$

is called a *stochastic distribution process* or an $(\mathcal{S})^N_{-1}$-*process.*

The process u is called continuous, differentiable, C^1, C^k, etc., if the $(\mathcal{S})^N_{-1}$-valued function u has these properties, respectively. For example, the partial derivative $\partial u / \partial x_k(x)$ of an $(\mathcal{S})_{-1}$-process u is defined by

$$\frac{\partial u}{\partial x_k}(x_1, \ldots, x_d)$$

$$= \lim_{\Delta x_k \to 0} \frac{u(x_1, \ldots, x_k + \Delta x_k, \ldots, x_d) - u(x_1, \ldots, x_k, \ldots, x_d)}{\Delta x_k}, \quad (2.8.4)$$

provided the limit exists in $(\mathcal{S})_{-1}$.

In terms of the Hermite transform $\widetilde{u}(x)(z) = \widetilde{u}(x; z)$, the limit on the right hand side of (2.8.4) exists if and only if there exists an element $Y \in (\mathcal{S})_{-1}$

such that

$$\frac{1}{\Delta x_k}[\tilde{u}(x_1, \ldots, x_k + \Delta x_k, \ldots, x_d; z) - \tilde{u}(x_1, \ldots, x_k, \ldots, x_d; z)] \to \tilde{Y}(z)$$

$$(2.8.5)$$

pointwise boundedly (or uniformly) in $\mathbb{K}_q(\delta)$ for some $q < \infty, \delta > 0$, according to Theorem 2.8.1. If this is the case, then Y is denoted by $\partial u/\partial x_k$.

When we apply the Hermite transform to solve stochastic differential equations the following observation is important.

For simplicity of notation, choose $N = d = 1$ and consider a differentiable $(\mathcal{S})_{-1}$-process $X(t, \omega)$. The statement that

$$\frac{dX(t, \omega)}{dt} = F(t, \omega) \quad \text{in} \quad (\mathcal{S})^{-1}$$

is then equivalent to saying that

$$\lim_{\Delta t \to 0} \frac{1}{\Delta t} \left(\tilde{X}(t + \Delta t; z) - \tilde{X}(t; z) \right) = \tilde{F}(t; z)$$

pointwise boundedly for $z \in \mathbb{K}_q(\delta)$ for some $q < \infty, \delta > 0$. For this it is clearly *necessary* that

$$\frac{d\tilde{X}(t; z)}{dt} = \tilde{F}(t; z) \quad \text{for each} \quad z \in \mathbb{K}_q(\delta),$$

but apparently not sufficient, because we also need that the pointwise convergence is *bounded* for $z \in \mathbb{K}_q(\delta)$. The following result is sufficient for our purposes.

Lemma 2.8.4 (Differentiation of $(\mathcal{S})_{-1}$-processes). *Suppose $X(t, \omega)$ and $F(t, \omega)$ are $(\mathcal{S})_{-1}$-processes such that*

$$\frac{d\tilde{X}(t; z)}{dt} = \tilde{F}(t; z) \quad \text{for each} \quad t \in (a, b), z \in \mathbb{K}_q(\delta) \qquad (2.8.6)$$

and that

$$\tilde{F}(t; z) \quad \text{is a bounded function of} \quad (t, z) \in (a, b) \times \mathbb{K}_q(\delta),$$
$$\text{continuous in} \quad t \in (a, b) \quad \text{for each} \quad z \in \mathbb{K}_q(\delta). \qquad (2.8.7)$$

Then $X(t, \omega)$ is a differentiable $(\mathcal{S})_{-1}$ process and

$$\frac{dX(t, \omega)}{dt} = F(t, \omega) \text{ for all } t \in (a, b). \qquad (2.8.8)$$

Proof By the mean value theorem we have

$$\frac{1}{\Delta t}\left(\widetilde{X}(t+\Delta t;z)-\widetilde{X}(t;z)\right)=\widetilde{F}(t+\theta\Delta t;z)$$

for some $\theta\in[0,1]$, for each $z\in\mathbb{K}_q(\delta)$. So if (2.8.6) and (2.8.7) hold, then

$$\frac{1}{\Delta t}\left(\widetilde{X}(t+\Delta t;z)-\widetilde{X}(t;z)\right)\to\widetilde{F}(t;z)\quad\text{as}\quad\Delta t\to 0,$$

pointwise boundedly for $z\in\mathbb{K}_q(\delta)$. $\square$

Similarly we can relate the integrability of an $(\mathcal{S})_{-1}$-process to the integrability of its $\mathcal{H}$-transform as follows:

We say that an $(\mathcal{S})_{-1}$-process $X(t)$ is (strongly) integrable in $(\mathcal{S})_{-1}$ over the interval $[a,b]$ if

$$\int_a^b X(t,\omega)dt:=\lim_{\Delta t_k\to 0}\sum_{k=0}^{n-1}X(t_k^*,\omega)\Delta t_k \qquad (2.8.9)$$

exists in $(\mathcal{S})_{-1}$, for all partitions $a=t_0<t_1<\cdots<t_n=b$ of $[a,b]$, $\Delta t_k=t_{k+1}-t_k$ and $t_k^*\in[t_k,t_{k+1}]$ for $k=1,\ldots,n-1$.

Taking $\mathcal{H}$-transforms and using Theorem 2.8.1, we get the following result:

Lemma 2.8.5. *Let $X(t)$ be an $(\mathcal{S})_{-1}$-process. Suppose there exist $q<\infty$, $\delta>0$ such that*

$$\sup\{\widetilde{X}(t;z);t\in[a,b],z\in\mathbb{K}_q(\delta)\}<\infty \qquad (2.8.10)$$

and

$$\widetilde{X}(t;z)\quad\text{is a continuous function}$$
$$\text{of } t\in[a,b]\quad\text{for each}\quad z\in\mathbb{K}_q(\delta). \qquad (2.8.11)$$

Then $X(t)$ is strongly integrable and

$$\mathcal{H}\left[\int_a^b X(t)dt\right]=\int_a^b\widetilde{X}(t)dt. \qquad (2.8.12)$$

Example 2.8.6. Choose $N=m=d=1$ and let $B(t,\omega)$ be Brownian motion. Then (see Example 2.2.5)

$$B(t;\omega)=\sum_{j=1}^{\infty}\int_0^t\xi_j(s)ds H_{\epsilon^{(j)}}(\omega),$$

and so

$$\widetilde{B}(t;z) = \sum_{j=1}^{\infty} \int_0^t \xi_j(s)ds z_j; \quad z \in (\mathbb{C}^N)_c.$$

Hence

$$\frac{d\widetilde{B}(t;z)}{dt} = \sum_{j=1}^{\infty} \xi_j(t)z_j, \quad \text{for each } z \in (\mathbb{C}^N)_c.$$

Moreover,

$$\left| \sum_{j=1}^{\infty} \xi_j(t)z_j \right|^2 \leq \left(\sum_{j=1}^{\infty} \xi_j^2(t)(2\mathbb{N})^{-2\epsilon^{(j)}} \right) \left(\sum_{j=1}^{\infty} |z^{\epsilon^{(j)}}|^2 (2\mathbb{N})^{2\epsilon^{(j)}} \right)$$

$$\leq \sup_{j,t} |\xi_j^2(t)| \sum_{j=1}^{\infty} (2j)^{-2} \sum_{\alpha} |z^\alpha|^2 (2\mathbb{N})^{2\alpha} \leq CR^2$$

for some constant C if $z \in \mathbb{K}_2(R)$. We also have that

$$t \to \sum_{j=1}^{\infty} \xi_j(t)z_j \quad \text{is continuous.}$$

Since $\sum_{j=1}^{\infty} \xi_j(t)z_j$ is the $\mathcal{H}$-transform of white noise $W(t,\omega)$ (see (2.6.8)), we conclude by Lemma 2.8.4 that

$$\frac{dB(t,\omega)}{dt} = W(t,\omega) \quad \text{in} \quad (\mathcal{S})_{-1}. \tag{2.8.13}$$

(Compare with (2.5.27).)

Example 2.8.7. Let us proceed one step further from the previous example and try to differentiate white noise $W(t,\omega)$. (Again we assume $m = d = 1$.) Since

$$\widetilde{W}(t;z) = \sum_{j=1}^{\infty} \xi_j(t)z_j; \quad z \in (\mathbb{C}^N)_c,$$

we get

$$\frac{d\widetilde{W}(t;z)}{dt} = \sum_{j=1}^{\infty} \xi_j'(t)z_j; \quad z \in (\mathbb{C}^N)_c.$$

Here the right hand side is clearly a continuous function of t for each z. It remains to prove boundedness for $z \in \mathbb{K}_q(\delta)$ for some $q < \infty, \delta > 0$. From the definition (2.2.1) of the Hermite functions ξ_j together with the estimate (2.2.5) we conclude that

$$\sup_{t \in [a,b]} |\xi_j'(t)| \leq Cj, \tag{2.8.14}$$

where $C = C_{a,b}$ is a constant depending only on a, b. Hence

$$\left| \sum_{j=1}^{\infty} \xi_j'(t) z_j \right|^2 \leq \left(\sum_{j=1}^{\infty} |\xi_j'(t)|^2 (2\mathbb{N})^{-4\epsilon^{(j)}} \right) \cdot \left(\sum_{j=1}^{\infty} |z_j|^2 (2\mathbb{N})^{4\epsilon^{(j)}} \right)$$

$$\leq C^2 \sum_{j=1}^{\infty} j^2 (2j)^{-4} \cdot \sum_{\alpha} |z^{\alpha}|^2 (2\mathbb{N})^{4\alpha} \leq C_1 R^2$$

if $z \in \mathbb{K}_4(R)$; $t \in [a, b]$. From Lemma 2.8.4 we conclude that

$$\frac{dW(t, \omega)}{dt} = \sum_{j=1}^{\infty} \xi_j'(t) H_{\epsilon^{(j)}}(\omega) \quad \text{in} \quad (\mathcal{S})_{-1}. \tag{2.8.15}$$

2.9 The $\mathcal{F}$-Transform and the Wick Product on $L^1(\mu)$

The $\mathcal{S}$-transform is closely related to the *Fourier transform* or $\mathcal{F}$-transform, which is defined on $L^1(\mu)$ as follows:

Definition 2.9.1. Let $g \in L^1(\mu_m), \phi \in \mathcal{S}(\mathbb{R}^d)$. Then the $\mathcal{F}$-transform, $\mathcal{F}[g](\phi)$, of g at ϕ, is defined by

$$\mathcal{F}[g](\phi) = \int_{\mathcal{S}'(\mathbb{R}^d)} e^{i\langle \omega, \phi \rangle} g(\omega) d\mu(\omega). \tag{2.9.1}$$

Note that if $g \in L^p(\mu_m)$ for some $p > 1$, then $g \in (\mathcal{S})^*$ (Corollary 2.3.8) and hence, with i denoting the imaginary unit,

$$(\mathcal{S}g)(i\phi) = \langle g, \exp^{\diamond}[w(i\phi, \cdot)] \rangle = \int_{\mathcal{S}'(\mathbb{R}^d)} \exp^{\diamond}[w(i\phi, \omega)] g(\omega) d\mu(\omega)$$

$$= \int_{\mathcal{S}'(\mathbb{R}^d)} \exp^{\diamond}[i\langle \omega, \phi \rangle] g(\omega) d\mu(\omega)$$

$$= e^{\frac{1}{2}\|\phi\|^2} \int_{\mathcal{S}'(\mathbb{R}^d)} \exp[i\langle \omega, \phi \rangle] g(\omega) d\mu(\omega)$$

$$= e^{\frac{1}{2}\|\phi\|^2} \mathcal{F}[g](\phi).$$

This gives

Lemma 2.9.2. a) *Suppose* $g \in L^1(\mu_m) \cap (\mathcal{S})_{-\rho}$ *for some* $\rho < 1$. *Then*

$$\mathcal{F}[g](\phi) = e^{-\frac{1}{2}\|\phi\|^2} (\mathcal{S}g)(i\phi) \tag{2.9.2}$$

for all $\phi \in \mathcal{S}(\mathbb{R}^d)$.

b) *Suppose $h \in L^1(\mu_m) \cap (\mathcal{S})_{-1}$. Then for all $\phi \in \mathcal{S}(\mathbb{R}^d)$ we have*

$$\mathcal{F}[h](\lambda\phi) = e^{-\frac{1}{2}\lambda^2 \|\phi\|^2}(\mathcal{S}h)(i\lambda\phi) \quad for \quad |\lambda| \quad small\ enough. \tag{2.9.3}$$

Proof **a)** We have proved that (2.9.2) holds if $g \in L^2(\mu_m)$. Since $L^2(\mu)$ is dense in both $(\mathcal{S})_{-\rho}$ and $L^1(\mu_m)$, the result follows.

b) Choose $h_n \in L^1(\mu_m) \cap (\mathcal{S})_{-\rho}$ (for some fixed $\rho < 1$) such that $h_n \to h$ in $L^1(\mu_m)$ and in $(\mathcal{S})_{-1}$. Then (2.9.2) holds for h_n for all n. Taking the limit as $n \to \infty$ we get (2.9.3). $\qquad\square$

This result gives the following connection between $\mathcal{F}$-transforms and Wick products.

Lemma 2.9.3. **a)** *Suppose X, Y and $X \diamond Y \in L^1(\mu_m) \cap (\mathcal{S})_{-\rho}$ for some $\rho < 1$.*
Then

$$\mathcal{F}[X \diamond Y](\phi) = e^{\frac{1}{2}\|\phi\|^2}\mathcal{F}[X](\phi) \cdot \mathcal{F}[Y](\phi); \ \phi \in \mathcal{S}(\mathbb{R}^d). \tag{2.9.4}$$

b) *Suppose X, Y and $X \diamond Y$ all belong to $L^1(\mu_m) \cap (\mathcal{S})_{-1}$. Then for all $\phi \in \mathcal{S}(\mathbb{R}^d)$ we have*

$$\mathcal{F}[X \diamond Y](\lambda\phi) = e^{\frac{1}{2}\lambda^2\|\phi\|^2}\mathcal{F}[X](\lambda\phi)\mathcal{F}[Y](\lambda\phi) \tag{2.9.5}$$

for $|\lambda|$ small enough.

Proof **a)** By Lemma 2.9.2 a) and Proposition 2.7.9 we have, for $\phi \in \mathcal{S}(\mathbb{R}^d)$,

$$\begin{aligned}
\mathcal{F}[X \diamond Y](\phi) &= e^{-\frac{1}{2}\|\phi\|^2}(\mathcal{S}(X \diamond Y))(i\phi) \\
&= e^{-\frac{1}{2}\|\phi\|^2}(\mathcal{S}X)(i\phi) \cdot (\mathcal{S}Y)(i\phi) \\
&= e^{-\frac{1}{2}\|\phi\|^2}e^{\frac{1}{2}\|\phi\|^2}\mathcal{F}[X](\phi)e^{\frac{1}{2}\|\phi\|^2}\mathcal{F}[Y](\phi) \\
&= e^{\frac{1}{2}\|\phi\|^2}\mathcal{F}[X](\phi)\mathcal{F}[Y](\phi).
\end{aligned}$$

b) This follows from Lemma 2.9.2 b) in the same way. $\qquad\square$

Using the $\mathcal{F}$-transform we can now (partially) extend the Wick product to $L^1(\mu_m)$ as follows:

Definition 2.9.4 (The Wick product on $L^1(\mu_m)$). Let $X, Y \in L^1(\mu_m)$. Suppose there exist $X_n, Y_n \in L^2(\mu_m)$ such that

$$X_n \to X \text{ in } L^1(\mu_m) \text{ and } Y_n \to Y \text{ in } L^1(\mu_m) \text{ as } n \to \infty \tag{2.9.6}$$

and such that

$$\lim_{n \to \infty} X_n \diamond Y_n \quad \text{exists in} \quad L^1(\mu_m). \tag{2.9.7}$$

Then we define the *Wick product of X and Y in* $L^1(\mu_m)$, denoted $X \hat{\diamond} Y$, by

$$X \hat{\diamond} Y = \lim_{n \to \infty} X_n \diamond Y_n. \tag{2.9.8}$$

We must show that $X \hat{\diamond} Y$ is well defined, i.e., we must show that $\lim_{n \to \infty} X_n \diamond Y_n$ does not depend on the actual sequences $\{X_n\}, \{Y_n\}$. This is done in the following lemma.

Lemma 2.9.5. *Let X_n, Y_n be as in Definition 2.9.4 and assume that X_n', Y_n' also satisfy*

$$X_n' \to X \text{ in } L^1(\mu_m) \text{ and } Y_n' \to Y \text{ in } L^1(\mu_m) \text{ as } n \to \infty \tag{2.9.9}$$

and

$$\lim_{n \to \infty} X_n' \diamond Y_n' \quad \text{exists in } L^1(\mu_m). \tag{2.9.10}$$

Then

$$\lim_{n \to \infty} X_n' \diamond Y_n' = \lim_{n \to \infty} X_n \diamond Y_n = X \hat{\diamond} Y. \tag{2.9.11}$$

Moreover, we have

$$\mathcal{F}[X \hat{\diamond} Y](\phi) = e^{\frac{1}{2}\|\phi\|^2} \mathcal{F}[X](\phi)\mathcal{F}[Y](\phi) \tag{2.9.12}$$

for all $\phi \in \mathcal{S}(\mathbb{R}^d)$.

Proof Set $Z = \lim_{n \to \infty} X_n \diamond Y_n$. Then by Lemma 2.9.3 we have

$$\begin{aligned}
\mathcal{F}[Z](\phi) &= \lim_{n \to \infty} \mathcal{F}[X_n \diamond Y_n](\phi) \\
&= \lim_{n \to \infty} e^{\frac{1}{2}\|\phi\|^2} \mathcal{F}[X_n](\phi)\mathcal{F}[Y_n](\phi) \\
&= e^{\frac{1}{2}\|\phi\|^2} \mathcal{F}[X](\phi)\mathcal{F}[Y](\phi), \ \phi \in \mathcal{S}(\mathbb{R}^d).
\end{aligned}$$

Similarly, we get, with $Z' = \lim_{n \to \infty} X_n' \diamond Y_n'$,

$$\mathcal{F}[Z'](\phi) = e^{\frac{1}{2}\|\phi\|^2} \mathcal{F}[X](\phi) \cdot \mathcal{F}[Y](\phi),$$

hence $\mathcal{F}[Z](\phi) = \mathcal{F}[Z'](\phi)$ for all $\phi \in \mathcal{S}(\mathbb{R}^d)$. Since the algebra $\mathcal{E}$ generated by the stochastic exponentials $\exp[i\langle \omega, \phi \rangle]$; $\phi \in \mathcal{S}(\mathbb{R}^d)$ is dense in $L^2(\mu_m)$ (see Theorem 2.1.3), a function in $L^1(\mu_m)$ is uniquely determined by its Fourier transform. Therefore $Z = Z'$. This proves (2.9.11) and also (2.9.12).   □

We can now verify that the two Wick products $\hat{\diamond}$ and $\diamond$ coincide on the intersection $L^1(\mu_m) \cap (\mathcal{S})_{-1}$.

Theorem 2.9.6. Let $X, Y \in L^1(\mu_m) \cap (\mathcal{S})_{-1}$. Assume that $X \hat{\diamond} Y$ exists in $L^1(\mu_m)$ and that $X \diamond Y$ (the Wick product in $(\mathcal{S})_{-1}$) belongs to $L^1(\mu_m)$. Then

$$X \hat{\diamond} Y = X \diamond Y.$$

Proof By Lemma 2.9.3 (ii) we have that

$$\mathcal{F}[X \diamond Y](\lambda \phi) = e^{\frac{1}{2}\lambda^2 \|\phi\|^2} \mathcal{F}[X](\lambda \phi) \cdot \mathcal{F}[Y](\lambda \phi)$$

for all $\phi \in \mathcal{S}(\mathbb{R}^d)$ if $|\lambda|$ is small enough. On the other hand, from Lemma 2.9.5 we have that

$$\mathcal{F}[X \hat{\diamond} Y](\psi) = e^{\frac{1}{2}\|\psi\|^2} \mathcal{F}[X](\psi) \cdot \mathcal{F}[Y](\psi)$$

for all $\psi \in \mathcal{S}(\mathbb{R}^d)$.
    This is sufficient to conclude that $X \hat{\diamond} Y = X \diamond Y$. $\square$

Remark In view of Theorem 2.9.6 we can – and will – from now on write $X \diamond Y$ for the Wick product in $L^1(\mu_m)$.

Corollary 2.9.7. Let $X, Y \in L^1(\mu_m)$, and assume that $X \diamond Y \in L^1(\mu_m)$ exists (in the sense of Definition 2.9.4). Then

$$E[X \diamond Y] = E[X] \cdot E[Y]. \qquad (2.9.13)$$

Proof Choose $\phi = 0$ in (2.9.12). $\square$

Functional Processes

As pointed out in the introduction, it is sometimes useful to smooth the singular white noise $\mathbf{W}(x, \omega)$ by a test function $\phi \in \mathcal{S}(\mathbb{R}^d)$, thereby obtaining the smoothed white noise process

$$\mathbf{W}_\phi(x, \omega) = \mathbf{w}(\phi_x, \omega), \qquad (2.9.14)$$

where $\phi_x(y) = \phi(y - x)$; $x, y \in \mathbb{R}^d$ (see (2.1.19)).
    The reason for doing this could be simply technical: By smoothing the white noise we get less singular equations to work with and therefore (we hope) less singular solutions.
    But the reason could also come from the model: In some cases the smoothed process (2.9.14) simply gives a more realistic model for the noise we consider. In these cases the choice of ϕ may have a physical significance. For example, in the modeling of fluid flow in a porous, random medium the smoothed positive noise

$$K_\phi(x, \omega) = \exp^\diamond[W_\phi(x, \omega)] \qquad (2.9.15)$$

will be a natural model for the (stochastic) permeability of the medium, and then the size of the support of ϕ will give the distance beyond which the permeability values at different points are independent. (See Chapter 4.)

In view of this, the following concept is useful:

Definition 2.9.8 (Functional processes). A *functional process* is a map

$$X : \mathcal{S}(\mathbb{R}^d) \times \mathbb{R}^d \to L^1(\mu_m).$$

If there exists $p \geq 1$ such that

$$X(\phi, x) \in L^p(\mu_m) \quad \text{for all} \quad \phi \in \mathcal{S}(\mathbb{R}^d), \ x \in \mathbb{R}^d,$$

then X is called an L^p-*functional process*.

Example 2.9.9. The processes $\mathbf{W}_\phi(x), K_\phi(x)$ given in (2.9.14) and (2.9.15) are both L^p-functional processes for all $p < \infty$.

In Chapters 3 and 4 we will give examples of smoothed stochastic differential equations with solutions $X(\phi, x)$ that are L^p-functional processes for $p = 1$ but not for any $p > 1$.

2.10 The Wick Product and Translation

There is a striking relation between Wick products, Wick exponentials of white noise and translation. This relation was first formulated on the Wiener space in Gjessing (1994), Theorem 2.10, and applied there to solve quasilinear anticipating stochastic differential equations. Subsequently the relation was generalized by Benth and Gjessing (2000), and applied to a class of nonlinear parabolic stochastic partial differential equations. The relation has also been applied to prove positivity of solutions of stochastic heat transport equations in Benth (1995).

In this section we will prove an $(\mathcal{S})_{-1}$-version of this relation (Theorem 2.10.2). Then in Chapter 3 we present a variation of the SDE application in Gjessing (1994), and in Chapter 4 we will look at some of the above-mentioned applications to SPDEs.

We first consider the translation on functions in $(\mathcal{S})_1$.

Theorem 2.10.1. *For $f \in (\mathcal{S})_1$ and $\omega_0 \in \mathcal{S}'(\mathbb{R}^d)$, define the function $T_{\omega_0} f : \mathcal{S}'(\mathbb{R}^d) \to \mathbb{R}$ by*

$$T_{\omega_0} f(\omega) = f(\omega + \omega_0); \omega \in \mathcal{S}'(\mathbb{R}^d). \tag{2.10.1}$$

Then the map $f \to T_{\omega_0} f$ is a continuous linear map from $(\mathcal{S})_1$ into $(\mathcal{S})_1$.

Proof Suppose $f \in (\mathcal{S})_1$ has the expansion

$$f(\omega) = \sum_{\beta} c_{\beta} H_{\beta}(\omega) = \sum_{\beta} c_{\beta} \langle \omega, \eta \rangle^{\diamond \beta},$$

where

$$\langle \omega, \eta \rangle^{\diamond \beta} = \langle \omega, \eta_1 \rangle^{\diamond \beta_1} \diamond \langle \omega, \eta_2 \rangle^{\diamond \beta_2} \diamond \cdots$$

(see (2.4.17)). Then

$$f(\omega + \omega_0) = \sum_{\beta} c_{\beta} \langle \omega + \omega_0, \eta \rangle^{\diamond \beta} = \sum_{\beta} c_{\beta} (\langle \omega, \eta \rangle + \langle \omega_0, \eta \rangle)^{\diamond \beta}$$

$$= \sum_{\beta} c_{\beta} \prod_{j=1}^{\infty} (\langle \omega, \eta_j \rangle + \langle \omega_0, \eta_j \rangle)^{\diamond \beta_j}$$

$$= \sum_{\beta} c_{\beta} \prod_{j=1}^{\infty} \left(\sum_{\gamma_j=0}^{\beta_j} \binom{\beta_j}{\gamma_j} \langle \omega, \eta_j \rangle^{\diamond \gamma_j} \langle \omega_0, \eta_j \rangle^{(\beta_j - \gamma_j)} \right)$$

$$= \sum_{\beta} c_{\beta} \sum_{0 \leq \gamma_k \leq \beta_k} \binom{\beta_1}{\gamma_1} \binom{\beta_2}{\gamma_2} \cdots \langle \omega, \eta_1 \rangle^{\diamond \gamma_1} \cdot \langle \omega, \eta_2 \rangle^{\diamond \gamma_2}$$

$$\cdots \langle \omega_0, \eta_1 \rangle^{(\beta_1 - \gamma_1)} \cdot \langle \omega_0, \eta_2 \rangle^{(\beta_2 - \gamma_2)} \cdots$$

$$= \sum_{\beta} c_{\beta} \sum_{0 \leq \gamma \leq \beta} \binom{\beta}{\gamma} \langle \omega, \eta \rangle^{\diamond \gamma} \langle \omega_0, \eta \rangle^{\beta - \gamma},$$

where we have used the multi-index notation

$$\binom{\beta}{\gamma} = \binom{\beta_1}{\gamma_1} \binom{\beta_2}{\gamma_2} \cdots = \frac{\beta_1!}{\gamma_1!(\beta_1 - \gamma_1)!} \cdot \frac{\beta_2!}{\gamma_2!(\beta_2 - \gamma_2)!} \cdots = \frac{\beta!}{\gamma!(\beta - \gamma)!}.$$

Hence the expansion of $f(\omega + \omega_0)$ is

$$f(\omega + \omega_0) = \sum_{\gamma} \sum_{\beta \geq \gamma} c_{\beta} \binom{\beta}{\gamma} \langle \omega_0, \eta \rangle^{\beta - \gamma} \langle \omega, \eta \rangle^{\diamond \gamma}.$$

Introduce

$$b_{\gamma} = \sum_{\beta \geq \gamma} c_{\beta} \binom{\beta}{\gamma} \langle \omega_0, \gamma \rangle^{\beta - \gamma}.$$

To show that $f(\omega + \omega_0) \in (\mathcal{S})_1$, we must verify that

$$J(q) := \sum_{\gamma} b_{\gamma}^2 (\gamma!)^2 (2\mathbb{N})^{q\gamma} < \infty \quad \text{for all} \quad q \in \mathbb{N}. \qquad (2.10.2)$$

Choose $q > 2$. Since we have $f \in (\mathcal{S})_1$, we know that for all $r \in \mathbb{N}$ there exists $M(r) \in (0, \infty)$ such that

$$c_\beta^2 (\beta!)^2 (2\mathbb{N})^{2r\beta} \leq M^2(r) \quad \text{for all } \beta,$$

i.e.,

$$|c_\beta| \leq M(r)(\beta!)^{-1}(2\mathbb{N})^{-r\beta} \quad \text{for all} \quad \beta.$$

Therefore, with r to be determined later,

$$J(q) \leq \sum_\gamma \left(\sum_{\beta \geq \gamma} M(r)(\beta!)^{-1}(2\mathbb{N})^{-r\beta} \binom{\beta}{\gamma} \langle \omega_0, \eta \rangle^{\beta-\gamma} \right)^2 (\gamma!)^2 (2\mathbb{N})^{q\gamma}$$

$$\leq M(r)^2 \sum_\gamma \left(\sum_{\beta \geq \gamma} \langle \omega_0, \eta \rangle^{\beta-\gamma} (2\mathbb{N})^{-r\beta} \right)^2 (2\mathbb{N})^{q\gamma}. \qquad (2.10.3)$$

By Theorem 2.3.1 we can write $\omega_0 = \sum_{j=1}^\infty b_j \eta_j$, where

$$\sum_{j=1}^\infty b_j^2 (\delta_1^{(j)})^{-\theta_1}(\delta_2^{(j)})^{-\theta_2} \cdots (\delta_d^{(j)})^{-\theta_d} < \infty$$

for some $\theta = (\theta_1, \ldots, \theta_d)$. Setting $\theta_0 = \max\{\theta_j; 1 \leq j \leq d\}$, we get

$$\sum_{j=1}^\infty b_j^2 (\delta_j^{(j)} \cdots \delta_d^{(j)})^{-\theta_0} < \infty.$$

By Lemma 2.3.4 this implies that

$$\sum_{j=1}^\infty b_j^2 j^{-\theta_0 d} < \infty.$$

In particular, there exists $K \in (1, \infty)$ such that

$$|\langle \omega_0, \eta_j \rangle| = |b_j| \leq K \cdot j^{\theta_0 d}. \qquad (2.10.4)$$

Using this in (2.10.3), we get

$$J(q) \leq M(r)^2 \sum_\gamma \left(\sum_{\beta \geq \gamma} (K\mathbb{N})^{\theta_0 d(\beta-\gamma)}(2\mathbb{N})^{-r\beta} \right)^2 (2\mathbb{N})^{q\gamma}.$$

Now choose

$$r = \theta_0 d(1 + \log_2 K) + q + 2. \qquad (2.10.5)$$

Then we get

$$
J(q) \le M(r)^2 \sum_{\gamma} \left(\sum_{\beta \ge \gamma} K^{\theta_0 d|\beta|} \mathrm{N}^{\theta_0 d\beta} 2^{-r|\beta|} \mathrm{N}^{-r\beta} \right)^2 (2\mathrm{N})^{q\gamma}
$$

$$
\le M(r)^2 \sum_{\gamma} \left(\sum_{\beta \ge \gamma} \mathrm{N}^{\theta_0 d\beta} 2^{-(q+2)|\beta|} \mathrm{N}^{-\theta_0 d\beta - (q+2)\beta} \right)^2 (2\mathrm{N})^{q\gamma}
$$

$$
= M(r)^2 \sum_{\gamma} \left(\sum_{\beta \ge \gamma} (2\mathrm{N})^{-q\beta - 2\beta} \right)^2 (2\mathrm{N})^{q\gamma}
$$

$$
\le M(r)^2 \sum_{\gamma} \left(\sum_{\beta \ge \gamma} (2\mathrm{N})^{-q\gamma - 2\beta} \right)^2 (2\mathrm{N})^{q\gamma}
$$

$$
\le M(r)^2 \sum_{\gamma} \left(\sum_{\beta \ge 0} (2\mathrm{N})^{-2\beta} \right) (2\mathrm{N})^{-q\gamma}
$$

$$
= M(r)^2 \sum_{\beta \ge 0} (2\mathrm{N})^{-2\beta} \sum_{\gamma \ge 0} (2\mathrm{N})^{-q\gamma} < \infty.
$$

This proves (2.10.2), and we conclude that $T_{\omega_0} f \in (\mathcal{S})_1$.

It is clear that the map $f \to T_{\omega_0} f$ is linear. Finally, to prove that $f \to T_{\omega_0} f$ is continuous from $(\mathcal{S})_1$ into $(\mathcal{S})_1$, note that the argument above actually shows that T_{ω_0} maps $(\mathcal{S})_{1,r}$ into $(\mathcal{S})_{1,q}$ when r is given by (2.10.5). This proves the continuity, for if f_n is a sequence in $(\mathcal{S})_1$ converging to 0 and

$$
N_{1,q,R} := \{ f \in (\mathcal{S})_{1,q}; \|f\|_{1,q} < R \}
$$

is a neighborhood of 0 in $(\mathcal{S})_1$, then $T_{\omega_0} f_n \in N_{1,q,R}$ if n is so large that we have $f_n \in (\mathcal{S})_{1,r}$. □

Remark It is proved in Hida et al. (1993), Theorem 4.15, that T_{ω_0} is a continuous linear map from $(\mathcal{S})(= (\mathcal{S})_0)$ into $(\mathcal{S})$. In Potthoff and Timpel (1995), the same is proved for the translation operator on the space $(\mathcal{G})$.

Definition 2.10.2. Fix $\omega_0 \in \mathcal{S}'(\mathbb{R}^d)$.

a) The map $T_{\omega_0} : (\mathcal{S})_1 \to (\mathcal{S})_1$ is called the *translation operator*.
b) The *adjoint translation operator* is the map

$$
T_{\omega_0}^* : (\mathcal{S})_{-1} \to (\mathcal{S})_{-1}
$$

defined by

$$
\langle T_{\omega_0}^* X, f \rangle = \langle X, T_{\omega_0} f \rangle; \quad f \in (\mathcal{S})_1, X \in (\mathcal{S})_{-1}. \tag{2.10.6}
$$

Remark Note that $T^*_{w_0}$ maps $(\mathcal{S})_{-1}$ into $(\mathcal{S})_{-1}$ because of Theorem 2.10.1.

The following result is the $(\mathcal{S})_{-1}$-version of Lemma 5.3 in Benth and Gjessing (2000) (see also Prop. 9.4 in Benth (1995)).

Theorem 2.10.3 Benth and Gjessing (2000). *Let* $w_0 \in \mathcal{S}'(\mathbb{R}^d)$ *and* $X \in (\mathcal{S})_{-1}$. *Then*

$$T^*_{w_0} X = X \diamond \exp^\diamond[w(w_0)], \qquad (2.10.7)$$

where

$$w(w_0, w) := \sum_{j=1}^{\infty} \langle w_0, \eta_j \rangle H_{\epsilon^{(j)}}(w) \in (\mathcal{S})^*. \qquad (2.10.8)$$

is the generalized smoothed white noise.

Proof First note that from (2.10.4) it follows that $w(w_0, \cdot) \in (\mathcal{S})^*$. We verify (2.10.7) by proving that $\mathcal{S}$-transforms of the two sides are equal: For $\phi \in \mathcal{S}(\mathbb{R}^d)$ and $|\lambda|$ small enough we have (see (2.7.11))

$$\begin{aligned}
(ST^*_{w_0}X)(\lambda\phi) &= \langle T^*_{w_0}X, \exp^\diamond[w(\lambda\phi, \cdot)]\rangle \\
&= \langle X, T_{w_0}(\exp^\diamond[w(\lambda\phi, \cdot)])\rangle \\
&= \langle X, \exp^\diamond[\langle w + w_0, \lambda\phi\rangle]\rangle \\
&= \langle X, \exp^\diamond[\langle w, \lambda\phi\rangle]\rangle \cdot \exp[\langle w_0, \lambda\phi\rangle] \\
&= (SX)(\lambda\phi) \cdot (Sw_{w_0})(\lambda\phi) \\
&= S(X \diamond w_{w_0})(\lambda\phi).
\end{aligned}$$

By Theorem 2.7.10 and the uniqueness of the Hermite transform on $(\mathcal{S})_{-1}$, the theorem is proved.   □

Corollary 2.10.4 Benth and Gjessing (2000). a) *If* $X \in (\mathcal{S})_{-1}$ *and* $w_0 \in \mathcal{S}'(\mathbb{R}^d)$, *then*

$$\langle \exp^\diamond[w(w_0)] \diamond X, f \rangle = \langle X, T_{w_0}f \rangle; f \in (\mathcal{S})_1. \qquad (2.10.9)$$

b) *If* $X \in (\mathcal{S})_1, f \in (\mathcal{S})_1$ *and* $w_0 = \phi \in L^2(\mathbb{R}^d)$, *then*

$$\int_{\mathcal{S}'} f(w) \cdot (\exp^\diamond[w(\phi)] \diamond X)(w)d\mu(w) = \int_{\mathcal{S}'} X(w)f(w + \phi)d\mu(w). \qquad (2.10.10)$$

Proof a) *follows directly from (2.10.7). Version* b) *is an* $(\mathcal{S})_1$-*version of* a).   □

In particular, as observed in Benth and Gjessing (2000), if we choose $X \equiv 1$, we recover a version of the *Girsanov formula*.

Corollary 2.10.5. *Let $f \in L^p(\mu_1)$ for some $p > 2$ and let $\phi \in L^2(\mathbb{R}^d)$. Then $f(\omega + \phi) \in L^2(\mu_1)$ and*

$$\int_{S'} f(\omega) \cdot \exp^\diamond[w(\phi)](\omega)d\mu_1(\omega) = \int_{S'} f(\omega + \phi)d\mu_1(\omega). \qquad (2.10.11)$$

Proof Fix $p > 2$ and $\varphi \in L^2(\mathbb{R}^d)$. Choose $f_n \in (\mathcal{S})_1$ such that $f_n \to f$ in $L^p(\mu_1)$. Then by (2.10.10) we get that (2.10.11) holds for each f_n and with $\phi = 1/2\varphi$, i.e.

$$\int_{S'} f_n(\omega) \cdot \exp^\diamond\left[w\left(\frac{1}{2}\varphi\right)\right](\omega)d\mu_1(\omega) = \int_{S'} f_n\left(\omega + \frac{1}{2}\varphi\right)d\mu_1(\omega); \quad n = 1, 2, \ldots$$

Since $\exp^\diamond[w(1/2\varphi)] = \exp[w(1/2\varphi) - 1/8\|\varphi\|^2_{L^2(\mathbb{R}^d)}]$ is in $L^q(\mu_1)$ for all $q < \infty$, we have by the Hölder inequality that $f_n \cdot \exp^\diamond[w(1/2\varphi)] \to f \cdot \exp^\diamond[w(1/2\varphi)]$ in $L^2(\mu_1)$. Therefore

$$\int_{S'} (f_n - f_m)^2 \left(\exp^\diamond\left[w\left(\frac{1}{2}\varphi\right)\right]\right)^2 d\mu_1 \to 0 \quad \text{as } m, n \to \infty$$

This is equivalent to

$$\int_{S'} (f_n - f_m)^2 \exp^\diamond[w(\varphi)]d\mu_1 \to 0 \quad \text{as } m, n \to \infty$$

By (2.10.10) this implies that

$$\int_{S'} (f_n(\omega + \varphi) - f_m(\omega + \varphi))^2 d\mu_1(\omega) \to 0 \quad \text{as } m, n \to \infty$$

and hence $\{f_n(\cdot + \varphi)\}_{n=1}^\infty$ is convergent in $L^2(\mu_1)$. Since a subsequence of $\{f_n\}$ converges to f a.e., we conclude that the $L^2(\mu_1)$ limit of $f_n(\cdot + \varphi)$ must be $f(\cdot + \varphi)$. $\qquad \square$

The following useful result first appeared in Gjessing (1994), Theorem 2.10, in the Wiener space setting, and subsequently in Benth and Gjessing (2000), Lemma 5.6, in a white noise setting (for the spaces $\mathcal{G}, \mathcal{G}^*$). We will here present the $L^2(\mu_1)$-version of their result.

Theorem 2.10.6 (Gjessing's Lemma). *Let $\phi \in L^2(\mathbb{R}^d)$ and $X \in L^p(\mu_1)$ for some $p > 1$. Then $X \diamond \exp^\diamond[w(\phi)] \in L^\rho(\mu_1)$ for all $\rho < p$, and almost surely we have*

$$(X \diamond \exp^\diamond[w(\phi)])(\omega) = T_{-\phi}X(\omega) \cdot \exp^\diamond[w(\phi)](\omega). \qquad (2.10.13)$$

Proof   First assume $X \in (\mathcal{S})_1$ and choose $f \in (\mathcal{S})_1$. Then, by (2.10.10) and (2.10.11),

$$\int_{\mathcal{S}'} f(\omega) \cdot (X \diamond \exp^{\diamond}[w(\phi)])(\omega) d\mu_1(\omega)$$

$$= \int_{\mathcal{S}'} X(\omega) f(\omega + \phi) d\mu_1(\omega)$$

$$= \int_{\mathcal{S}'} X(\omega - \phi) f(\omega) \exp^{\diamond}[w(\phi)](\omega) d\mu_1(\omega). \qquad (2.10.14)$$

By Corollary 2.10.5 we know that $X(\cdot - \phi) \in L^{\rho}(\mu_1)$ for all $\rho < p$ and hence the same is true for $X(\cdot - \phi) \cdot \exp^{\diamond}[w(\phi)]$. Since (2.10.14) holds for all $f \in (\mathcal{S})_1$ and $(\mathcal{S})_1$ is dense in $L^q(\mu_1)$ for all $q < \infty$, we conclude that

$$X \diamond \exp^{\diamond}[w(\phi)] = X(\omega - \phi) \cdot \exp^{\diamond}[w(\phi)], \text{ almost surely,}$$

as claimed. $\square$

2.11 Positivity

In many applications the noise that occurs is not white. The following example illustrates this.

If we consider fluid flow in a porous rock, we often lack exact information about the permeability of the rock at each point. The lack of information makes it natural to model the permeability as a (multiparameter) noise (see Chapter 1). This noise will, of course, not be white, but positive, since permeability is always a non-negative quantity. In this section we will discuss the positivity in the case of distributions and also in the case of functional processes. Let $(\mathcal{S})_1$ and $(\mathcal{S})_{-1}$ be the spaces defined in Section 2.3.

Definition 2.11.1.   An element $\Phi \in (\mathcal{S})_{-1}$ is called *positive* if for any positive $\phi \in (\mathcal{S})_1$ we have $\langle \Phi, \phi \rangle \geq 0$. The collection of positive elements in $(\mathcal{S})_{-1}$ is denoted by $(\mathcal{S})_{-1}^+$.

Before we state an important characterization of positive distributions, we must provide some preparatory results. For simplicity we assume $d = 1$. Let A be an operator on $L^2(\mathbb{R})$ given by

$$A = -\left(\frac{d}{dx}\right)^2 + (x^2 + 1). \qquad (2.11.1)$$

Then the Hermite function $\xi_n, n \geq 1$ is an eigenfunction of A with eigenvalue $2n$. Let $\mathcal{S}_p(\mathbb{R})$ be the completion of $\mathcal{S}(\mathbb{R})$ under the norm $|\cdot|_{2,p} := \|A^p \cdot\|_{L^2(\mathbb{R})}$.

Denote by $\mathcal{S}_{-p}(\mathbb{R})$ the dual space of $\mathcal{S}_p(\mathbb{R})$, with the norm $|\cdot|_{2,-p}$. It is well known that $\mathcal{S}(\mathbb{R})$ is the projective limit of $\mathcal{S}_p(\mathbb{R}), p > 0$ and $\mathcal{S}'(\mathbb{R})$ is the union of $\mathcal{S}_{-p}(\mathbb{R}), p > 0$, with inductive topology.

Lemma 2.11.2 Kondratiev et al. (1995a), Corollary 1.   *Let $p > 0$ be a constant such that the embedding $\mathcal{S}_p(\mathbb{R}) \to L^2(\mathbb{R})$ is Hilbert–Schmidt. Assume that $\phi \in (\mathcal{S})_1$. Then for any $\varepsilon > 0$, there exists a constant $C_{\varepsilon,p}$ such that*

$$|\phi(x)| \le C_{\varepsilon,p}\|\phi\|_{1,p}e^{\varepsilon|x|_{2},-p}; \ x \in \mathcal{S}_{-p}(\mathbb{R}). \tag{2.11.2}$$

Proof   See Corollary 1 in Kondratiev et al. (1995a). $\square$

Theorem 2.11.3 Kondratiev et al. (1995a), Theorem 2.   *Let $\Phi \in (\mathcal{S})_{-1}^{+}$. Then there exists a unique positive measure ν on $(\mathcal{S}'(\mathbb{R}), \mathcal{B}(\mathcal{S}'(\mathbb{R})))$ such that for all $\phi \in (\mathcal{S})_1$,*

$$\langle \Phi, \phi \rangle = \int_{\mathcal{S}'(\mathbb{R})} \phi(x)\nu(dx). \tag{2.11.3}$$

Proof   We will construct the measure ν by estimating the moments of the distributions. Since the polynomials $\mathcal{P} \subset (\mathcal{S})_1$, we can define the moments of the distribution Φ as

$$M_n(\zeta_1, \zeta_2, \ldots, \zeta_n) = \langle \Phi, \prod_{j=1}^{n} \langle \cdot, \zeta_j \rangle \rangle; \ n \in \mathbb{N}, 1 \le j \le n, \ \zeta_j \in \mathcal{S}(\mathbb{R})$$

$$M_0 = \langle \Phi, 1 \rangle. \tag{2.11.4}$$

First assume $\zeta_1 = \zeta_2 = \cdots = \zeta_n = \zeta \in \mathcal{S}(\mathbb{R})$. Since $\Phi \in (\mathcal{S})_{-p}$ for some $p > 0$, we have

$$|\langle \Phi, \langle \cdot, \zeta \rangle^n \rangle| \le \|\Phi\|_{-1,-p}\|\langle \cdot, \zeta \rangle^n\|_{1,p}. \tag{2.11.5}$$

To obtain a bound of $\|\langle \cdot, \zeta \rangle^n\|_{1,p}$, we use the well-known Hermite decomposition (see Appendix C)

$$\langle \cdot, \zeta \rangle^n = \sum_{k=0}^{[\frac{n}{2}]} \frac{n!}{k!(n-2k)!}\left(-\frac{1}{2}\|\zeta\|_{L^2(\mathbb{R})}^2\right)^k \int_{\mathbb{R}^{n-2k}} \zeta^{\otimes n-2k}dB^{\otimes n-2k}. \tag{2.11.6}$$

But for any integer $n \ge 1$,

$$\int_{\mathbb{R}^n} \zeta^{\otimes n}dB^{\otimes n} = \sum_{|\alpha|=n} \langle \zeta^{\otimes n}, \xi^{\otimes \alpha} \rangle \int_{\mathbb{R}^n} \xi^{\otimes \alpha}dB^{\otimes|\alpha|} = \sum_{|\alpha|=n} \langle \zeta^{\otimes n}, \xi^{\otimes \alpha} \rangle H_\alpha(\omega).$$

$$\tag{2.11.7}$$

Hence,

$$
\left\| \int_{\mathbb{R}^n} \zeta^{\otimes n} dB^{\otimes n} \right\|_{1,p}^2 = \sum_{|\alpha|=n} \langle \zeta^{\otimes n}, \xi^{\otimes \alpha} \rangle^2 (\alpha!)^2 (2\mathbb{N})^{p\alpha}
$$

$$
\leq \left(\sum_{|\alpha|=n} \langle \zeta^{\otimes n}, \xi^{\otimes \alpha} \rangle^2 (2\mathbb{N})^{p\alpha} \right) (n!)^2
$$

$$
= (n!)^2 \left(\sum_{|\alpha|=n} \langle (A^{\frac{p}{2}})^{\otimes n} \zeta^{\otimes n}, \xi^{\otimes \alpha} \rangle^2 \right) = (n!)^2 |\zeta|_{2,\frac{p}{2}}^{2n}.
$$

$$(2.11.8)$$

Observe that $\int_{\mathbb{R}^n} \zeta^{\otimes n} dB^{\otimes n}, n \geq 1$, are orthogonal in $(\mathcal{S})_{-p}$. Thus we obtain from (2.11.6) and (2.11.8) that

$$
\| \langle \cdot, \zeta \rangle^n \|_{1,p}^2 = \sum_{k=0}^{[\frac{n}{2}]} \left(\frac{n!}{k!(n-2k)!2^k} \right)^2 \| \zeta \|_{L^2(\mathbb{R})}^{4k} \left\| \int_{\mathbb{R}^n} \zeta^{\otimes n-2k} dB^{\otimes n-2k} \right\|_{1,p}^2
$$

$$
\leq (n!)^2 |\zeta|_{2,\frac{p}{2}}^{2n} \sum_{k=0}^{[\frac{n}{2}]} \frac{2^{2n}}{(k!)^2} \cdot 2^{-2k} \leq (n!)^2 4^n |\zeta|_{2,\frac{p}{2}}^{2n} C, \qquad (2.11.9)
$$

where $C = \sum_{k=1}^{\infty} 2^{-2k}/(k!)^2$. By the polarization formula, this implies

$$
\left\| \prod_{j=1}^n \langle \cdot, \zeta_j \rangle \right\|_{1,p} \leq \sqrt{C} 2^n (n!) \prod_{j=1}^n |\zeta_j|_{2,\frac{p}{2}}. \qquad (2.11.10)
$$

Hence we obtain from (2.11.4) that

$$
|M_n(\zeta_1, \ldots, \zeta_n)| \leq \| \Phi \|_{-1,-p} \sqrt{C} 2^n (n!) \prod_{j=1}^n |\zeta_j|_{2,\frac{p}{2}}. \qquad (2.11.11)
$$

Due to the kernel theorem – see Gelfand and Vilenkin (1964) – the following decomposition holds:

$$
M_n(\zeta_1, \ldots, \zeta_n) = \langle M^{(n)}, \zeta_1 \otimes \zeta_2 \otimes \cdots \otimes \zeta_n \rangle, \qquad (2.11.12)
$$

where $M^{(n)} \in \mathcal{S}'(\mathbb{R})^{\hat{\otimes} n}$. The sequence $\{M^{(n)}, n \in \mathbb{N}\}$ has the following property of positivity: For any finite sequence of smooth kernels $\{f^{(n)}, n \in \mathbb{N}\}$, i.e., that $f^{(n)} \in \mathcal{S}(\mathbb{R})^{\otimes n}, f^{(n)} = 0, n \geq n_0$

$$
\sum_{k,j}^{n_0} \langle M^{(k+j)}, f^{(k)} \otimes \overline{f^{(j)}} \rangle = \langle \Phi, |\phi|^2 \rangle \geq 0, \qquad (2.11.13)
$$

where $\phi(x) = \sum_{n=0}^{n_0} \langle x^{\otimes n}, f^{(n)} \rangle, x \in \mathcal{S}'(\mathbb{R})$.

By the result in Berezansky and Kondratiev (1988), (2.11.11) and (2.11.13) are sufficient to ensure the existence of a uniquely defined measure ν on the probability space $(\mathcal{S}'(\mathbb{R}), \mathcal{B}(\mathcal{S}'(\mathbb{R})))$, such that for any $\phi \in \mathcal{P}$

$$\langle \Phi, \phi \rangle = \int_{\mathcal{S}'(\mathbb{R})} \phi(\omega)\nu(d\omega). \tag{2.11.14}$$

By Corollary 2.4 in Zhang (1992), it is known that any element $\phi \in (\mathcal{S})_1$ is defined pointwise and continuous. Thus to show (2.11.14) also holds for any $\phi \in (\mathcal{S})_1$, by Lemma 2.11.2 it suffices to prove that there exists $p' > 0$ and $\varepsilon > 0$ such that $\exp[\varepsilon |x|_{2,-p'}]$ is integrable with respect to ν. Choose $p' > p/2$ such that the embedding $i^{p'} : \mathcal{S}_{p'}(\mathbb{R}) \to \mathcal{S}_{p/2}(\mathbb{R})$ is of the Hilbert–Schmidt type. Then we let $\{e_k, k \in \mathbb{N}\} \subset \mathcal{S}(\mathbb{R})$ be an orthonormal basis in $\mathcal{S}_{p'}(\mathbb{R})$. This gives

$$|x|_{2,-p'}^2 = \sum_{k=1}^{\infty} \langle x, e_k \rangle^2, x \in \mathcal{S}_{-p'}(\mathbb{R}) \tag{2.11.15}$$

and

$$\int_{\mathcal{S}'(\mathbb{R})} |\omega|_{2,-p'}^{2n} \nu(d\omega) = \sum_{k_1=1}^{\infty} \cdots \sum_{k_n=1}^{\infty} \int_{\mathcal{S}'(\mathbb{R})} \langle \omega, e_{k_1} \rangle^2 \cdots \langle \omega, e_{k_n} \rangle^2 \nu(d\omega).$$

Using the bound (2.11.11), we have

$$\int_{\mathcal{S}'(\mathbb{R})} |\omega|_{2,-p'}^{2n} \nu(d\omega) \leq \|\Phi\|_{-1,-p} \sqrt{C} 2^{2n} (2n)! \sum_{k_1=1}^{\infty} \cdots \sum_{k_n=1}^{\infty} |e_{k_1}|_{2,\frac{p}{2}}^2 \cdots |e_{k_n}|_{2,\frac{p}{2}}^2$$

$$= \|\Phi\|_{-1,-p} \sqrt{C} 2^{2n} (2n)! (\|i^{p'}\|_{HS})^{2n},$$

because

$$\sum_{k=1}^{\infty} |e_k|_{2,\frac{p}{2}}^2 = \|i^{p'}\|_{HS}^2.$$

For an arbitrary integer $n \geq 1$,

$$\int_{\mathcal{S}'(\mathbb{R})} |\omega|_{2,-p'}^n \nu(d\omega) \leq \left(\int_{\mathcal{S}'(\mathbb{R})} |\omega|_{2,-p'}^{2n} \nu(dx) \right)^{\frac{1}{2}} \nu(\mathcal{S}'(\mathbb{R}))^{\frac{1}{2}}$$

$$\leq \sqrt{\|\Phi\|_{-1,-p}} C^{\frac{1}{4}} 2^n \cdot 2^n n! (\|i^{p'}\|_{HS})^n M_0^{\frac{1}{2}}, \tag{2.11.16}$$

where we have used that $(2n)! \leq 4^n (n!)^2$ and $\nu(\mathcal{S}'(\mathbb{R})) = M_0 < +\infty$. Choose $\varepsilon < 1/4 \|i^{p'}\|_{HS}^{-1}$. Then

$$\int_{\mathcal{S}'(\mathbb{R})} \exp[\varepsilon|\omega|_{2,-p'}]\nu(d\omega) = \sum_{n=0}^{\infty} \frac{\varepsilon^n}{n!} \int_{\mathcal{S}'(\mathbb{R})} |\omega|_{2,-p'}^n \nu(d\omega)$$

$$\leq M_0^{\frac{1}{2}} C^{\frac{1}{4}} \sqrt{\|\Phi\|_{-1,-p}} \sum_{n=0}^{\infty} (\varepsilon 4 \||i^{p'}\|_{HS})^n < +\infty.$$

$$\tag{2.11.17}$$

$\square$

Let $X = \sum_{\alpha} c_{\alpha} H_{\alpha} \in (\mathcal{S})^*$ (the Hida distribution space defined in Section 2.3). As we know, the Hermite transform of X is given by

$$\mathcal{H}X(z) = \widetilde{X}(z) = \sum_{\alpha} c_{\alpha} z^{\alpha}, \; z = (z_1, z_2, \dots, z_n, \dots) \in \mathbb{C}^{\mathbb{N}}. \tag{2.11.18}$$

By Lindstrøm et al. (1991), Lemma 5.3, and Zhang (1992), $\widetilde{X}(z)$ converges absolutely for $z = (z_1, \dots, z_n, 0, 0, \dots)$ for each integer n. Therefore, the function $\widetilde{X}^{(n)}(z_1, \dots, z_n) := \widetilde{X}(z_1, z_2, \dots, z_n, 0 \cdots 0)$ is analytic on $\mathbb{C}^n$ for each n. Following Definition 2.11.1, we can define the positivity in $(\mathcal{S})^*$. The following characterization is sometimes useful.

Theorem 2.11.4 Lindstrøm et al. (1991a). *Let $X \in (\mathcal{S})^*$. Then X is positive if and only if*

$$g_n(y) := \widetilde{X}^{(n)}(iy)e^{-\frac{1}{2}|y|^2}; \; y \in \mathbb{R}^n \tag{2.11.19}$$

is positive definite for all n.

Before giving the proof, let us recall the definition of positive definiteness. A function $g(y), y \in \mathbb{R}^n$, is called positive definite if for all positive integers m and all $y^{(1)}, \dots, y^{(m)} \in \mathbb{R}^n, a = (a_1, \dots, a_m) \in \mathbb{C}^m$,

$$\sum_{j,k}^{m} a_j \bar{a}_k g(y^{(j)} - y^{(k)}) \geq 0. \tag{2.11.20}$$

Proof Let $d\lambda(x)$ be the standard Gaussian measure on $\mathbb{R}^{\infty}$, i.e., the direct product of infinitely many copies of the normalized Gaussian measure on $\mathbb{R}$. Set $F(z) = \widetilde{X}^{(n)}(z)$, for $z = (z_1, \dots, z_n) \in \mathbb{C}^n$. Define

$$J_n(x) = J_n(x_1, \dots, x_n) = \int \widetilde{X}^{(n)}(x + iy)d\lambda(y)$$

$$= \int F(x + iy)e^{-\frac{1}{2}|y|^2}(2\pi)^{-\frac{n}{2}} dy, \tag{2.11.21}$$

where $y = (y_1, \dots, y_n), dy = dy_1 \cdots dy_n$.

We write this as

$$J_n(x) = e^{-\frac{1}{2}|x|^2} \int F(z) e^{\frac{1}{2}z^2} \cdot e^{-i(x,y)} (2\pi)^{-\frac{n}{2}} dy$$

$$= e^{-\frac{1}{2}|x|^2} (2\pi)^{-\frac{n}{2}} \int G(z) e^{-i(x,y)} dy, \qquad (2.11.22)$$

where $z = (z_1, \ldots, z_n)$, $z_k = x_k + iy_k$, $z^2 = z_1^2 + \cdots + z_n^2$, $(x,y) = \sum_{k=1}^{n} x_k y_k$, and $G(z) := F(z) e^{\frac{1}{2}z^2}$ is analytic. Consider the function

$$f(x, \eta) = \int G(x + iy) e^{-i(\eta, y)} dy, x, \eta \in \mathbb{R}^n. \qquad (2.11.23)$$

Using the Cauchy–Riemann equations, we have

$$\frac{\partial f}{\partial x_1} = \int \frac{\partial G}{\partial x_1} \cdot e^{-i(\eta, y)} dy = \int (-i) \frac{\partial G}{\partial y_1} e^{-i(\eta, y)} dy.$$

But

$$\int_{-\infty}^{+\infty} (-i) \frac{\partial G}{\partial y_1} e^{-i\eta_1 y_1} dy_1 = i \int_{-\infty}^{+\infty} G(z) e^{-i\eta_1, y_1} (-i\eta_1) dy_1.$$

This gives

$$\frac{\partial f(x, \eta)}{\partial x_1} = \eta_1 f(x, \eta). \qquad (2.11.24)$$

Hence we have $f(x_1, x_2, \ldots, x_n; \eta) = f(0, x_2, \ldots, x_n; \eta) e^{\eta_1 x_1}$, and so on for $x_2, \ldots, x_n$. Therefore,

$$f(x, \eta) = f(0, \eta) e^{\eta x} = e^{(\eta, x)} \int G(iy) e^{-i(\eta, y)} dy. \qquad (2.11.25)$$

We conclude from (2.11.21)–(2.11.25) that

$$J_n(x) = e^{\frac{1}{2}|x|^2} (2\pi)^{-\frac{n}{2}} \int \widetilde{X}^{(n)}(iy) e^{-\frac{1}{2}|y|^2} e^{-i(x,y)} dy$$

$$= e^{\frac{1}{2}|x|^2} \hat{g}_n(x), \qquad (2.11.26)$$

where $\hat{g}_n(x) = (2\pi)^{-n/2} \int g_n(y) e^{-i(x,y)} dy$ is the Fourier transform of g_n.

Note that $g_n \in \mathcal{S}(\mathbb{R}^n)$, and hence $\hat{g}_n \in \mathcal{S}(\mathbb{R}^n)$. Therefore, we can apply the Fourier inversion to obtain

$$g_n(y) = (2\pi)^{-\frac{n}{2}} \int \hat{g}_n(-x) e^{i(x,y)} dx = (2\pi)^{-\frac{n}{2}} \int J_n(-x) e^{-\frac{1}{2}|x|^2} e^{i(x,y)} dx$$

$$= \int J_n(-x) e^{i(x,y)} d\lambda(x) = \int J_n(x) e^{-i(x,y)} d\lambda(x), \qquad (2.11.27)$$

so if $y^{(1)}, \ldots, y^{(m)} \in \mathbb{R}^n$ and $a = (a_1, \ldots, a_m) \in \mathbb{C}^m$, then

$$\sum_{j,k}^{m} a_j \bar{a}_k g_n(y^{(j)} - y^{(k)}) = \int |\gamma(x)|^2 J_n(x) d\lambda(x), \qquad (2.11.28)$$

where $\gamma(x) = \sum_{j=1}^{m} a_j e^{-ixy^{(j)}}$. Since

$$\int \eta dB = \left(\int \eta_1(t) dB(t), \int \eta_2(t) dB(t), \ldots, \int \eta_n(t) dB(t), \ldots \right)$$

has distribution $d\lambda(x)$, we can rewrite (2.11.28) as

$$\sum_{j,k}^{m} a_j \bar{a}_k g_n(y^{(j)} - y^{(k)}) = E\left[\left| \gamma \left(\int \eta dB \right) \right|^2 J_n \left(\int \eta dB \right) \right], \qquad (2.11.29)$$

since $(\mathcal{S})$ is an algebra (see, e.g., Zhang (1992)), we have that $|\gamma(\int \eta dB)|^2 \in (\mathcal{S})$. Since $J_n(\int \eta dB) \to X$ in $(\mathcal{S})^*$,

$$\sum_{j,k}^{m} a_j \bar{a}_k g_n(y^{(j)} - y^{(k)}) \to < X, \left| \gamma \left(\int \eta dB < \right) \right|^2 >, \quad \text{as} \quad n \to +\infty.$$
$$(2.11.30)$$

So if X is positive for almost all ω, we deduce that

$$\lim_{n \to +\infty} \sum_{j,k}^{m} a_j \bar{a}_k g_n(y^{(j)} - y^{(k)}) \geq 0. \qquad (2.11.31)$$

But with $y^{(1)}, \ldots, y^{(m)} \in \mathbb{R}^n$ fixed, $g_n(y^{(j)} - y^{(k)})$ eventually becomes constant as $n \to +\infty$, so (2.11.31) implies that $g_n(y)$ is positive definite.

Conversely, if g_n is positive definite, then, by (2.11.27), $J_n(x) \geq 0$ for almost all x with respect to $d\lambda$, and if this is true for all $n \geq 1$, we have that

$$X(\omega) = \lim_n J_n \left(\int \eta dB \right) \text{ is positive.} \qquad \square$$

Remark Let $X \in L^2(\mu_1) \subset (\mathcal{S})^*$. Then X is positive in $(\mathcal{S})^*$ if and only if X is a non-negative random variable.

Definition 2.11.5. A functional process $X(\phi, x, \omega)$ is called *positive* or a *positive noise* if

$$X(\phi, x, \omega) \geq 0 \quad \text{for} \quad \text{almost all } \omega \qquad (2.11.32)$$

for all $\phi \in \mathcal{S}(\mathbb{R}^d), x \in \mathbb{R}^d$.

Example 2.11.6. Let $w(\phi)$ be the smoothed white noise defined as in Section 2.6. Then the Wick exponential $\exp^\diamond[w(\phi, \omega)]$ is a positive noise.

This follows from the identity (see 2.6.55)

$$\exp^{\diamond}[w(\phi,\omega)] = \exp\left[w(\phi,\omega) - \frac{1}{2}\|\phi\|^2\right].$$

Corollary 2.11.7. *Let $X = X(\phi,\omega)$ and $Y = Y(\phi,\omega)$ be positive L^2-functional processes of the following form:*

$$X(\phi,\omega) = \sum_{\alpha} a_{\alpha}(\phi^{\otimes|\alpha|})H_{\alpha}(\omega),$$
$$Y(\phi,\omega) = \sum_{\alpha} b_{\alpha}(\phi^{\otimes|\alpha|})H_{\alpha}(\omega). \tag{2.11.33}$$

where $a_{\alpha}, b_{\alpha} \in H^{-s}(\mathbb{R}^{nd})$ for some s. If $X \diamond Y$ is well defined, then $X \diamond Y$ is also positive.

Proof For $\phi \in \mathcal{S}(\mathbb{R}^d)$, consider $\widetilde{X}^{(n)}(\phi, iy)e^{-1/2|y|^2}$ as before and, similarly, $\widetilde{Y}^{(n)}(\phi, iy)e^{-1/2|y|^2}$. Replacing ϕ by $\rho\phi$ where $\rho > 0$, Theorem 2.11.3 yields that

$$g_n^{(\rho)}(y) = \widetilde{X}^{(n)}(\phi, i\rho y)e^{-\frac{1}{2}|y|^2} \quad \text{is positive definite,}$$

hence

$$\sigma_n(y) := \widetilde{X}^{(n)}(\phi, iy)e^{-\frac{1}{2}|\frac{y}{\rho}|^2} \quad \text{is positive definite,} \tag{2.11.34}$$

and, similarly,

$$\gamma_n(y) = \widetilde{Y}^{(n)}(\phi, iy)e^{-\frac{1}{2}|\frac{y}{\rho}|^2} \quad \text{is positive definite.} \tag{2.11.35}$$

Therefore the product $\sigma_n\gamma_n(y) = (\widetilde{X}^{(n)}(\phi)\widetilde{Y}^{(n)}(\phi))(iy)e^{-|\frac{y}{\rho}|^2}$ is positive definite. If we choose $\rho = \sqrt{2}$, this gives that

$$\mathcal{H}(X \diamond Y)^{(n)}(\phi, iy)e^{-\frac{1}{2}|y|^2} \quad \text{is positive definite.}$$

So from Theorem 2.11.4, we have $X \diamond Y \geq 0$.   □

Exercises

2.1 To obtain a formula for $E[\langle \cdot, \phi \rangle^n]$, replace ϕ by $\alpha\phi$ with $\alpha \in \mathbb{R}$ in equation (2.1.3), and compute the nth derivative with respect to α at $\alpha = 0$. Then use polarization to show that we have $E[\langle \cdot, \phi \rangle\langle \cdot, \psi \rangle] = (\phi, \psi)$ for functions $\phi, \psi \in \mathcal{S}(\mathbb{R}^d)$.

2.2 Extend Lemma 2.1.2 to functions that are not necessarily orthogonal in $L^2(\mathbb{R}^d)$.

2.3 Show that $E[|\tilde{B}(x_1) - \tilde{B}(x_2)|^4] = 3|x_1 - x_2|^2$.

2.4 Prove formula (2.1.7). (Hint: Set $F(\alpha, \beta) = \int_{\mathbb{R}} e^{i\alpha t - \beta t^2} dt$ for $\beta > 0$. Verify that $\partial F/\partial \beta = \partial^2 F/\partial \alpha^2$ and $F(0, \beta) = (\pi/\beta)^{1/2}$, and use this to conclude that F must coincide with the right-hand side of (2.1.7).)

2.5 Give an alternative proof of Lemma 2.1.2. (Hint: Use (2.1.3) to prove that the characteristic function of the random variable $(\langle \omega, \xi_1 \rangle, \langle \omega, \xi_2 \rangle, \ldots, \langle \omega, \xi_n \rangle)$ coincides with that of the Gaussian measure λ_n on $\mathbb{R}^n$.)

2.6 Prove statement (2.1.9): If $\phi \in L^2(\mathbb{R}^d)$ and we choose $\phi_n \in \mathcal{S}(\mathbb{R}^d)$ such that $\phi_n \to \phi$ in $L^2(\mathbb{R}^d)$, then

$$\langle \omega, \phi \rangle := \lim_{n \to \infty} \langle \omega, \phi_n \rangle \text{ exists in } L^2(\mu_1)$$

and is independent of the choice of $\{\phi_n\}$. (Hint: From Lemma 2.1.2 (or from Exercise 2.1), we get $E[\langle \omega, \phi \rangle^2] = \|\phi\|^2$ for all $\phi \in \mathcal{S}(\mathbb{R}^d)$. Hence $\{\langle \cdot, \phi_n \rangle\}_{n=1}^{\infty}$ is a Cauchy sequence in $L^2(\mu_1)$ and therefore convergent.)

2.7 Use the Kolmogorov's continuity theorem (see, e.g., Stroock and Varadhan (1979), Theorem 2.1.6) to prove that the process $\tilde{B}(x) := \langle \omega, \chi_{[0,x_1] \times \cdots \times [0,x_d]} \rangle$ defined in (2.1.10) has a continuous version.

2.8 Find the Wiener–Itô chaos expansion (2.2.21),

$$f(\omega) = \sum_{\alpha \in \mathcal{J}} c_\alpha H_\alpha(\omega); \quad c_\alpha \in \mathbb{R}^N,$$

for the following $f \in L^2(\mu_m)$ (when nothing else is said, assume $N = m = 1$):

a) $f(\omega) = w^{\diamond 2}(\phi, \omega)$, $\phi \in \mathcal{S}(\mathbb{R}^d)$. (Hint: Use (2.2.23) and (2.4.2).)
b) $f(\omega) = B^{\diamond 2}(x, \omega)$; $x \in \mathbb{R}^d$. (Hint: Use (2.2.24) and (2.4.2).)
c) $f(\omega) = B^2(x, \omega)$; $x \in \mathbb{R}^d$. (Hint: Use b) and (2.4.14).)
d) $f(\omega) = B^3(x, \omega)$; $x \in \mathbb{R}^d$. (Hint: Use (2.4.17).)
e) $f(\omega) = \exp^{\diamond}[w(\eta_1, \omega)]$. (Hint: Use (2.6.48).)
f) $f(\omega) = \exp[w(\eta_1, \omega)]$. (Hint: Use (2.6.55).)
g) $m \geq 1$, $f(\omega) = B_1(x, \omega) + \cdots + B_m(x, \omega)$; $x \in \mathbb{R}^d$. (Hint: Use (2.2.26).)
h) $m \geq 1$, $f(\omega) = B_1^2(x, \omega) + \cdots + B_m^2(x, \omega)$; $x \in \mathbb{R}^d$. (Hint: Use (2.2.26) and c).)
i) $N \geq 1$, $f(\omega) = \left(2B(x, \omega) + 1, B^2(x, \omega)\right)$.

2.9 Prove (2.4.15):

$$w(\phi) \diamond w(\psi) = w(\phi) \cdot w(\psi) - (\phi, \psi)$$

for all $\phi, \psi \in L^2(\mathbb{R}^d)$. (Hint: Use (2.4.10) and (2.2.23).)

2.10 Let $F \in L^1(\mu) \cap (\mathcal{S})_{-1}$, with chaos expansion

$$F(\omega) = \sum_\alpha c_\alpha H_\alpha(\omega).$$

Prove that $E[F] = c_0$. (Hint: Combine (2.9.1) and (2.9.3) when $\phi = 0$).

2.11 Let $K_\phi(x, \omega) = \exp^\circ[W_\phi(x, \omega)]$ be as in (2.6.56). Prove that

$$E[K_\phi(x, \cdot)] = 1 \quad \text{and} \quad \mathrm{Var}[K_\phi(x, \cdot)] = \exp[\|\phi\|^2] - 1.$$

(Hint: Use (2.6.54) and (2.6.55).)

2.12 Let ϕ be normally distributed with mean 0 and variance σ^2. Prove that

$$E[\phi^{2k}] = (2k - 1)(2k - 3) \cdots 3 \cdot 1 \cdot \sigma^{2k} \text{ for } k \in \mathbb{N}.$$

(Hint: We may assume that $\sigma = 1$. Use integration by parts and induction:

$$E[\phi^{2k}] = \int_{\mathbb{R}} x^{2k} d\lambda(x) = \int_{-\mathbb{R}} x \cdot x^{2k-1} d\lambda(x)$$

$$= \left. -x^{2k-1} \cdot e^{-\frac{1}{2}x^2} \cdot \frac{1}{\sqrt{2\pi}} \right|_{-\infty}^{\infty} + \int_{\mathbb{R}} (2k-1) x^{2k-2} d\lambda(x),$$

with $d\lambda(x) = d\lambda_1(x)$ as in (2.1.4).)

2.13 For $X \in (\mathcal{S})_{-1}$ we define the *Wick-cosine of X*, $\cos^\circ[X]$, and the *Wick-sine of X*, $\sin^\circ[X]$, by

$$\cos^\circ[X] = \sum_{n=0}^{\infty} \frac{(-1)^n}{(2n)!} X^{\circ(2n)}$$

and

$$\sin^\circ[X] = \sum_{n=1}^{\infty} \frac{(-1)^{n-1}}{(2n-1)!} X^{\circ(2n-1)},$$

respectively. Prove that

a) $\cos^\circ[w(\phi)] = \exp\left[\frac{1}{2}\|\phi\|^2\right] \cdot \cos[w(\phi)]$

b) $\sin^\circ[w(\phi)] = \exp\left[\frac{1}{2}\|\phi\|^2\right] \cdot \sin[w(\phi)]$.

(Hint: Use Lemma 2.6.16 and the formulas

$$\cos^\circ[w(\phi)] = \frac{1}{2}(\exp^\circ[w(i\phi)] + \exp^\circ[w(-i\phi)])$$

$$\sin^\circ[w(\phi)] = \frac{1}{2i}(\exp^\circ[w(i\phi)] - \exp^\circ[w(-i\phi)]),$$

where $i = \sqrt{-1}$ is the imaginary unit.)

2.14

a) Prove that

$$\exp^\circ[-\exp^\circ[w(\phi)]] = \sum_{n=0}^{\infty} \frac{(-1)^n}{n!} \exp\left[nw(\phi) - \frac{n^2}{2}\|\phi\|^2\right],$$

where the right-hand side converges in $L^1(\mu)$.

b) Give an example to show that the Wick exponential $\exp^\circ[X]$ need not in general be positive. In fact, it may not even be bounded below.
(Hint: Consider $f(x,y) = \sum_{n=0}^{\infty} (-1)^n/n! \exp[nx - n^2 y^2]$. If we have that $x = 2y^2 > 2\ln(3 + M)$ for $M > 0$, then $f(x,y) < -M$.)

2.15

a) Show the following generating formula for the Hermite polynomials:

$$\exp\left[tx - \frac{1}{2}t^2\right] = \sum_{n=0}^{\infty} \frac{t^n}{n!} h_n(x).$$

(Hint: Write $\exp[tx - 1/2t^2] = \exp[1/2x^2] \cdot \exp[-1/2(x - t)^2]$ and use Taylor's Theorem at $t = 0$ on the last factor. Then combine with Definition (C.1).)

b) Show that

$$\exp\left[w(\phi) - \frac{1}{2}\|\phi\|^2\right] = \sum_{n=0}^{\infty} \frac{\|\phi\|^n}{n!} h_n\left(\frac{w(\phi)}{\|\phi\|}\right)$$

for all $\phi \in L^2(\mathbb{R}^d)$, where $\|\phi\| = \|\phi\|_{L^2(\mathbb{R}^d)}$.

c) Deduce that

$$\exp\left[B(t) - \frac{1}{2}t\right] = \sum_{n=0}^{\infty} \frac{t^{\frac{n}{2}}}{n!} h_n\left(\frac{B(t)}{\sqrt{t}}\right) \text{ for all } t \geq 0.$$

d) Combine b) with Lemma 2.6.16 and (2.6.48) to give an alternative proof of (2.4.17):

$$w(\phi)^{\circ n} = \|\phi\|^n h_n\left(\frac{w(\phi)}{\|\phi\|}\right),$$

for all $n \in \mathbb{N}$ and all $\phi \in L^2(\mathbb{R}^d)$.

2.16 Show that $\exp^\circ[w(\phi)^{\circ 2}]$ is *not* positive.

2.17 Show that

$$\|\exp^\circ[nw(\phi)]\|_{L^p(\mu)} = \exp\left[(p-1)n^2\|\phi\|_{L^2(\mathbb{R}^d)}^2\right]$$

for all $n \in \mathbb{N}, \phi \in L^2(\mathbb{R}^d), p \geq 1$.

2.18

a) Show that $\exp^{\circ}[\exp^{\circ}[w(\phi)]] \in (\mathcal{S})_{-1} \cap L^1(\mu)$.
b) Show that if $p > 1$, then $\exp^{\circ}[\exp^{\circ}(w(\phi))] \notin L^p(\mu)$.

2.19 Let $\psi = \chi_{[0,t]}$. Use Itô's formula to show that

a) $\int_{\mathbb{R}} \int_{\mathbb{R}} \psi^{\hat{\otimes}2} dB^{\otimes 2} = B^2(t) - t$.
b) $\int_{\mathbb{R}^3} \psi^{\hat{\otimes}3} dB^{\otimes 3} = B^3(t) - 3tB(t)$.
   (Compare with (2.2.29).)

2.20 Let $\phi \in (\mathcal{S}_1)$. Prove that ϕ is pointwise defined and continuous on $\mathcal{S}'(\mathbb{R})$. (Hint: Use Definition 2.3.2 and formula (C.2).)

2.21 Consider the space $L^2(\mathbb{R}, \lambda)$ where

$$d\lambda(x) = \frac{1}{\sqrt{2\pi}} e^{-\frac{1}{2}x^2} dx.$$

Let $h_n(x); n = 0, 1, 2, \ldots$ be the Hermite polynomials defined in (2.2.1). In this 1-dimensional situation we can construct $(\mathcal{S})$ and $(\mathcal{S})^*$ as follows: For $p \geq 1$, define

$$(\mathcal{S})_p = \left\{ u(x) = \sum_{n=0}^{\infty} c_n h_n(x) \in L^2(\mathbb{R}, \lambda); \sum_{n=0}^{\infty} c_n^2 n! 2^{np} < \infty \right\};$$

set $(\mathcal{S}) = \bigcap_{p \geq 1} (\mathcal{S})_p$ and $(\mathcal{S})^* = (\mathcal{S})'$, the dual of $(\mathcal{S})$. Prove that if $f \in (\mathcal{S})_1$, then

$$f(x) \exp\left[-\frac{1}{2}x^2\right] \in \mathcal{S}(\mathbb{R}).$$

2.22 Let G be a Borel subset of $\mathbb{R}$. Let $\mathcal{F}_G$ be the σ-algebra generated by all random variables of the form

$$\int_{\mathbb{R}} \chi_A(t) dB(t) = \int_A dB(t); \quad A \subset G \text{ Borel set.}$$

Thus if $G = [0, t]$, we have, with $\mathcal{F}_t$ as in Appendix B, $\mathcal{F}_{[0,t]} = \mathcal{F}_t$ for $t \geq 0$.

a) Let $g \in L^2(\mathbb{R})$ be deterministic. Show that

$$E\left[\int_{\mathbb{R}} g(t) dB(t) | \mathcal{F}_G\right] = \int_{\mathbb{R}} \chi_G(t) g(t) dB(t).$$

b) Let $v(t, \omega) \in \mathbb{R}$ be a stochastic process such that $v(t, \cdot)$ is $\mathcal{F}_t \cap \mathcal{F}_G$-measurable for all t and

$$E\left[\int_{\mathbb{R}} v^2(t, \omega) dt\right] < \infty.$$

Show that $\int_G v(t,\omega)dB(t)$ is $\mathcal{F}_G$-measurable. (Hint: We can assume that $v(t,\omega)$ is a step function $v(t,\omega) = \sum_i v_i(\omega)\chi_{[t_i,t_{i+1})}(t)$ where $v_i(\omega)$ is $\mathcal{F}_{t_i} \cap \mathcal{F}_G$-measurable. Then

$$\int_G v(t,\omega)dB(t) = \sum_i \int_{G\cap[t_i,t_{i+1})} v_i(\omega)dB(t)$$

$$= \sum_i v_i(\omega) \int_{G\cap[t_i,t_{i+1})} dB(t).)$$

c) Let $u(t,\omega)$ be an $\mathcal{F}_t$-adapted process such that

$$E\left[\int_{\mathbb{R}} u^2(t,\omega)dt\right] < \infty.$$

Show that

$$E\left[\int_{\mathbb{R}} u(t,\omega)dB(t)|\mathcal{F}_G\right] = \int_G E[u(t,\omega)|\mathcal{F}_G]dB(t).$$

(Hint: By b) it suffices to verify that

$$E\left[f(\omega) \cdot \int_{\mathbb{R}} u(t,\omega)dB(t)\right] = E\left[f(\omega) \cdot \int_G E[u(t,\omega)|\mathcal{F}_G]dB(t)\right]$$

for all $f(\omega) = \int_A dB(t), A \subset G$.)

d) Let $f_n \in \hat{L}^2(\mathbb{R}^n)$. Show that

$$E\left[\int_{\mathbb{R}^n} f_n dB^{\otimes n}|\mathcal{F}_G\right] = \int_{\mathbb{R}^n} f_n(t_1,\ldots,t_n)\chi_G(t_1)\cdots\chi_G(t_n)dB^{\otimes n}(t_1,\ldots,t_n).$$

(Hint: Apply induction to c).)

e) We say that two random variables $\phi_1, \phi_2 \in L^2(\mu)$ are *strongly independent* if there exist two Borel sets $G_1, G_2 \subset \mathbb{R}$ such that ϕ_i is $\mathcal{F}_{G_i}$-measurable for $i = 1,2$ and $G_1 \cap G_2$ has Lebesgue measure 0. Suppose ϕ_1, ϕ_2 are strongly independent. Show that $\phi_1 \diamond \phi_2 = \phi_1 \cdot \phi_2$. (See Example 2.4.9.) (Hint: Use Proposition 2.4.2 and d).)

2.23 Find the Wiener–Itô expansion (2.2.35)

$$f(\omega) = \sum_{n=0}^{\infty} \int_{\mathbb{R}^n} f_n dB^{\otimes n}; \ f_n \in \hat{L}^2(\mathbb{R}^n)$$

of the following random variables $f(\omega) \in L^2(\mu)$ ($N = m = d = 1$):

a) $f(\omega) = B(t_0,\omega); \ t_0 > 0$

b) $f(\omega) = B^2(t_0, \omega); \ t_0 > 0$

c) $f(\omega) = \exp[\int_{\mathbb{R}} g(s)dB(s, \omega)]; \ g \in L^2(\mathbb{R})$ deterministic

d) $f(\omega) = B(t, \omega)(B(T, \omega) - B(t, \omega)); \ 0 \le t \le T.$

(Answers:

a) $f_0 = 0, \ f_1 = \chi_{[0,t_0]}, \ f_n = 0$ for $n \ge 2.$

b) $f_0 = t_0, \ f_1 = 0, \ f_2(t_1, t_2) = \chi_{[0,t_0]}(t_1) \cdot \chi_{[0,t_0]}(t_2), \ f_n = 0$ for $n \ge 3.$

c) $f_n = \frac{1}{n!} \exp\left[\frac{1}{2}\|g\|^2_{L^2(\mathbb{R})}\right] g^{\otimes n}$ for all $n \ge 0.$

d) $f_0 = 0, \ f_1 = 0, \ f_2(t_1, t_2) = \frac{1}{2}(\chi_{\{t_1 < t < t_2 < T\}} + \chi_{\{t_2 < t < t_1 < T\}}), \ f_n = 0$ for $n \ge 3.$)

2.24 Find the following Skorohod integrals using Definition 2.5.1:

a) $\int_0^T B(t_0, \omega)\delta B(t); \ 0 \le t_0 \le T$

b) $\int_0^T \int_0^T g(s)dB(s)\delta B(t)$, where $g \in L^2(\mathbb{R})$ is deterministic.

c) $\int_0^T B^2(t_0, \omega)\delta B(t); \ 0 \le t_0 \le T.$

d) $\int_0^T \exp[B(T, \omega)]\delta B(t).$

e) $\int_0^T B(t, \omega)(B(T, \omega) - B(t, \omega))\delta B(t).$

(Hint: Use the expansions you found in Exercise 2.23.)

(Answers:

a) $B(t_0)B(T) - t_0.$

b) $B(T) \cdot \int_0^T g(s)dB(s) - \int_0^T g(s)ds.$

c) $B^2(t_0)B(T) - 2t_0 B(t_0).$

d) $\exp\left[\frac{1}{2}T\right] \sum_{n=0}^{\infty} \frac{1}{n!} T^{\frac{n+1}{2}} h_{n+1}\left(\frac{B(T)}{\sqrt{T}}\right)$

e) $\frac{1}{6}(B(T)^3 - 3T B(T)).$)

2.25 Compute the Skorohod integrals in Exercise 2.24 by using the Wick product representation in Theorem 2.5.9. (Hint: In e) apply Exercise 2.12 e).) Remark: Note how much easier the calculation is with Wick products!

2.26 Let
$$\mathbf{w}(\phi, \omega) = (\langle \omega_1, \phi_1 \rangle, \langle \omega_2, \phi_2 \rangle)$$
be the 2-dimensional smoothed white noise vector defined by (2.1.27). Define
$$\mathbf{w}_c(\phi, \omega) = \langle \omega_1, \phi_2 \rangle + i\langle \omega_2, \phi_2 \rangle,$$

where $i = \sqrt{-1}$ is the imaginary unit. We call $\mathbf{w}_c$ the *complex smoothed white noise.*

Prove that

$$\mathbf{w}_c^{\diamond 2}(\phi, \omega) = \mathbf{w}_c^2(\phi, \omega).$$

For generalizations of this curious result, see Benth et al. (1996).

2.27 Let

$$w(\omega_1) = w(\omega_1, \omega) = \sum_{j=1}^{\infty} \langle \omega_1, \eta_j \rangle H_{\varepsilon_j}(\omega) \in (\mathcal{S})^*; \quad \omega_1 \in \mathcal{S}'(\mathbb{R}^d)$$

be the generalized smoothed white noise defined in (2.10.8). Prove that

$$\exp^{\diamond}[w(\omega_1 + \omega_2)] = \exp^{\diamond}[w(\omega_1)] \diamond \exp^{\diamond}[w(\omega_2)].$$

(Note that both sides are functions of $\omega \in \mathcal{S}'(\mathbb{R}^d)$.)

2.28 Let $\omega_1, \omega_2 \in \mathcal{S}'(\mathbb{R}^d)$. Prove that

$$T_{\omega_1 + \omega_2}^* = T_{\omega_1}^* T_{\omega_2}^* = T_{\omega_2}^* T_{\omega_1}^*.$$

(Hint: See Theorem 2.10.3 and use Exercise 2.24.)

2.29 In this exercise, we let ϕ denote the Hermite function of order $k \in \mathbb{N}$, i.e., $\phi(x) = \xi_k(x)$ where $\xi_k(x)$ is given by (2.2.2).

a) Define $X(\omega) = \sum_{n=0}^{\infty} a_n w(\phi)^{\diamond n}$, where

$$\sum_{n=0}^{\infty} (n!)^2 a_n^2 < \infty.$$

Show that

$$\psi_X(x) := \sum_{n=0}^{\infty} a_n h_n(x) \in L^2(\mathbb{R}, d\lambda),$$

where

$$d\lambda(x) = \frac{1}{\sqrt{2\pi}} e^{-\frac{1}{2}x^2} dx$$

and that

$$X(\omega) = \psi_X(w(\phi)).$$

(Hint: See Exercise 2.15.)

b) Define

$$\delta(w(\phi)) := \sum_{n=0}^{\infty} \frac{(-1)^n}{2^n n! \sqrt{2\pi}} w(\phi)^{\diamond 2n} = \frac{1}{\sqrt{2\pi}} \exp^{\diamond}\left(-\frac{1}{2} w(\phi)^{\diamond 2}\right).$$

This is called the *Donsker delta function*. Note that $\delta(w(\phi)) \in (\mathcal{S})^*$. Show that $X \in (\mathcal{S})_1$ and that

$$\langle \delta(w(\phi)), X \rangle = \psi_X(0).$$

c) Show that no element $Z \in (\mathcal{S})_{-1}$ can satisfy the relation

$$Z \diamond w(\phi) = 1.$$

d) In spite of the result in c), we can come close to a *Wick inverse of $w(\phi)$* proceeding as follows:
   With a slight abuse of notation, define

$$w(\phi)^{-\diamond 1} = \sum_{k=0}^{\infty} \frac{(-1)^k}{(2k+2)2^k k!} w(\phi)^{\diamond (2k+1)}.$$

Show that $w(\phi)^{-\diamond 1} \in (\mathcal{S})_{-1}$ and that

$$w(\phi)^{-\diamond 1} \diamond w(\phi) = 1 - \sqrt{2\pi}\delta(w(\phi)).$$

$w(\phi)^{-\diamond 1}$ is called the *Wick inverse of white noise*. See Hu et al. (1995) for more details.

2.30 Prove (2.3.38), i.e., that $W(t) = \frac{d}{dt}B(t)$ in $(\mathcal{S})^*$.

Chapter 3
Applications to Stochastic Ordinary Differential Equations

As mentioned in the introduction, the framework that we developed in Chapter 2 for the main purpose of solving stochastic partial differential equations, can also be used to obtain new results – as well as new proofs of old results – for stochastic ordinary differential equations. In this chapter we will illustrate this by discussing some important examples.

3.1 Linear Equations

3.1.1 Linear 1-Dimensional Equations

In this section we consider the general 1-dimensional linear Wick type Skorohod stochastic differential equation in $X(t) = X(t, \omega)$

$$\begin{cases} dX(t) = g(t,\omega)dt + r(t,\omega) \diamond X(t)dt + \sum_{i=1}^{m} \alpha_i(t,\omega) \diamond X(t)\delta B_i(t) \\ X(0) = X_0(\omega), \end{cases} \tag{3.1.1}$$

where $g(t,\omega), r(t,\omega), X_0(\omega)$ and $\alpha_i(t,\omega); 1 \leq i \leq m$, are random functions, possibly anticipating. $B(t) = (B_1(t), \ldots, B_m(t))$ is m-dimensional Brownian motion.

In view of the relation given by Theorem 2.5.9 between Skorohod integrals and Wick products with white noise, we rewrite equation (3.1.1) as

$$\begin{cases} \frac{dX(t)}{dt} = g(t,\omega) + r(t,\omega) \diamond X(t) + \sum_{i=1}^{m} \alpha_i(t,\omega) \diamond X(t) \diamond W_i(t) \\ X(0) = X_0(\omega), \end{cases} \tag{3.1.2}$$

H. Holden et al., *Stochastic Partial Differential Equations*, 2nd ed., Universitext,   115
DOI 10.1007/978-0-387-89488-1_3, © Springer Science+Business Media, LLC 2010

where $W(t) = (W_1(t), \ldots, W_m(t))$ is m-dimensional white noise. In this setting it is natural to assume that X_0 and the coefficients involved are $(\mathcal{S})_{-1}$-valued, as well as $X(t)$.

Theorem 3.1.1.   *Suppose $g(t, \omega), r(t, \omega)$ and $\alpha_i(t, \omega); \; 1 \le i \le m$ are continuous $(\mathcal{S})_{-1}$-processes and that $X_0 \in (\mathcal{S})_{-1}$. Then equation (3.1.2) has a unique continuously differentiable $(\mathcal{S})_{-1}$-valued solution given by*

$$X(t, \omega) = X_0(\omega) \diamond \exp^\diamond \left[\int_0^t \left(r(u, \omega) + \sum_{i=1}^m \alpha_i(u, \omega) \diamond W_i(u) \right) du \right]$$

$$+ \int_0^t \exp^\diamond \left[\int_s^t \left(r(u, \omega) + \sum_{i=1}^m \alpha_i(u, \omega) \diamond W_i(u) \right) du \right] \diamond g(s, \omega) ds.$$

$$(3.1.3)$$

The generalized expectation of $X(t)$ is

$$E[X(t)] = E[X_0] \exp \left[\int_0^t E[r(u)] du \right]$$

$$+ \int_0^t \exp \left[\int_s^t E[r(u)] du \right] E[g(s)] ds. \qquad (3.1.4)$$

Proof   Taking $\mathcal{H}$-transforms we get the equation

$$\begin{cases} \frac{d\widetilde{X}(t)}{dt} = \widetilde{g}(t) + \left[\widetilde{r}(t) + \sum_{i=1}^m \widetilde{\alpha}_i(t) \widetilde{W}_i(t) \right] \widetilde{X}(t) \\ \widetilde{X}(0) = \widetilde{X}_0, \end{cases} \qquad (3.1.5)$$

where we have suppressed the z in the notation: $\widetilde{X}(t) = \widetilde{X}(t; z)$, etc.
   For each $z \in (\mathbb{C}^{\mathbb{N}})_c$ we now use ordinary calculus to obtain that

$$\widetilde{X}(t) = \widetilde{X}_0 \exp \left[\int_0^t \left(\widetilde{r}(u) + \sum_{i=1}^m \widetilde{\alpha}_i(u) \widetilde{W}_i(u) \right) du \right]$$

$$+ \int_0^t \exp \left[\int_s^t \left(\widetilde{r}(u) + \sum_{i=1}^m \widetilde{\alpha}_i(u) \widetilde{W}_i(u) \right) du \right] \widetilde{g}(s) ds \qquad (3.1.6)$$

is the unique solution of (3.1.5). Moreover, $\widetilde{X}(t; z)$ is clearly a continuous function of t for each $z \in (\mathbb{C}^{\mathbb{N}})_c$. Hence by (3.1.5) and our assumptions on g, r and α the same is true for $d\widetilde{X}/dt(t; z)$. Moreover, by (3.1.6) and (3.1.5) we also have that $d\widetilde{X}/dt(t;z)$ is a bounded function of $(t,z) \in [0,T] \times \mathbb{K}_q(R)$ for any $T < \infty$ and suitable $q, R < \infty$ (depending on g, r and α).

We conclude from Lemma 2.8.4 that

$$
X(t,\omega) = X_0(\omega) \diamond \exp^\diamond \left[\int_0^t \left(r(u,\omega) + \sum_{i=1}^m \alpha_i(u,\omega) \diamond W_i(u) \right) du \right]
$$
$$
+ \int_0^t \exp^\diamond \left[\int_s^t \left(r(u,\omega) + \sum_{i=1}^m \alpha_i(u,\omega) \diamond W_i(u) \right) du \right] \diamond g(s,\omega) ds
$$

solves (3.1.3), as claimed. The last statement of the theorem, (3.1.4), follows from (2.6.45) and (2.6.54). □

Note that (3.1.2) is a special case of the *general linear 1-dimensional Wick equation*

$$
\begin{cases}
\frac{dX(t)}{dt} = g(t,\omega) + h(t,\omega) \diamond X(t) \\
X(0) = X_0(\omega),
\end{cases} \tag{3.1.7}
$$

where $g(t,\omega), h(t,\omega)$ are continuous $(\mathcal{S})_{-1}$ processes, $X_0 \in (\mathcal{S})_{-1}$. In (3.1.2) we have $h(t,\omega) = r(t,\omega) + \sum_{i=1}^m \alpha_i(t,\omega) \diamond W_i(t)$. By the same method as above, we obtain

Theorem 3.1.2. *The unique solution $X(t) \in (\mathcal{S})_{-1}$ of (3.1.7) is given by*

$$
X(t,\omega) = X_0(\omega) \diamond \exp^\diamond \left[\int_0^t h(u,\omega) du \right]
$$
$$
+ \int_0^t \exp^\diamond \left[\int_s^t h(u,\omega) du \right] \diamond g(s,\omega) ds. \tag{3.1.8}
$$

Corollary 3.1.3. *Suppose $g(t,\omega)$ is a continuous $(\mathcal{S})_{-1}$-process, that $r(t)$ and $\alpha_i(t); 1 \leq i \leq m$ are real (deterministic) continuous functions and that $X_0 \in (\mathcal{S})_{-1}$. Then the unique solution of (3.1.2) is given by*

$$
X(t) = X_0 \diamond \exp \left[\int_0^t r(u) - \frac{1}{2} \sum_{i=1}^m \alpha_i^2(u) du + \sum_{i=1}^m \int_0^t \alpha_i(u) dB_i(u) \right]
$$
$$
+ \int_0^t \exp \left[\int_s^t r(u) - \frac{1}{2} \sum_{i=1}^m \alpha_i^2(u) du + \sum_{i=1}^m \int_s^t \alpha_i(u) dB_i(u) \right] \diamond g(s) ds.
$$
$$
\tag{3.1.9}
$$

Proof This follows from Theorem 3.1.1 combined with the relation (2.6.56) between Wick exponentials ($\exp^\diamond$) and ordinary exponentials ($\exp$) of smoothed white noise. □

Corollary 3.1.4. *Let r and α be real constants and assume that X_0 is independent of $\{B(s); s \geq 0\}$. Then the unique solution of the Itô equation*

$$\begin{cases} dX(t) = rX(t)dt + \alpha X(t)dB(t) \\ X(0) = X_0 \end{cases} \tag{3.1.10}$$

is given by

$$X(t) = X_0 \cdot \exp\left[\left(r - \frac{1}{2}\alpha^2\right)t + \alpha B(t)\right]. \tag{3.1.11}$$

If $|X_0|$ has a finite expectation, then the expectation of $X(t)$ is

$$E[X] = E[X_0] \cdot \exp[rt]. \tag{3.1.12}$$

Proof This follows from (3.1.9) and (3.1.4). Note that the Wick product with X_0 in (3.1.9) coincides with the ordinary product, since X_0 is independent of $\{B(s); s \geq 0\}$. □

Remark The solution (3.1.11), usually called *geometric Brownian motion*, is well known from Itô calculus (see, e.g., Øksendal (1995), equation (5.5)). Note that our method does not involve the use of Itô's formula, just ordinary calculus rules (with Wick products).

3.1.2 Some Remarks on Numerical Simulations

Expressions like the one in (3.1.11) are quite easy to handle with respect to numerical simulations. We have an explicit formula defined in terms of a function applied to Brownian motion. This can, of course, be done in several different ways. The construction we have found most useful to exploit is that Brownian motion has independent increments. We thus have

$$B\left(\frac{k}{n}\right) = B\left(\frac{1}{n}\right) - B(0) + B\left(\frac{2}{n}\right) - B\left(\frac{1}{n}\right) + \cdots + B\left(\frac{k}{n}\right) - B\left(\frac{k-1}{n}\right)$$

$$= \sum_{j=0}^{k} \Delta B_j \tag{3.1.13}$$

where $\Delta B_j = B(j+1/n) - B(j/n)$. Here all the $\Delta B_j, j = 0, 1, \ldots$ are independent normally distributed $\mathcal{N}[0, 1/n]$ random variables. Many computer programs have ready–made random number generators to sample pseudo-random numbers of the form $\mathcal{N}[0, \sigma^2]$. We then produce a sample path of Brownian motion simply by adding these numbers together. The figure below shows three different sample paths of geometric Brownian motion, generated from (3.1.11) using the scheme above.

The figure also shows the average value $\bar{X}(t) = E[X(t)] = E[X_0] \cdot e^{rt}$ (see (2.6.54)). As parameter values we have used $r = 1, \alpha = 0.6$ and $X_0 = 1$. Note that $\bar{X}(t)$ coincides with the solution of the no-noise equation, i.e., the case when $\alpha = 0$.

3.1.3 Some Linear Multidimensional Equations

We now consider the multidimensional case. In this section we will assume that all the coefficients except one are constant with respect to time. We will, however, allow them to be random, possibly anticipating. The general linear case will be considered in Section 3.3.

Theorem 3.1.5 (1-dimensional noise). *Suppose that $G(t, \omega) \in (\mathcal{S})^N_{-1}$ is t-continuous and that $X_0 \in (\mathcal{S})^N_{-1}$, $R \in (\mathcal{S})^{N \times N}_{-1}$ and $A \in (\mathcal{S})^{N \times N}_{-1}$ are constant with respect to t. Assume that R and A commute with respect to matrix Wick multiplication, i.e., $R \diamond A = A \diamond R$. Let $W(t)$ be 1-dimensional. Then the unique $(\mathcal{S})^N_{-1}$-valued process $X(t) = X(t, \omega)$ solving*

$$\begin{cases} \frac{dX(t)}{dt} = G(t, \omega) + R(\omega) \diamond X(t) + A(\omega) \diamond X(t) \diamond W(t) \\ X(0) = X_0 \end{cases} \quad (3.1.14)$$

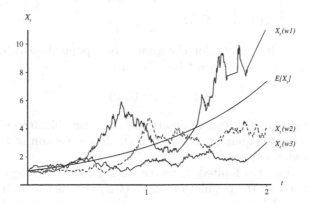

Fig. 3.1 Three sample paths of geometric Brownian motion.

is given by

$$X(t,\omega) = X_0(\omega) \diamond \exp^\diamond[tR(\omega) + A(\omega)B(t,\omega)]$$

$$+ \int_0^t \exp^\diamond[(t-s)R(\omega) + A(\omega)(B(t,\omega) - B(s,\omega))] \diamond G(s,\omega)ds,$$

(3.1.15)

where $B(t,\omega)$ is 1-dimensional Brownian motion.

Here the Wick matrix exponential $\exp^\diamond[K(\omega)]$, with $K \in (\mathcal{S})_{-1}^{N \times N}$, is defined by

$$\exp^\diamond[K(\omega)] = \sum_{n=0}^\infty \frac{1}{n!} K(\omega)^{\diamond n}, \qquad (3.1.16)$$

where $K^{\diamond n}$ is the Wick matrix power of order n, i.e., $K^{\diamond n} = K \diamond K \diamond \cdots \diamond K$ (n factors). The convergence of (3.1.16) is easily shown by taking $\mathcal{H}$-transforms, or by using Theorem 2.6.12.

The proof of Theorem 3.1.5 follows the same lines as the proof of Theorem 3.1.1 and is omitted. Unfortunately this method does not extend to the case when R and A vary with t.

3.2 A Model for Population Growth in a Crowded, Stochastic Environment

If the environment of a population has an infinite carrying capacity, then the equation (3.1.2) with deterministic coefficients, rewritten as

$$\begin{cases} \frac{dX(t,\omega)}{dt} = [r(t) + \alpha(t)W(t,\omega)] \diamond X(t,\omega) \\ X(0,\omega) = X_0(\omega), \end{cases} \qquad (3.2.1)$$

may be regarded as a model for the growth of a population $X(t,\omega)$, where the relative growth rate at time t has the form

$$r(t) + \alpha(t)W(t,\omega). \qquad (3.2.2)$$

This is a mathematical way of describing unpredictable, irregular changes in the environment. Equation (3.2.1) is a stochastic version of the Malthus model for population growth.

If the environment is limited, however, with a finite carrying capacity K, then the equation for the population size $X(t,\omega)$ at time t must be modified. The Verhulst model assumes that the relation growth rate in this case is proportional to the "free life space" $K - X(t,\omega)$. This gives two possible models, in which the population sizes are denoted by Y and X, respectively. The first model is

MODEL A $\qquad \begin{cases} \frac{dY(t)}{dt} = Y(t)(K - Y(t))[r(t) + \alpha(t)W(t)] \\ Y(0) = x_0 > 0 \quad \text{(constant)}, \end{cases}$

which can be interpreted in the usual Itô sense

$$\begin{cases} dY(t) = r(t)Y(t)[K - Y(t)]dt + \alpha(t)Y(t)[K - Y(t)]dB(t) \\ Y(0) = x_0. \end{cases} \tag{3.2.3}$$

The second model is

MODEL B $\qquad \begin{cases} \frac{dX(t)}{dt} = X(t) \diamond (K - X(t)) \diamond [r(t) + \alpha(t)W(t)] \\ X(0) = x_0, \end{cases}$

or, in Skorohod integral interpretation,

$$\begin{cases} dX(t) = r(t)X(t) \diamond (K - X(t))dt + \alpha(t)X(t) \diamond (K - X(t))\delta B(t) \\ X(0) = x_0. \end{cases}$$
$$\tag{3.2.4}$$

Note that the two models coincide in the deterministic case ($\alpha = 0$). It is therefore a relevant question which of the two models gives the best mathematical model in the stochastic case. We emphasize that there is no reason to assume a priori that the pointwise product $Y(t, \omega) \cdot [K - Y(t, \omega)]$ is better than the Wick product $(X(t, \cdot) \diamond [K - X(t, \cdot)])(\omega)$. In this section we will show how to solve (3.2.4) explicitly and compare the stochastic properties of this solution $X(t)$ with the solution $Y(t)$ of (3.2.3).

Model B was first discussed in Lindstrøm et al. (1992) (as in equation in $L^2(\mu)$) and subsequently in Benth (1996), as an equation in $(\mathcal{S})_{-1}$. Here we first present the approach in Benth (1996), and then compare it with the solution in Lindstrøm et al. (1992). For simplicity, we assume that the units are chosen such that $K = 1$.

3.2.1 The General $(\mathcal{S})_{-1}$ Solution

Theorem 3.2.1 Benth (1996). *Suppose $X_0 \in (\mathcal{S})_{-1}$ with $E[X_0] > 0$. Then the stochastic distribution process*

$$X(t, \omega) = [1 + \theta_0 \diamond \exp^\diamond[-rt - \alpha B(t)]]^{\diamond(-1)}, \tag{3.2.5}$$

with

$$\theta_0 = (X_0)^{\diamond(-1)} - 1, \tag{3.2.6}$$

is the unique continuously differentiable $(S)_{-1}$*-process solving the equation*

$$X(t) = X_0 + r \int_0^t X(s) \diamond (1 - X(s)) ds + \alpha \int_0^t X(s) \diamond (1 - X(s)) \diamond W(s) ds; t \geq 0. \tag{3.2.7}$$

Proof Taking the $\mathcal{H}$-transform of (3.2.7) gives us the equation (where we have $\widetilde{X}(t) = \widetilde{X}(t; z)$)

$$\begin{cases} \frac{d\widetilde{X}(t)}{dt} = (r + \alpha \widetilde{W}(t))\widetilde{X}(t)(1 - \widetilde{X}(t)) \\ \widetilde{X}(0) = \widetilde{X}_0; z \in (\mathbb{C}^{\mathbb{N}})_c, \end{cases} \tag{3.2.8}$$

which has the solution

$$\widetilde{X}(t) = \frac{1}{1 + \widetilde{\theta}_0 \exp[-rt - \alpha \widetilde{B}(t)]}, \tag{3.2.9}$$

where

$$\widetilde{\theta}_0 = \widetilde{\theta}_0(z) = \frac{1}{\widetilde{X}_0(z)} - 1. \tag{3.2.10}$$

Since $\widetilde{X}_0(0) = E[X_0] > 0$, there exist $\epsilon > 0$ and a neighborhood $\mathbb{K}_q(\delta)$ such that

$$|\widetilde{X}_0(z)| \geq \epsilon > 0 \quad \text{for all} \quad z \in \mathbb{K}_q(\delta).$$

Hence $\widetilde{\theta}_0(z)$ is a bounded analytic function in $\mathbb{K}_q(\delta)$. Moreover, since $\widetilde{\theta}_0(0) > 0$ and $\widetilde{B}(t; 0) = 0$, we see that for all $t \geq 0$ there exist $q(t), \delta(t)$ such that $\widetilde{X}(t; z)$ is a bounded analytic function in z, for $z \in \mathbb{K}_{q(t)}(\delta(t))$. Moreover, for given $T < \infty$, the numbers $q_1 = q(t), \delta_1 = \delta(t)$ can be chosen to work for all $t \leq T$. Therefore, by (3.2.8) the derivative $d\widetilde{X}(t; z)/dt$ is analytic in z and bounded for $(t, z) \in [0, T] \times \mathbb{K}_{q_1}(\delta_1)$. From (3.2.8) we also see that $d\widetilde{X}(t; z)/dt$ is a continuous function of t when defined. We conclude from Lemma 2.8.4 that

$$X(t, \omega) = \mathcal{H}^{-1}(\widetilde{X}(t; z)) = [1 + \theta_0 \diamond \exp^{\diamond}[-rt - \alpha B(t)]]^{\diamond(-1)}$$

with

$$\theta_0 = (X_0)^{\diamond(-1)} - 1$$

solves equation (3.2.7). The uniqueness follows from the uniqueness of the solution of (3.2.8). □

3.2.2 A Solution in $L^1(\mu)$

For simplicity, let us from now on assume that $X_0 = x_0 > 0$ is a constant. Moreover, assume $r > 0$. First consider the case

$$x_0 > \frac{1}{2}, \text{ i.e., } \theta_0 := \frac{1}{x_0} - 1 \in (-1, 1). \qquad (3.2.11)$$

Then formula (3.2.5) can be written

$$
\begin{aligned}
X(t) = X_1(t) &= \left(1 + \theta_0 \exp^\diamond\left[-rt - \alpha B(t)\right]\right)^{\diamond(-1)} \\
&= \sum_{m=0}^{\infty} (-1)^m \theta_0^m \exp^\diamond[-rmt - \alpha m B(t)] \\
&= \sum_{m=0}^{\infty} (-1)^m \theta_0^m \exp\left[-\left(rm + \frac{1}{2}\alpha^2 m^2\right)t - \alpha m B(t)\right].
\end{aligned}
$$

$$\qquad (3.2.12)$$

Since $E\left[\exp^\diamond[-\alpha m B(t)]\right] = 1$, the sum (3.2.12) converges in $L^1(\mu)$ for all $t \geq 0$. Moreover, the $L^1(\mu)$ process $X_1(t)$ defined by (3.2.12) satisfies equation (3.2.7), when the Wick product is interpreted in the $L^1(\mu)$ sense as described in Section 2.9. To see this, note that

$$
\begin{aligned}
X_1(t) \diamond (1 - X_1(t)) &= \lim_{N \to \infty} \left(\sum_{m=0}^{N} (-1)^m \theta_0^m \exp^\diamond[-rmt - \alpha m B(t)] \right) \\
&\quad \diamond \left(1 - \sum_{n=0}^{N} (-1)^n \theta_0^n \exp^\diamond[-rnt - \alpha n B(t)] \right) \\
&= \lim_{N \to \infty} \sum_{\substack{m=0 \\ n=1}}^{N} (-1)^{m+n+1} \theta_0^{m+n} \exp^\diamond[-r(m + n)t \\
&\quad - \alpha(m + n)B(t)] \\
&= \lim_{N \to \infty} \sum_{k=1}^{2N} (-1)^{k+1} \theta_0^k k \exp^\diamond[-rkt - \alpha k B(t)] \\
&= \sum_{k=1}^{\infty} (-1)^{k+1} \theta_0^k k \exp^\diamond[-rkt - \alpha k B(t)] \in L^1(\mu).
\end{aligned}
$$

$$\qquad (3.2.13)$$

Hence $X_1(t) \diamond (1 - X_1(t))$ exists in $L^1(\mu)$ and is given by (3.2.13). Let

$$Y_k(t) = \exp^\diamond[-rkt - \alpha k B(t)]; \quad k = 0, 1, 2, \dots.$$

Then by Itô's formula

$$dY_k(t) = d\left(\exp\left[-\left(rk + \frac{1}{2}\alpha^2 k^2\right)t - \alpha k B(t)\right]\right)$$

$$= Y_k(t)\left(-\left(rk + \frac{1}{2}\alpha^2 k^2\right)dt - \alpha k dB(t) + \frac{1}{2}\alpha^2 k^2 dt\right)$$

$$= kY_k(t)(-rdt - \alpha dB(t)).$$

Hence

$$\int_0^t X_1(s) \diamond (1 - X_1(s))(rds + \alpha dB(s))$$

$$= \sum_{k=1}^{\infty}(-1)^k \theta_0^k \int_0^t kY_k(s)(-rds - \alpha dB(s))$$

$$= \sum_{k=1}^{\infty}(-1)^k \theta_0^k (Y_k(t) - 1) = \sum_{k=0}^{\infty}(-1)^k \theta_0^k (Y_k(t) - 1)$$

$$= X_1(t) - \frac{1}{1 + \theta_0} = X_1(t) - x_0,$$

as claimed. We conclude that $X_1(t)$ satisfies equation (3.2.7).
Next consider the case

$$0 < x_0 < \frac{1}{2}, \text{ i.e., } \theta_0 = \frac{1}{x_0} - 1 > 1. \tag{3.2.14}$$

Then formula (3.2.5) can be written

$$X(t,\omega) = X_2(t,\omega)$$

$$= \theta_0^{-1} \exp^{\diamond}[rt + \alpha B(t)] \diamond (1 + \theta_0^{-1} \exp[rt + \alpha B(t)])^{\diamond(-1)}$$

$$= \theta_0^{-1} \exp^{\diamond}[rt + \alpha B(t)] \diamond \left(\sum_{m=0}^{\infty}(-1)^m \theta_0^{-m} \exp^{\diamond}[rmt + \alpha m B(t)]\right)$$

$$= \sum_{m=1}^{\infty}(-1)^{m+1}\theta_0^{-m} \exp\left[\left(rm - \frac{1}{2}\alpha^2 m^2\right)t + \alpha m B(t)\right].$$

$$\tag{3.2.15}$$

This converges in $L^1(\mu)$ for all t such that $\exp[rt] < \theta_0$, i.e., for

$$t < T_0 := \frac{1}{r}\ln\theta_0.$$

A similar calculation as above shows that $X_2(t)$ is an $L^1(\mu)$ solution of (3.2.7) for $t < T_0$. Note that the series (3.2.15) in fact converges pointwise for almost all ω for all values of t, but we can only deduce $L^1(\mu)$ convergence when $t < T_0$. In the following we let $X_2(t)$ denote the pointwise a.e. limit of (3.2.15) for all $t \geq 0$.

We summarize this as follows:

Theorem 3.2.2 Lindstrøm et al. (1992).   *Assume that $r > 0, \alpha \in \mathbb{R}$ are constants.*

a) *Suppose $x_0 > 1/2$, i.e., $\theta_0 := 1/x_0 - 1 \in (-1, 1)$. Then the process*

$$X(t) = X_1(t) = \sum_{m=0}^{\infty} (-1)^m \theta_0^m \exp^\diamond[-rmt - \alpha m B(t)] \qquad (3.2.16)$$

is an $L^1(\mu)$ solution of the equation

$$\begin{cases} X(t) = x_0 + \int_0^t rX(s) \diamond (1 - X(s))ds + \int_0^t \alpha X(s) \diamond (1 - X(s))dB(s) \\ X(0) = x_0 \end{cases}$$

$$(3.2.17)$$

for all $t \geq 0$.

b) *Suppose $0 < x_0 < 1/2$, i.e., $\theta_0 := 1/x_0 - 1 > 1$.*
Then the process

$$X(t) = X_2(t) = \sum_{m=1}^{\infty} (-1)^{m+1} \theta_0^{-m} \exp^\diamond[rmt + \alpha m B(t)] \qquad (3.2.18)$$

converges for almost all ω for all t and is an $L^1(\mu)$ solution of (3.2.17) for

$$t < T_0 := \frac{1}{r} \ln \theta_0. \qquad (3.2.19)$$

Some interesting properties of these solutions are the following:

Corollary 3.2.3.   *Let $X_1(t), X_2(t)$ be as in Theorem 3.2.2.*

a) *Let $x_0 > 1/2$. Then we have*

$$E^{x_0}[X_1(t)] = x(t), \quad \text{for all } t, \qquad (3.2.20)$$

where $x(t)$ is the solution of (3.2.17) in the deterministic case $(\alpha = 0)$, i.e.,

$$\begin{cases} \frac{dx(t)}{dt} = rx(t)(1 - x(t)) \\ x(0) = x_0. \end{cases} \qquad (3.2.21)$$

Moreover,

$$\lim_{t \to \infty} X_1(t) = 1 \quad a.s., \qquad (3.2.22)$$

and for all $t > 0$ we have

$$P^{x_0}[X_1(t) > 1] > 0 \quad and \quad P^{x_0}[X_1(t) < 1] > 0. \tag{3.2.23}$$

b) *Let $0 < x_0 < 1/2$. Then, with T_0 as in (3.2.19),*

$$E^{x_0}[X_2(t)] = x(t) \quad for \quad t < T_0. \tag{3.2.24}$$

Moreover, for all $t < T_0$ we have

$$P^{x_0}[X_2(t) > 1] > 0 \quad and \quad P^{x_0}[X_2(t) < 1] > 0. \tag{3.2.25}$$

Proof   Properties (3.2.20), (3.2.24) follow immediately from the fact that

$$E\left[\exp^\diamond\left[\int \phi(t)dB(t)\right]\right]$$

$$= E\left[\exp\left[\int \phi(t)dB_t - \frac{1}{2}\|\phi\|^2\right]\right] = 1 \tag{3.2.26}$$

for all $\phi \in L^2(\mathbb{R})$. Property (3.2.22) follows from the expression (3.2.16) combined with the *law of iterated logarithm* for Brownian motion:

$$\limsup_{t\to\infty} \frac{B_t}{\sqrt{2t\ln(\ln t)}} = 1 \quad \text{almost surely,} \tag{3.2.27}$$

(see, e.g., Lamperti (1966), Section 22). Statements (3.2.23) and (3.2.25) are consequences of formulas (3.2.16), (3.2.18) for $X_1(t), X_2(t)$ plus the fact that for any $t > 0$ Brownian motion obtains arbitrary large or small values with positive probability (the density function for $B(t)$ is positive on the whole of $\mathbb{R}$).   □

Computer simulations of some paths of $X_1(t), X_2(t)$ are shown on the figure below.

Remark   For $x_0 = 1/2$ the $(\mathcal{S})_{-1}$ solution $X(t)$ does not seem to allow as simple a representation as in (3.2.16) or (3.2.18). For this reason the point $x_0 = 1/2$ is called a *stochastic bifurcation point* in Lindstrøm et al. (1992). Note, however, that the $(\mathcal{S})_{-1}$ solution exists for all initial values $x_0 > 0$ and for all $t \geq 0$. But the $\mathcal{H}$-transform $\widetilde{X}(t; z)$ given by (3.2.9) cannot be extended to an analytic function of z on $\mathbb{C}^n$ for any $n \geq 1$ (the function is meromorphic). Therefore, by Proposition 2.6.5 we see that $X(t)$ does not belong to $(\mathcal{S})_{-\rho}$ for any $\rho < 1$. In particular, $X(t)$ *is not in* $(\mathcal{S})^*$ *for any* $t > 0$ *and any* $x_0 > 0$.

Logistic paths

The same sample with $r = 1$, $\alpha = 1$. Starting points: 0.75, 0.6.

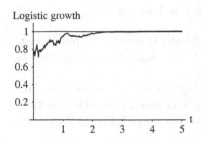

Logistic growth

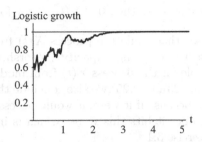

Logistic growth

Different sample paths with $r = 1/5$, $\alpha = 1/2$. Starting point: 0.6.

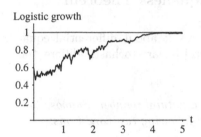

Logistic growth

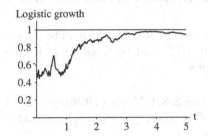

Logistic growth

Different sample paths with $r = 1/5$, $\alpha = 1$. Starting point: 0.25.

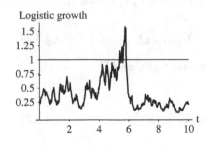

Logistic growth

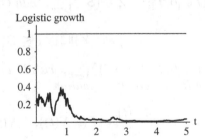

Logistic growth

3.2.3 A Comparison of Model A and Model B

It is known (see, e.g., Lungu and Øksendal (1997)) that the Itô stochastic differential equation (3.2.3) of Model A has a unique continuous, $\mathcal{F}_t$-adapted solution $Y(t, \omega)$ for all $t \geq 0$ (still assuming $r > 0$ and $x_0 > 0$). Moreover, the solution is a strong Markov process, and it is easily seen that it has the following properties:

If $x_0 > 1$, then $Y(t, \omega) > 1$ for all t and almost all ω,   (3.2.28)

If $0 < x_0 < 1$, then $0 < Y(t, \omega) < 1$ for all t and almost all ω.   (3.2.29)

We see that while the process $X(t)$ from Model B allows the population to cross the carrying capacity ("overshoot"), by (3.2.23), (3.2.25), this is impossible for the process $Y(t)$ from model A.

By (3.2.23), (3.2.25) we also conclude that $X(t)$ cannot be a strong Markov process, because if it were, it would necessarily continue to have the constant value 1 once it hits this value, which is impossible by (3.2.16), (3.2.18). See also Exercise 3.1.

3.3 A General Existence and Uniqueness Theorem

In this section we first formulate a general result about differential equations in $(\mathcal{S})_{-1}^N$ and then we apply it to general linear stochastic differential equations.

Theorem 3.3.1 Våge (1996b).   *Let k be a natural number. Suppose that $F : [0, T] \times (\mathcal{S})_{-1,-k}^N \to (\mathcal{S})_{-1,-k}^N$ satisfies the following two conditions:*

$$\|F(t, Y) - F(t, Z)\|_{-1,-k} \le C\|Y - Z\|_{-1,-k}   (3.3.1)$$

for all $t \in [0, T]$; $Y, Z \in (\mathcal{S})_{-1,-k}^N$, with C independent of t, Y and Z;

$$\|F(t, Y)\|_{-1,-k} \le D(1 + \|Y\|_{-1,-k})   (3.3.2)$$

for all $t \in [0, T], Y \in (\mathcal{S})_{-1,-k}^N$, with D independent of t and Y.
Then the differential equation

$$\frac{dX(t)}{dt} = F(t, X(t)); \quad X(0) = X_0 \in (\mathcal{S})_{-1,-k}^N   (3.3.3)$$

has a unique t-continuous solution $X(t) : [0, T] \to (\mathcal{S})_{-1,-k}^N$.

Proof   The result follows by standard methods for differential equations of this type. The details are omitted. See Exercise 3.2.   □

We wish to apply this to the general linear Wick stochastic differential equation. For this we need the following useful estimate for Wick products:

Proposition 3.3.2 Våge's inequality, Våge (1996a). *Suppose*

$$F = \sum_\alpha a_\alpha H_\alpha \in (\mathcal{S})_{-1,-l}, \ G = \sum_\beta b_\beta H_\beta \in (\mathcal{S})_{-1,-k},$$

where $l, k \in \mathbb{Z}$ *with*

$$k > l + 1. \tag{3.3.4}$$

Then

$$\|F \diamond G\|_{-1,-k} \le A(k-l) \cdot \|F\|_{-1,-l} \cdot \|G\|_{-1,-k}, \tag{3.3.5}$$

where

$$A(k-l) = \left[\sum_\alpha (2\mathbb{N})^{(l-k)\alpha} \right]^{\frac{1}{2}} < \infty \tag{3.3.6}$$

by Proposition 2.3.3.

Proof

$$\|F \diamond G\|_{-1,-k}^2 = \lim_{n \to \infty} \sum_{\gamma \in \Gamma_n} \left(\sum_{\alpha+\beta=\gamma} a_\alpha b_\beta \right)^2 (2\mathbb{N})^{-k\gamma}$$

$$= \lim_{n \to \infty} \sum_{\gamma \in \Gamma_n} \left(\sum_{\alpha+\beta=\gamma} a_\alpha (2\mathbb{N})^{-\frac{k}{2}\alpha} b_\beta (2\mathbb{N})^{-\frac{k}{2}\beta} \right)^2$$

where

$$\Gamma_n = \{\gamma = (\gamma_1, \ldots, \gamma_n) \in \mathbb{N}^n; \gamma_i \le n \ \text{ for } \ i = 1, \ldots, n\}.$$

Define $f(\alpha) = a_\alpha (2\mathbb{N})^{-\frac{k}{2}\alpha}$, $g(\beta) = b_\beta (2\mathbb{N})^{-\frac{k}{2}\beta}$ and write $\alpha \le \gamma$ if there exists a multi-index β such that $\alpha + \beta = \gamma$. Then from the above we get

$$\|F \diamond G\|_{-1,-k}^2 = \lim_{n \to \infty} \sum_{\gamma \in \Gamma_n} \left(\sum_{\alpha \le \gamma} f(\alpha) g(\gamma - \alpha) \right)^2$$

$$= \lim_{n \to \infty} \sum_{\gamma \in \Gamma_n} \sum_{\alpha, \alpha' \le \gamma} f(\alpha) f(\alpha') g(\gamma - \alpha) g(\gamma - \alpha')$$

$$= \lim_{n \to \infty} \sum_{\alpha, \alpha' \in \Gamma_n} f(\alpha) f(\alpha') \sum_{\gamma \ge \alpha, \alpha', \gamma \in \Gamma_n} g(\gamma - \alpha) g(\gamma - \alpha')$$

$$\le \limsup_{n \to \infty} \sum_{\alpha, \alpha' \in \Gamma_n} f(\alpha) f(\alpha') \left(\sum_{\beta \in \Gamma_n} g(\beta)^2 \right)^{\frac{1}{2}} \left(\sum_{\beta \in \Gamma_n} g(\beta)^2 \right)^{\frac{1}{2}}$$

$$\le \limsup_{n \to \infty} \sum_{\alpha, \alpha' \in \Gamma_n} |f(\alpha)| |f(\alpha')| \sum_\beta g(\beta)^2$$

$$= \left(\sum_\alpha |f(\alpha)| \right)^2 \sum_\beta g(\beta)^2$$

$$= \left(\sum_{\alpha} |a_{\alpha}|(2\mathbb{N})^{-\frac{k}{2}\alpha} \right)^2 \|G\|_{-1,-k}^2$$

$$= \left(\sum_{\alpha} |a_{\alpha}|(2\mathbb{N})^{-\frac{l}{2}\alpha}(2\mathbb{N})^{\frac{l-k}{2}\alpha} \right)^2 \|G\|_{-1,-k}^2$$

$$\leq \sum_{\alpha} a_{\alpha}^2 (2\mathbb{N})^{-l\alpha} \sum_{\alpha} (2\mathbb{N})^{(l-k)\alpha} \|G\|_{-1,-k}^2$$

$$= \sum_{\alpha} (2\mathbb{N})^{(l-k)\alpha} \|F\|_{-1,-l} \|G\|_{-1,-k}^2. \qquad \square$$

Theorem 3.3.5 (The general linear multi-dimensional Wick stochastic differential equation). *Let $T > 0$ and $l \in \mathbb{Z}$. Suppose $X_0 \in (\mathcal{S})_{-1}^N$, $G : [0,T] \to (\mathcal{S})_{-1}^N$ and $H : [0,T] \to (\mathcal{S})_{-1}^{N \times N}$ for $1 \leq i, j \leq N$. Moreover, suppose there exists $M < \infty$ such that*

$$\|G_i(t)\|_{-1,-l} + \|H_{ij}(t)\|_{-1,-l} \leq M \qquad (3.3.7)$$

for all $t \in [0,T]; 1 \leq i, j \leq N$.

Then there is a unique solution $X : [0,T] \to (\mathcal{S})_{-1}^N$ of the general linear system of equations

$$\frac{dX_i(t)}{dt} = G_i(t) + \sum_{j=1}^{N} H_{ij}(t) \diamond X_j(t); 1 \leq i \leq N \qquad (3.3.8)$$

with initial condition

$$X(0) = X_0. \qquad (3.3.9)$$

Proof Define $F = (F_1, \ldots, F_n) : [0,T] \times (\mathcal{S})_{-1}^N \to (\mathcal{S})_{-1}^N$ by

$$F_i(t, Y) = G_i(t) + \sum_{j=1}^{N} H_{ij}(t) \diamond Y_j$$

for $Y = (Y_1, \ldots, Y_N) \in (\mathcal{S})_{-1}^N$.

Choose $k > l + 1$, $Y, Z \in (\mathcal{S})_{-1}^N$. Then by Proposition 3.3.2 we have

$$\|F_i(t, Y) - F_i(t, Z)\|_{-1,-k} \leq \left\| \sum_{j=1}^{N} H_{ij}(t) \diamond (Y_j - Z_j) \right\|_{-1,-k}$$

$$\leq \sum_{j=1}^{N} A(k - l) \|H_{ij}\|_{-1,-l} \|Y_j - Z_j\|_{-1,-k}$$

$$\leq A(k - l)M \sum_{j=1}^{N} \|Y_j - Z_j\|_{-1,-k}$$

$$\leq A(k - l)M \cdot N \|Y - Z\|_{-1,-k}$$

and
$$\|F_i(t, Y)\|_{-1,-k} \le NM(1 + A(k - l))\|Y\|_{-1,-k}$$

for $1 \le i \le N$.

Hence the result follows from Theorem 3.3.1. □

As in Section 3.1 we note that (3.3.8) is a generalization of the linear stochastic differential equation

$$\frac{dX_i(t)}{dt} = G_i(t) + \sum_{j=1}^{N} R_{ij} \diamond X_j(t)$$

$$+ \sum_{\substack{1 \le j \le N \\ 1 \le k \le m}} W_k(t) \diamond A_{ij}^{(k)}(t) \diamond X_j(t); 1 \le i \le N \qquad (3.3.10)$$

with initial condition
$$X(0) = X_0 \in (\mathcal{S})_{-1}^N, \qquad (3.3.11)$$

where $G_i : [0, T] \to (\mathcal{S})_{-1}, R_{ij} : [0, T] \to (\mathcal{S})_{-1}$ and $A_{ij}^{(k)} : [0, T] \to (\mathcal{S})_{-1}$ satisfy

$$\|G_i(t)\|_{-1,-l} + \|R_{ij}(t)\|_{-1,-l} + \|A_{ij}^{(k)}(t)\|_{-1,-l} \le M \qquad (3.3.12)$$

for $1 \le i, j \le N$, $1 \le k \le m$.

Corollary 3.3.6. *There is a unique solution $X(t) \in (\mathcal{S})_{-1}^N$; $0 \le t \le T$ of the equation (3.3.10).*

Proof We apply Theorem 3.3.5 to the case when

$$H_{ij}(t) = R_{ij}(t) + \sum_{k=1}^{m} W_k(t) \diamond A_{ij}^{(k)}(t). \qquad (3.3.13)$$

We must verify that (3.3.7) holds: Using the Proposition 2.3.10 we have that $W_k(t) \in (\mathcal{S})_{0,-l} \subset (\mathcal{S})_{-1,-l}$ for all $l > 1$. Therefore, combining (3.3.12) with Proposition 3.3.2, we obtain (3.3.7). □

3.4 The Stochastic Volterra Equation

The classical stochastic (linear) Volterra equation has the form

$$X(t) = \theta(t) + \int_0^t b(t, s)X(s)ds + \int_0^t \sigma(t, s)X(s)dB(s); 0 \le t \le T \qquad (3.4.1)$$

where $\theta(s) = \theta(s, \omega) : \mathbb{R} \times \Omega \to \mathbb{R}, b(t, s) = b(t, s, \omega) : \mathbb{R}^2 \times \Omega \to \mathbb{R}$ and where we have that $\sigma(t, s) = \sigma(t, s, \omega) : \mathbb{R}^2 \times \Omega \to \mathbb{R}$ are given $\mathcal{F}_s$-adapted processes, $b(t, s) = \sigma(t, s) = 0$ for $s \geq t$, and $T > 0$ is a constant. The solution $X(t)$ is then required to be $\mathcal{F}_t$-adapted also, and the integral on the right of (3.4.1) is an Itô integral.

If θ, b or σ is not adapted, i.e., if some of them are *anticipating*, then the integral on the right is interpreted as a *Skorohod* integral (see Section 2.5).

Skorohod Volterra equations with anticipating kernel (but adapted initial condition $\theta(t)$) have been studied in Pardoux and Protter (1990). See also Berger and Mizel (1982), and the survey in Pardoux (1990). In Ogawa (1986), the stochastic Volterra equation is studied in the setting of Ogawa integrals. In Cochran et al. (1995), existence and uniqueness of solution of (3.4.1) is proved when θ is (possibly) anticipating, $b = 0$ and $\sigma(t, s)$ is deterministic, measurable with a possible singularity at $s = t$ of the type

$$|\sigma(t, s)| \leq A(t - s)^{\alpha}; \ 0 \leq s \leq t$$

for some constants $A \geq 0$ and $1/2 \leq \alpha \leq 0$.

There are several situations that can be modeled by stochastic Volterra equations. The following economic example is from Øksendal and Zhang (1993):

Example 3.4.1. An investment in an economic production, for example, the purchase of new production equipment, will usually have effects over a long period of time. Let $X(t, u)$ denote the capital density with respect to u at time t resulting from investments that were made u units of time ago (i.e., have age u). More precisely, let $\int_U X(t, u) du$ denote the total capital gained at time t from all investments with age $u \in U$. Assume that

$$\frac{\partial X(t, u)}{\partial t} + \frac{\partial X(t, u)}{\partial u} = -m(u)X(t, u), \tag{3.4.2}$$

where $m(u) \geq 0$ is the age-dependent relative "death" rate of the equipment or of the machines involved in the production. Since the left-hand side of (3.4.2) can be expressed as

$$\lim_{\Delta t \to 0} \frac{X(t + \Delta t, u + \Delta t) - X(t, u)}{\Delta t},$$

the interpretation of this equation is simply that such a rate of change of $X(t, u)$ is proportional to $X(t, u)$.

Moreover, assume that the amount of new capital $X(t, 0)$ at time t is described by the equation

$$X(t,0) = \int_0^\infty X(t,u)p(u)du, \qquad (3.4.3)$$

where $p(u)$ is the production at age u per capital unit. So in this model we assume that all the produced capital is reinvested into the production process.

If the initial capital density $X(0,u) = \phi(u)$ is known, then the solution $X(t,u)$ of (3.4.2) is given by

$$X(t,u) = \begin{cases} \phi(u-t) \cdot \exp\left[-\int_0^t m(r+u-t)dr\right]; & 0 \le t < u \\[2ex] X(t-u,0) \cdot \exp\left[-\int_0^u m(r)dr\right]; & t \ge u. \end{cases} \qquad (3.4.4)$$

Substituting this in (3.4.3) we get the Volterra equation

$$X(t,0) = Y(t) + \int_0^t K(t-s)X(s,0)ds, \qquad (3.4.5)$$

where

$$Y(t) = \int_0^\infty \phi(s) \exp\left[-\int_0^t m(s+r)dr\right]p(t+s)ds \qquad (3.4.6)$$

and

$$K(t) = p(t) \exp\left[-\int_0^t m(r)dr\right]. \qquad (3.4.7)$$

If the productivity function $p(u)$ is subject to random fluctuations, we could model $p(u)$ by

$$p(u,\omega) = p_0(u) + \epsilon W(u,\omega), \qquad (3.4.8)$$

where $\epsilon > 0, p_0(u) = E[p(u,\omega)]$ is deterministic and $W(u,\omega)$ is white noise. This leads to a *stochastic Volterra equation* of the form

$$X(t,\omega) = a(t,\omega) + \int_0^t b(t,s)X(s,\omega)ds$$

$$+ \int_0^t \sigma(t,s)X(s,\omega) \diamond W(s,\omega)ds, \qquad (3.4.9)$$

where

$$a(t,\omega) = \int_0^\infty \phi(s) \exp\left[-\int_0^t m(s+r)dr\right] p_0(t+s)ds$$

$$+ \epsilon \int_t^\infty \phi(v-t) \exp\left[-\int_0^t m(r+v-t)dr\right] dB_v, \qquad (3.4.10)$$

$$b(t,s) = p_0(t-s) \exp\left[-\int_0^{t-s} m(r)dr\right]; \ 0 \le s \le t \qquad (3.4.11)$$

and

$$\sigma(t,s) = \epsilon \exp\left[-\int_0^{t-s} m(r)dr\right]; \ 0 \le s \le t. \qquad (3.4.12)$$

Note that $a(t,\omega)$ is not adapted in this case. Then there is no reason to expect that $X(t,\omega)$ will be adapted either. An alternative formulation of (3.4.9) would be in terms of the *Skorohod* integral

$$X(t,\omega) = a(t,\omega) + \int_0^t b(t,s)X(s,\omega)ds + \int_0^t \sigma(t,s)X(s,\omega)\delta B(s) \qquad (3.4.13)$$

(see Section 2.5).

We may regard (3.4.9) and (3.4.13) as special cases of the following general linear stochastic Volterra equation:

$$X(t) = J(t) + \int_0^t K(t,s) \diamond X(s)ds, \quad 0 \le t \le T, \qquad (3.4.14)$$

where $T > 0$ is a given number and

$$J : [0,T] \to (\mathcal{S})_{-1} \quad \text{and} \quad K : [0,T] \times [0,T] \to (\mathcal{S})_{-1}$$

are given stochastic distribution processes. Using Wick calculus we can solve this equation explicitly.

Theorem 3.4.2 Øksendal and Zhang (1993), Øksendal and Zhang (1996). *Let* $J : [0,T] \to (\mathcal{S})_{-1}, K : [0,T] \times [0,T] \to (\mathcal{S})_{-1}$ *be continuous stochastic distribution processes. Suppose there exists* $l < \infty, M < \infty$ *such that*

$$\|K(t,s)\|_{-1,-l} \le M \quad \text{for} \quad 0 \le s \le t \le T. \qquad (3.4.15)$$

Then there exists a unique continuous stochastic distribution process $X(t)$ that solves the stochastic Volterra equation

$$X(t) = J(t) + \int_0^t K(t,s) \diamond X(s)ds; \ 0 \le t \le T. \qquad (3.4.16)$$

The solution is given by

$$X(t) = J(t) + \int_0^t H(t,s) \diamond J(s)ds, \qquad (3.4.17)$$

where

$$H(t,s) = \sum_{n=1}^{\infty} K_n(t,s), \qquad (3.4.18)$$

with K_n given inductively by

$$K_{n+1}(t,s) = \int_s^t K_n(t,u) \diamond K(u,s)du; \quad n \ge 1 \qquad (3.4.19)$$

$$K_1(t,s) = K(t,s). \qquad (3.4.20)$$

Proof With K_n defined as in (3.4.19)–(3.4.20), note that

$$K_2(t,s) = \int_s^t K(t,u) \diamond K(u,s)du$$

and

$$K_3(t,s) = \int_s^t K_2(t,u_2) \diamond K(u_2,s)du_2$$

$$= \int_s^t \left(\int_{u_2}^t K(t,u_1) \diamond K(u_1,u_2)du_1 \right) \diamond K(u_2,s)du_2$$

$$= \int\int_{s \le u_2 \le u_1 \le t} K(t,u_1) \diamond K(u_1,u_2) \diamond K(u_2,s)du_1 du_2.$$

By induction we see that

$$K_n(t,s) = \int_{s \le u_{n-1} \le \cdots \le u_1 \le t} \cdots \int \prod_{0 \le k \le n-1}^{\diamond} K(u_k, u_{k+1})du_1 \cdots du_{n-1}. \qquad (3.4.21)$$

Here $\prod_{0\leq k\leq n-1}^{\diamond}$ denotes the Wick product from $k=0$ to $k=n-1$, and we set $u_0 = t, u_n = s$. Choose $k > l+1$, and apply Proposition 3.3.2 to (3.4.21):

$$\|K_n(t,s)\|_{-1,-k} \leq A^n M^n \int \cdots \int\limits_{s\leq u_{n-1}\leq\cdots\leq u_1\leq t} du_1 \cdots du_{n-1}$$

$$= \frac{A^n M^n (t-s)^{n-1}}{(n-1)!}, \tag{3.4.22}$$

where $A = A(k-l)$. This shows that

$$H(t,s) := \sum_{n=1}^{\infty} K_n(t,s)$$

converges absolutely in $(\mathcal{S})_{-1,-k}$. So we can define $X(t)$ by (3.4.17), i.e.,

$$X(t) = J(t) + \int_0^t H(t,s) \diamond J(s)ds; \ 0 \leq t \leq T.$$

We verify that this is a solution of (3.4.16):

$$J(t) + \int_0^t K(t,r) \diamond X(r)dr$$

$$= J(t) + \int_0^t K(t,r) \diamond \left(J(r) + \int_0^r H(r,u) \diamond J(u)du \right) dr$$

$$= J(t) + \int_0^t K(t,r) \diamond J(r)dr + \int_0^t \int_0^r K(t,r) \diamond \sum_{n=1}^{\infty} K_n(r,u) \diamond J(u)dudr$$

$$= J(t) + \int_0^t K(t,r) \diamond J(r)dr + \sum_{n=1}^{\infty} \int_0^t \left(\int_u^t K(t,r) \diamond K_n(r,u)dr \right) \diamond J(u)du$$

$$= J(t) + \int_0^t K(t,u) \diamond J(u)du + \sum_{n=1}^{\infty} \int_0^t K_{n+1}(t,u) \diamond J(u)du$$

$$= J(t) + \int_0^t \sum_{m=1}^{\infty} K_m(t,u) \diamond J(u)du$$

$$= J(t) + \int_0^t H(t,u) \diamond J(u)du = X(t).$$

(Note that (3.4.21) implies that $K_{n+1}(t,u) = \int_u^t K(t,r) \diamond K_n(r,u)dr$.) This shows that $X(t)$ is indeed a solution of (3.4.16).

It remains to prove uniqueness. Suppose $Y(t)$ is another continuous solution of (3.4.16), so that

$$Y(t) = J(t) + \int_0^t K(t,s) \diamond Y(s)ds. \qquad (3.4.23)$$

Subtracting (3.4.23) from (3.4.16), we get

$$Z(t) = \int_0^t K(t,s) \diamond Z(s)ds; \ 0 \leq t \leq T, \qquad (3.4.24)$$

where $Z(t) = X(t) - Y(t)$. This, together with Proposition 3.3.2, gives that

$$\|Z(t)\|_{-1,-k} \leq M \int_0^t \|Z(s)\|_{-1,-k}ds$$

for some constant $M < \infty$. Applying the Gronwall inequality, we conclude that $Z(t) = 0$ for all t. $\qquad \square$

Example 3.4.3. We verify that the conditions of Theorem 3.4.2 are satisfied for the equation (3.4.9). Here $J(t) = a(t,\omega)$ is clearly continuous, even as a mapping from $[0,T]$ into $L^2(\mu)$. In this case we have,

$$K(t,s) = b(t,s) + \sigma(t,s) \diamond W(s).$$

Since b, σ are bounded, deterministic and continuous, it suffices to consider $W(s)$. By Example 2.7.3 we have $W(s) \in (\mathcal{S})_{0,-q}$ for all $q > 1$. Moreover, for $s, t \in [0,T]$, we have, by (2.8.14),

$$\|W(s) - W(t)\|_{0,-q}^2 = \sum_{k=1}^{\infty}(\eta_k(s) - \eta_k(t))^2(2k)^{-q}$$

$$\leq \sum_{k=1}^{\infty}C_{0,T}k^2|s - t|^2(2k)^{-q}$$

$$\leq C_{0,T}|s - t|^2 \sum_{k=1}^{\infty}(2k)^{2-q} < \infty,$$

for $q > 3$. Hence $W(s)$, and consequently $K(t,s)$, is continuous in $(\mathcal{S})_{0,-3}$ and therefore in $(\mathcal{S})_{-1,-3}$.

Example 3.4.4.   In Grue and Øksendal (1997), a stochastic Volterra equation is deduced from a second–order ordinary (Wick type) stochastic differential equation modeling the slow-drift motions of offshore structures. Such constructions are well known from the deterministic case, and the Wick calculus allows us to perform a similar procedure when the coefficients are stochastic distribution processes.

Consider a linear second–order stochastic differential equation of the form

$$\ddot{X}(t) + \alpha(t) \diamond \dot{X}(t) + \beta(t) \diamond X(t) + \gamma(t) = 0, \qquad (3.4.25)$$

where the coefficients $\alpha(t), \beta(t)$ and $\gamma(t) : \mathbb{R} \to (\mathcal{S})_{-1}$ are (possibly anticipating) continuous stochastic distribution processes. If we Wick multiply this equation by

$$M(t) := \exp^{\diamond}\left[\int_0^t \alpha(u)du\right], \qquad (3.4.26)$$

we get

$$\frac{d}{dt}(M(t) \diamond \dot{X}(t)) = -M(t) \diamond \gamma(t) - M(t) \diamond \beta(t) \diamond X(t).$$

Hence

$$M(t) \diamond \dot{X}(t) = \dot{X}(0) - \int_0^t M(s) \diamond \gamma(s)ds - \int_0^t M(s) \diamond \beta(s) \diamond X(s)ds$$

or

$$\dot{X}(t) = \dot{X}(0) \diamond \exp^{\diamond}\left[-\int_0^t \alpha(u)du\right] - \int_0^t \exp^{\diamond}\left[-\int_s^t \alpha(u)du\right] \diamond \gamma(s)ds$$

$$- \int_0^t \exp^{\diamond}\left[-\int_s^t \alpha(u)du\right] \diamond \beta(s) \diamond X(s)ds.$$

From this we get

$$X(t) = X(0) + \dot{X}(0) \diamond \int_0^t \exp^{\diamond}\left[-\int_0^v \alpha(u)du\right] dv$$

$$- \int_0^t \int_0^v \exp^{\diamond}\left[-\int_s^v \alpha(u)du\right] \diamond \gamma(s)dsdv$$

$$- \int_0^t \int_0^v \exp^{\diamond}\left[-\int_s^v \alpha(u)du\right] \diamond \beta(s) \diamond X(s)dsdv. \qquad (3.4.27)$$

Now

$$\int_0^t \int_0^v \exp^\diamond \left[-\int_s^v \alpha(u)du \right] \diamond \beta(s) \diamond X(s)dsdv$$

$$= \int_0^t \int_s^t \exp^\diamond \left[-\int_s^v \alpha(u)du \right] dv \diamond \beta(s) \diamond X(s)ds.$$

Therefore (3.4.27) is a stochastic Volterra equation of the form

$$X(t) = J(t) + \int_0^t K(t,s) \diamond X(s)ds; \quad t \geq 0 \tag{3.4.28}$$

with

$$J(t) = X(0) + \dot{X}(0) \diamond \int_0^t \exp^\diamond \left[-\int_0^v \alpha(u)du \right] dv$$

$$- \int_0^t \int_0^v \exp^\diamond \left[-\int_s^v \alpha(u)du \right] \diamond \gamma(s)dsdv \tag{3.4.29}$$

and

$$K(t,s) - - \int_s^t \exp^\diamond \left[-\int_s^v \alpha(u)du \right] dv \diamond \beta(s); \quad 0 \leq s \leq t. \tag{3.4.30}$$

Example 3.4.5 (Oscillations in a stochastic medium). Let us consider the motion of an object attached to an oscillating string with a stochastic force constant (Hooke's constant) k. If we represent k by a positive noise process of the form

$$k = k(t,\omega) = \exp^\diamond[W_{\phi_t}(\omega)] \tag{3.4.31}$$

for a suitable test function $\phi \in \mathcal{S}(\mathbb{R})$, then this motion can be modeled by the stochastic differential equation

$$\ddot{X}(t) + \exp^\diamond[W_{\phi_t}] \diamond X(t) = 0; \quad X(0) = a, \dot{X}(0) = 0. \tag{3.4.32}$$

According to (3.4.28)–(3.4.30) this can be transformed into a stochastic Volterra equation

$$X(t) = a + \int_0^t K^\phi(t,s) \diamond X(s)ds, \tag{3.4.33}$$

where

$$K^\phi(t, s) = -(t - s) \exp^\circ[W_{\phi_s}]; \quad 0 \leq s \leq t. \qquad (3.4.34)$$

Hence by Theorem 3.4.2 the solution is

$$X(t) = a \left(1 + \sum_{n=1}^{\infty} \int_0^t K_n^\phi(t, s) ds \right), \qquad (3.4.35)$$

where by (3.4.21) K_n^ϕ is given by

$$K_n^\phi(t, s)$$

$$= (-1)^n \int \cdots \int_{s \leq u_{n-1} \leq \cdots \leq u_1 \leq t} \prod_{k=0}^{n-1} (u_k - u_{k+1}) \exp^\circ \left[\sum_{k=1}^{n} W_{\phi_{u_k}} \right] du_1 \cdots du_{n-1},$$

where $u_0 = t, u_n = s$. Therefore

$$\|K_n^\phi(t, s)\|_{L^1(\mu)}$$

$$= \int \cdots \int_{s \leq u_{n-1} \leq \cdots \leq u_1 \leq t} \prod_{k=0}^{n-1} (u_k - u_{k+1}) du_1 \cdots du_{n-1} \leq \frac{(t - s)^{2n-1}}{(n - 1)!}.$$

It follows that $\sum_{n=1}^{\infty} K_n^\phi(t, s)$ converges in $L^1(\mu)$, uniformly on compacts in (t, s). We conclude that $X(t)$ given by (3.4.35) belongs to $L^1(\mu)$ (as well as $(\mathcal{S})_{-1}$). Moreover, we see that $\exp^\circ[W_{\phi_t}] \diamond X(t) \in L^1(\mu)$, also. Therefore, if we define

$$x(t) = E[X(t)],$$

then by taking the expectation of (3.4.32) we get

$$\ddot{x}(t) + x(t) = 0; \quad x(0) = a, \quad \dot{x}(0) = 0$$

and hence

$$E[X(t)] = a \cos t; \quad t \geq 0,$$

which is the solution when $\phi = 0$, i.e., when there is no noise. It is natural to ask what can be said about other probabilistic properties of $X(t)$.

3.5 Wick Products Versus Ordinary Products: a Comparison Experiment

The presentation in this section is based on the discussion in Holden et al. (1993b). In Chapter 1 and in Section 3.2 we discussed the use of Wick products versus ordinary products when modeling stochastic phenomena. A natural and important question is:

Which type of product gives the best model?

This question is not as easy to answer as one might think. How does one test a stochastic dynamic model? The problem is that it is usually difficult to "re-run" a stochastic dynamic system in real life. Here the random parameter ω can be regarded as one particular realization of the "experiment" or of "the world". How do we re-run the price development of a stock? How do we re-run the population growth in a random environment?

There is, however, an example where it should be possible to test the model: fluid flow in a random medium. Here each ω can be regarded as a sample of the medium, so different experiments are obtained by choosing independent samples of the medium. Here we discuss one aspect of such flow: The pressure equation, described in the introduction. This equation will be considered in arbitrary dimension in Chapter 4. We now only look at the 1-dimensional case, modeling the fluid flow in a long, thin (heterogeneous) cylinder:

$$\frac{d}{dx}\left(K(x) \cdot \frac{d}{dx}p(x)\right) = 0 \; ; \; x \geq 0 \qquad (3.5.1)$$

with initial conditions

$$p(0) = 0, \quad K(0)p'(0) = a. \qquad (3.5.2)$$

Here $K(x) \geq 0$ is the permeability of the medium at x, $p(x)$ is the pressure of the fluid at x, and $a > 0$ is a constant. Condition (3.5.2) states that at the left endpoint of the cylinder the pressure of the fluid is 0 and the flux is a. If the medium is heterogeneous, then the permeability function $K(x)$ may vary in an irregular and unpredictable way. As argued in the introduction it is therefore natural to represent this quantity by the positive noise process

$$K(x) = \exp^\diamond[W_\phi(x)] = \exp\left[W_\phi(x) - \frac{1}{2}\|\phi\|_2^2\right] \qquad (3.5.3)$$

(see (2.6.56)), where $\phi \geq 0$ is a (deterministic) test function with compact support in $[0, \infty)$. The diameter of the support of ϕ indicates the maximal distance within which there is a correlation between the permeability values (depending on the sizes of the pores and other geometrical properties of the medium). The L^1 norm of ϕ, $\|\phi\|_1 = \int_\mathbb{R} |\phi(x)|dx$, reflects the size of the noise.

The figure below shows some typical sample paths of the Wick exponential $K(x) = K(x, \omega)$. In the figure we have used

$$\phi(x) = \frac{1}{h}\chi_{[0,h]}(x),$$

with $h = 1, 3, 5, 7, 9$ and 11.

Let us now consider the solutions of (3.5.1)–(3.5.2) in the two cases.

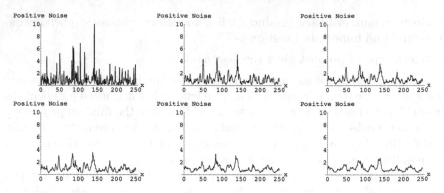

a) Ordinary product. In this case the equation is

$$(K(x,\omega) \cdot p'(x,\omega))' = 0; \ x \geq 0 \tag{3.5.4}$$

$$p(0,\omega) = 0, \quad K(0,\omega) \cdot p'(0,\omega) = a, \tag{3.5.5}$$

which is solved for each ω to give the solution

$$p(x,\omega) = p_1(x,\omega) = a \int_0^x \exp\left[-W_\phi(t) + \frac{1}{2}\|\phi\|_2^2\right] dt. \tag{3.5.6}$$

To find the expected value of p_1, we note that

$$p_1(x,\omega) = a \int_0^x \exp\left[W_{-\phi}(t) - \frac{1}{2}\|\phi\|_2^2\right] dt \cdot \exp[\|\phi\|_2^2]$$

$$= a \cdot \exp[\|\phi\|_2^2] \cdot \int_0^x \exp^\diamond[-W_\phi(t)] dt.$$

Hence, by (2.6.59) we conclude that

$$E[p_1(x)] = ax \cdot \exp[\|\phi\|_2^2]. \tag{3.5.7}$$

b) Wick product version. In this case the equation is

$$(K(x,\cdot) \diamond p'(x,\cdot))'(\omega) = 0; \ x \geq 0 \tag{3.5.8}$$

$$p(0,\omega) = 0, \quad (K(0,\cdot) \diamond p'(0,\cdot))(\omega) = a. \tag{3.5.9}$$

Straightforward Wick calculus gives the solution

$$p(x,\omega) = p_2(x,\omega) = a \int_0^x \exp^\diamond[-W_\phi(t)] dt. \tag{3.5.10}$$

In other words, the relation between the solutions is

$$p_1(x, \omega) = p_2(x, \omega) \exp[\|\phi\|_2^2], \qquad (3.5.11)$$

and we have

$$E[p_2(x)] = ax. \qquad (3.5.12)$$

Note that $E[p_2(x)] = ax$ coincides with the solution $\bar{p}(x)$ of the equation obtained by taking the average of the coefficients:

$$(1 \cdot \bar{p}'(x))' = 0; \; x \geq 0, \qquad (3.5.13)$$
$$\bar{p}(0) = 0, \quad \bar{p}'(0) = a. \qquad (3.5.14)$$

This property will hold for solutions of Wick type stochastic differential equations in general, basically because of (2.6.44).

If we let $\phi = \phi_n$ approach the Dirac delta function δ in the sense that

$$\int_{\mathbb{R}} g(x)\phi_n(x)dx \to g(0) \quad \text{as} \quad n \to \infty, g \in C_0(\mathbb{R}), \qquad (3.5.15)$$

then $\int_{\mathbb{R}} \phi_n(x)dx \to 1$ and $\mathrm{supp}\phi_n \to \{0\}$ as $n \to \infty$.

It follows that $\|\phi_n\|_2 \to \infty$ as $n \to \infty$. Hence we see that there are substantial differences between the two solutions $p_1(x, \omega)$ and $p_2(x, \omega)$ as $\phi \to \delta$:

$$\lim_{\phi \to \delta} p_1(x, \omega) = +\infty \qquad (3.5.16)$$

while

$$\lim_{\phi \to \delta} p_2(x, \omega) = \int_0^x \exp^\diamond[-W(t)]dt \in (\mathcal{S})^*. \qquad (3.5.17)$$

See also the solution in Potthoff (1992).

Although (3.5.17) only makes sense as a generalized stochastic process, this means that there are certain stability properties attached to the solution p_2. For a further discussion of this, see Lindstrøm et al. (1995).

3.5.1 Variance Properties

In reservoir simulation one often finds that Monte Carlo simulated solutions do not behave in the same manner as the solution of the averaged equation. The simple calculation (with the ordinary product) in the previous section sheds some light on this point of view. There may, however, be other explanations for this phenomenon. In the renormalized case, i.e., with the Wick product, it may happen that the typical sample path behaves very differently from the average solution. In this case the correlation width controls much of the behavior. It also suggests that certain scaling ratios are more favorable than others. To investigate this, we estimate the variance in some special cases.

To simplify the formulas, we use the function

$$\phi(x) = \frac{\epsilon}{h}\chi_{[0,h)}(x). \tag{3.5.18}$$

The parameter h is the correlation width, and ϵ controls the size of the noise.

With this choice of ϕ we get (see Lindstrøm et al. (1991a), p. 300, for the case $a = 1$ and see Exercise 3.3)

$$a^2 \max\left\{\epsilon^2\left(x + \frac{h}{3}\right), \frac{xh}{2}\left(e^{\frac{\epsilon^2}{2h}} - 1\right)\right\}$$

$$\leq E[(p_2(x,\omega) - ax)^2] \leq a^2 \frac{h(x+h)}{2}\left(e^{\frac{2\epsilon^2}{h}} - 1\right). \tag{3.5.19}$$

From this we can easily deduce the following:
For all $x > 0$

$$\lim_{h \to 0} \text{Var}[p_2(x,\omega)] = \infty; \tag{3.5.20}$$

For all $x > 0$

$$\lim_{h \to \infty} \text{Var}[p_2(x,\omega)] = \infty. \tag{3.5.21}$$

Hence if the correlation width is very small or very large, we can expect that typical sample paths differ significantly from the averages value. In these circumstances there is little point in estimating the average values from Monte Carlo experiments.

On the other hand it can be seen from the estimates that the variance (as a function of the correlation width) has a lower point. Around this point a Monte Carlo approach might be more favorable. For this to be true, the noise parameter ϵ must not be too large. More precisely we can see that
If $\epsilon^2 \ll h \ll x$, then

$$\text{Var}[p_2(x,\omega)] \approx \epsilon a\sqrt{x}. \tag{3.5.22}$$

When the parameters can be adjusted to conform with these scaling ratios, a Monte Carlo approach will give relevant information about the average value. Below we show some sample paths of the solution according to various choices of parameters.

In the figures below we used the value $a = 1$ and

$$\begin{array}{ll} \text{i) } h = 10, \ \epsilon = 1 & \text{ii) } h = 0.5, \ \epsilon = 1 \\ \text{iii) } h = 0.3, \ \epsilon = 1 & \text{iv) } h = 0.2, \ \epsilon = 1. \end{array}$$

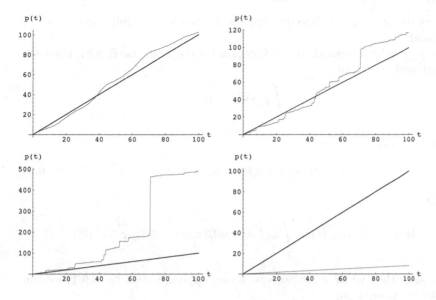

In the first two cases the variance is reasonably small. In the two last cases we are outside the favorable region, and the typical sample path is very much different from the average value.

The variance estimates are essentially the same in the case where we use the usual product. The two solutions differ by the constant factor $e^{\|\phi\|^2_{L^2}} = e^{\epsilon^2/h}$. Multiplying both sides of (3.5.19) by $e^{2\epsilon^2/h}$, we get

$$a^2 e^{\frac{2\epsilon^2}{h}} \max\left\{\epsilon^2\left(x+\frac{h}{3}\right), \frac{xh}{2}\left(e^{\frac{\epsilon^2}{2h}}-1\right)\right\}$$
$$\leq E[(p_1(x,\omega) - E[p_1])^2] \leq a^2 e^{\frac{2\epsilon^2}{h}} \frac{h(x+h)}{2}\left(e^{\frac{2\epsilon^2}{h}}-1\right).$$

If we examine the relations above, it is not hard to see that the properties (3.5.20–3.5.22) also apply in the case of the usual product. The stability region will, however, be somewhat smaller than in the Wick product case.

3.6 Solution and Wick Approximation of Quasilinear SDE

Consider an Itô stochastic differential equation of the form

$$dX(t) = b(t, X(t))dt + \sigma(t, X(t))dB(t), \ t > 0; \ X_0 = x \in \mathbb{R}, \qquad (3.6.1)$$

where $b(t, x) : \mathbb{R}^2 \to \mathbb{R}$ and $\sigma(t, x) : \mathbb{R}^2 \to \mathbb{R}$ are Lipschitz continuous of at most linear growth. Then we know that a unique, strong solution X_t exists.

If we try to approximate this equation and its solution, the following approach is natural:

Let $\rho \geq 0$ be a smooth (C^∞) function on the real line $\mathbb{R}$ with compact support and such that

$$\int_{\mathbb{R}} \rho(s)ds = 1.$$

For $k = 1, 2, \ldots$ define

$$\phi_k(s) = k\rho(ks) \quad \text{for } s \in \mathbb{R} \tag{3.6.2}$$

and let

$$W_{(k)}(t) := W_{\phi_k}(t) = \int_{\mathbb{R}} \phi_k(s - t)dB(s, \omega); \ t \in \mathbb{R}, \omega \in \mathcal{S}'(\mathbb{R})$$

be the smoothed white noise process. As an approximation to (3.6.1) we can now solve the equation

$$\frac{dY_k(t)}{dt} = b(t, Y_k(t)) + \sigma(t, Y_k(t)) \cdot W_{(k)}(t), t > 0; \ Y_k(0) = x \tag{3.6.3}$$

as an ordinary differential equation in t for each ω. Then, by the Wong–Zakai theorem, Wong and Zakai (1965), we know that $Y_k(t) \to Y(t)$ as $k \to \infty$, uniformly on bounded t-intervals for each ω, where $Y(t)$ is the solution of the *Stratonovich equation*

$$dY(t) = b(t, Y(t))dt + \sigma(t, Y(t)) \circ dB(t, \omega), t > 0; \ Y(0) = x. \tag{3.6.4}$$

So, perhaps surprisingly, we missed the solution $X(t)$ of our original Itô equation (3.6.1).

However, as conjectured in Hu and Øksendal (1996), we may perhaps recover $X(t)$ if we replace the ordinary product by the Wick product in the approximation procedure (3.6.3) above. Such a conjecture is supported by the relation between Wick products and Itô/Skorohod integration in general (see Theorem 2.5.9). Thus we consider the equation

$$\frac{dX_k(t)}{dt} = b(t, X_k(t)) + \sigma(t, X_k(t)) \diamond W_{(k)}(t), t > 0; X_k(0) = x \tag{3.6.5}$$

for each k and we ask

Does (3.6.4) have a unique solution for each k? If so, does $X_k(t) \to X(t)$ as $k \to \infty$?

The answer to these questions appears in general to be unknown. In this section we will apply the results from Section 2.10 to give a positive answer to these questions in the quasilinear case.

Following Gjessing (1994), we will consider more general (anticipating) quasilinear equations and first establish existence and uniqueness of solutions of such equations.

Theorem 3.6.1 Gjessing (1994).   *Suppose that the function $b(t, x, \omega)$:*
$\mathbb{R} \times \mathbb{R} \times \mathcal{S}'(\mathbb{R}) \rightarrow \mathbb{R}$ *satisfies the following condition:*

There exists a constant C such that
$$|b(t, x, \omega)| \leq C(1 + |x|) \quad \text{for all} \quad t, x, \omega \tag{3.6.6}$$

and
$$|b(t, x, \omega) - b(t, y, \omega)| \leq C|x - y| \quad \text{for all} \quad t, x, y, \omega. \tag{3.6.7}$$

Moreover, suppose that $\sigma(t)$ is a deterministic function, bounded on bounded intervals. Then the quasilinear, anticipating (Skorohod-type) stochastic differential equation

$$\frac{dX(t)}{dt} = b(t, X(t), \omega) + \sigma(t)X(t) \diamond W(t), t > 0; \ X(0) = x \tag{3.6.8}$$

has a unique (global) solution $X(t) = X(t, \omega); t \geq 0$. Moreover, we have

$$X(t, \cdot) \in L^p(\mu) \quad \text{for all} \quad p < \infty, t \geq 0. \tag{3.6.9}$$

Proof   Put $\sigma^{(t)}(s) = \sigma(s)\chi_{[0,t]}(s)$ and define

$$J_\sigma(t) = \exp^\diamond \left[-\int_0^t \sigma(s)dB(s) \right] = \exp^\diamond \left[-\int_\mathbb{R} \sigma^{(t)}(s)dB(s) \right]. \tag{3.6.10}$$

Regarding (3.6.8) as an equation in $(\mathcal{S})_{-1}$, we can Wick-multiply both sides by $J_\sigma(t)$ and this gives, after rearranging,

$$J_\sigma(t) \diamond \frac{dX(t)}{dt} - \sigma(t)J_\sigma(t) \diamond W(t) \diamond X(t) = J_\sigma(t) \diamond b(t, X(t), \omega)$$

or

$$\frac{dZ(t)}{dt} = J_\sigma(t) \diamond b(t, X(t), \omega), \tag{3.6.11}$$

where

$$Z(t) = J_\sigma(t) \diamond X(t). \tag{3.6.12}$$

By Theorem 2.10.6 we have, if $X(t) \in L^p(\mu)$ for some $p > 1$,

$$J_\sigma(t) \diamond b(t, X(t), \omega) = J_\sigma(t) \cdot b(t, T_{\sigma^{(t)}}X(t), \omega + \sigma^{(t)}) \tag{3.6.13}$$

and
$$Z(t) = J_\sigma(t) \cdot T_{\sigma(t)} X(t). \tag{3.6.14}$$

Substituting this into (3.6.11) we get the equation

$$\frac{dZ(t)}{dt} = J_\sigma(t) \cdot b(t, J_\sigma^{-1}(t)Z(t), \omega - \sigma^{(t)}), t > 0; \ Z_0 = x. \tag{3.6.15}$$

This equation can be solved for ω as an ordinary differential equation in t.

Because of our assumptions on b we get a unique solution $Z(t) = Z(t, \omega)$ for all ω. Moreover, from (3.6.15) we have

$$|Z(t)| \leq |x| + \left| \int_0^t J_\sigma(s) \cdot b(s, J_\sigma^{-1}(s)Z(s), \omega - \sigma^{(s)})ds \right|$$

$$\leq |x| + \int_0^t J_\sigma(s)C(1 + J_\sigma^{-1}(s)|Z(s)|)ds$$

$$= |x| + C \int_0^t J_\sigma(s)ds + C \int_0^t |Z(s)|ds.$$

Hence, by the Gronwall inequality,

$$|Z(t)| \leq \left(|x| + C \int_0^T J_\sigma(s)ds \right) \exp[Ct] \quad \text{for} \quad t \leq T. \tag{3.6.16}$$

Then, for $t \leq T$, we have

$$E[|Z(t)|^p] \leq \exp[pCt] \left(2^p |x|^p + 2^p C^p E \left[\left(\int_0^T J_\sigma(s)ds \right)^p \right] \right)$$

$$\leq C_1 + C_2 E \left[\left(\int_0^T 1 ds \right)^{\frac{p}{q}} \cdot \int_0^T J_\sigma(s)^p ds \right]$$

$$\leq C_1 + C_3 \int_0^T E[|J_\sigma(s)|^p] ds$$

$$\leq C_1 + C_3 T \exp \left[\frac{1}{2} p^2 \|\sigma\|_2^2 \right] < \infty.$$

We conclude that $Z(t) \in L^p(\mu)$ for all $t \geq 0, p < \infty$. It follows from (3.6.14) that the same is true for $T_{-\sigma(t)} X(t)$. From Corollary 2.10.5 we get that this is also true for $X(t)$. $\qquad\square$

Next, we consider the approximation question stated earlier in this section.

Theorem 3.6.2 Hu and Øksendal (1996).   *Let $b(t, x, \omega)$ and $\sigma(t)$ be as in Theorem 3.6.1, and let $W_{(k)}(t)$ be the ϕ_k-smoothed white noise process as defined in (3.6.2). Moreover, suppose that it is possible to find a map $D(\omega, \theta) : \mathcal{S}'(\mathbb{R}) \times \mathcal{S}(\mathbb{R}) \to (0, \infty)$ such that*

$$|b(t, x, \omega + \theta) - b(t, x, \omega)| \le D(\omega, \theta) \quad \text{for all} \quad t, x, \omega, \theta \qquad (3.6.18)$$

and

$$E_\mu[D^p(\cdot, \theta)] \to 0 \quad \text{as} \quad \theta \to 0 \quad \text{in} \quad \mathcal{S}'(\mathbb{R}) \qquad (3.6.19)$$

for some $p > 1$. Then for each $k \in \mathbb{N}$ there is a unique solution $X_k(t) \in L^p(\mu)$ for all $p < \infty$ of the equation

$$\frac{dX_k(t)}{dt} = b(t, X_k(t), \omega) + \sigma(t) X_k(t) \diamond W_{(k)}(t), t > 0; \; X_k(0) = x. \quad (3.6.20)$$

Moreover, for all $q < p$, we have

$$E_\mu[|Z_k(t) - Z(t)|^q] \to 0 \quad \text{as} \quad k \to \infty \qquad (3.6.21)$$

uniformly for t in bounded intervals.

Proof   Note that

$$\int_0^t \sigma(s) W_{(k)}(s) ds = \int_0^t \sigma(s) \int_\mathbb{R} \phi_k(r - s) dB(r) ds = \int_\mathbb{R} \sigma_k^{(t)}(r) dB(r),$$

where

$$\sigma_k^{(t)}(r) = \int_0^t \sigma(s) \phi_k(r - s) ds; \; t \ge 0.$$

Set

$$J_{\sigma_k}(t) = \exp^\diamond \left[-\int_0^t \sigma(s) W_{(k)}(s) ds \right] = \exp^\diamond \left[-\int_\mathbb{R} \sigma_k^{(t)}(r) dB(r) \right].$$

From now on we proceed exactly as in the proof of Theorem 3.6.1, except that σ is replaced by σ_k. Thus, with

$$Z_k(t) = J_{\sigma_k}(t) \diamond X_k(t) \qquad (3.6.22)$$

we get that

$$\frac{dZ_k(t)}{dt} = J_{\sigma_k}(t) \cdot b(t, J_{\sigma_k}^{-1}(t) Z_k(t), \omega - \sigma_k^{(t)}), t > 0; \; Z_k(0) = x, \qquad (3.6.23)$$

which has a solution $Z_k(t) \in L^p(\mu)$ for all $p < \infty$ just as equation (3.6.15). Finally, to prove (3.6.21) we use (3.6.18)–(3.6.19) to get

$$
|Z_k(t) - Z(t)| \leq \left| \int\limits_0^t \left(J_{\sigma_k}(s) b(s, J_{\sigma_k}^{-1}(s) Z_k(s), \omega - \sigma_k^{(s)}) \right. \right.
$$

$$
\left. \left. - J_\sigma(s) b(s, J_\sigma^{-1}(s) Z(s), \omega - \sigma^{(s)}) \right) ds \right|
$$

$$
\leq \int\limits_0^t \left(J_{\sigma_k} [C |Z_k(s) - Z(s)| \right.
$$

$$
\left. + D(\omega, \sigma_k^{(s)} - \sigma^{(s)})] + |J_{\sigma_k} - J_\sigma| C (1 + |Z(s)|) \right) ds.
$$

By the Gronwall inequality, this leads to

$$
|Z_k(t) - Z(t)| \leq F \cdot \exp\left[C \int\limits_0^t J_{\sigma_k}(s) ds \right] \quad \text{for} \quad t \leq T,
$$

where

$$
F = \int\limits_0^T \left(J_{\sigma_k}(s) D(\omega, \sigma_k^{(s)} - \sigma^{(s)}) + |J_{\sigma_k}(s) - J_\sigma(s)| C (1 + |Z_s|) \right) ds.
$$

From this we see that (3.6.21) follows. $\qquad\qquad\qquad\qquad\qquad\qquad\qquad$ □

3.7 Using White Noise Analysis to Solve General Nonlinear SDEs

From the previous sections one might get the impression that white noise analysis and the Wick product can only be used to solve linear, and some quasi-linear, SDEs. However, this is not the case. In a remarkable paper by Lanconelli and Proske (2004) the authors give an explicit solution formula for a general SDE. The formula and its proof uses the machinery of white noise analysis and the Wick calculus. We now present their result in more detail. For simplicity we only deal with the 1-dimensional case.

Consider an Itô stochastic differential equation of the form

$$
dX(t) = b(t, X(t))dt + \sigma(t, X(t))dB(t); \; X(0) = x, \; 0 \leq t \leq T \qquad (3.7.1)
$$

We assume that the functions $b : [0, T] \times \mathbb{R} \to \mathbb{R}$ and $\sigma : [0, T] \times \mathbb{R} \to \mathbb{R}$ are Lipschitz continuous and have at most linear growth. This is sufficient to guarantee that a unique, strong solution $X(t)$ exists. Moreover, we know that

$$\mathrm{E}\left[\int_0^T X^2(t)dt\right] < \infty \tag{3.7.2}$$

(See, e.g., Øksendal (2003), Chapter 5.)

For notational convenience we will in the following assume that the equation (3.7.1) is time-homogeneous, in the sense that $b(t, x) = b(x)$ and $\sigma(t, x) = \sigma(x)$. Moreover, we impose the conditions that

$$\sigma(x) > 0 \quad \text{for all } x \in \mathbb{R} \text{ and } \sigma \in C^1(\mathbb{R}) \tag{3.7.3}$$

and

$$\frac{b(x)}{\sigma(x)} \text{ is bounded on } \mathbb{R} \tag{3.7.4}$$

In the following we need to introduce the stocastic integral of an $(\mathcal{S})^*$-valued process $\phi(t)$ with respect to a Brownian motion $\hat{B}(t)$ on a filtered probability space $(\hat{\Omega}, \hat{\mathcal{F}}, \hat{\mu}), \{\hat{\mathcal{F}}_t\}_{t \geq 0}$, where $\hat{\mathcal{F}}_t$ is the filtration generated by $\hat{B}(s); s \leq t$. We only give the construction in the case when

$$\phi(t) = W(t) \in (\mathcal{S})^*$$

is white noise, which is the case we are interested in.

First consider a step function approximation $W^{(n)}(t)$ to $W(t)$, defined as follows

$$W^{(n)}(t) = \sum_{i=1}^{k_n - 1} W(t_i^{(n)}) \mathcal{X}_{[t_i^{(n)}, t_{i+1}^{(n)})}(t),$$

where $0 = t_1^{(n)} < \cdots < t_{k_n}^{(n)} = T$ is a partition of the interval $[0, T]$ such that $\max_i |t_i^{(n)} - t_{i+1}^{(n)}| \to 0$ as $n \to \infty$. Note that there exists a $q > 1$ such that

$$||W^{(n)}(t) - W(t)||_{-0,-q} \to 0 \text{ as } n \to \infty,$$

uniformly in $t \in [0, T]$. We define the stochastic integral of $W^{(n)}(t)$ with respect to $\hat{B}(t)$ by

$$\int_0^T W^{(n)}(t)d\hat{B}(t) = \sum_{i=1}^{k_n - 1} W(t_i^{(n)})(\hat{B}(t_{i+1}^{(n)}) - \hat{B}(t_i^{(n)}))$$

It can be verified that the following Itô isometry holds

$$\mathrm{E}_{\hat{\mu}}\left[\left|\left|\int_0^T W^{(n)}(t)d\hat{B}(t) - \int_0^T W^{(m)}(t)d\hat{B}(t)\right|\right|_{-0,-q}^2\right]$$

$$= \int_0^T ||W^{(n)}(t) - W^{(m)}(t)||_{-0,-q}^2 dt \qquad (3.7.5)$$

Hence we can define the stochastic integral

$$\int_0^T W(t)d\hat{B}(t) \in (\mathcal{S})^*$$

as the limit in $L^2(\hat{\mu}, (\mathcal{S})_{-0,-q})$ of the Cauchy sequence $\{\int_0^T W^{(n)}(t)d\hat{B}(t)\}_{n=1}^\infty$, as follows:

$$\int_0^T W(t)d\hat{B}(t) = \lim_{n\to\infty} \int_0^T W^{(n)}(t)d\hat{B}(t) \text{ in } L^2(\hat{\mu}, (\mathcal{S})_{-0,-q}). \qquad (3.7.6)$$

It is easy to see that the limit does not depend on the partition chosen.

The following result is needed in the proof of the main theorem:

Lemma 3.7.1.   *The map* $\Phi : \hat{\Omega} \to (\mathcal{S})_{-1}$ *defined by*

$$\hat{\omega} \to \exp^\diamond\left[\int_0^T W(s,\omega)d\hat{B}(s,\hat{\omega})\right]; \; \omega \in \Omega, \hat{\omega} \in \hat{\Omega}$$

is Bochner integrable in $(\mathcal{S})_{-1}$ *with respect to the measure* $\hat{\mu}$.

Proof   By Theorem 2.6.11 for all $R > 0$ there exist constants $C < \infty$ and $r \geq q$ such that if $F \in (\mathcal{S})_{-1,-q}$, then

$$\sup_{z\in\mathbb{K}_q(R)} |\mathcal{H}F(z)| \leq R||F||_{-1,-q} \leq R||F||_{-1,-2r} \leq C \sup_{z\in\mathbb{K}_q(R)} |\mathcal{H}F(z)| \quad (3.7.7)$$

Therefore it suffices to verify that

$$\mathrm{E}_{\hat{\mu}}\left[\sup_{z\in\mathbb{K}_q(R)} |\mathcal{H}(\Phi(\hat{\omega}))(z)|\right] < \infty.$$

By our construction of the stochastic integral we have that if l is large enough, then

$$\int_0^T W(s,\omega)d\hat{B}(s,\hat{\omega}) \in (\mathcal{S})_{-1,-l}$$

for a.a. $\hat{\omega}$ with respect to $\hat{\mu}$. Combined with Theorem 2.6.11 this gives the estimate

$$
\begin{aligned}
&\mathrm{E}_{\hat{\mu}}\left[\sup_{z \in \mathbb{K}_q(R)}|\mathcal{H}(\varPhi(\hat{\omega}))(z)|\right] \\
&\leq \mathrm{E}_{\hat{\mu}}\left[\exp\left[\sup_{z \in \mathbb{K}_q(R)}\left|\mathcal{H}\left(\int_0^T W(s)d\hat{B}(s)\right)(z)\right|\right]\right] \\
&\leq \mathrm{E}_{\hat{\mu}}\left[\exp\left[R\left\|\int_0^T W(s)d\hat{B}(s)\right\|_{-1,-l}\right]\right] \\
&\leq C_1 + \mathrm{E}_{\hat{\mu}}\left[\exp\left[R\left\|\int_0^T W(s)d\hat{B}(s)\right\|_{-1,-l}^2\right]\right]
\end{aligned}
$$

Since $\int_0^T W(s)d\hat{B}(s)$ is a zero mean Gaussian random variable in $(\mathcal{S})_{-l}^*$, it follows by Fernique's theorem (see, e.g., Da Prato and Zabczyk (1992)) that the last expectation is finite. Hence the conclusion follows from (3.7.7). $\square$

We are now ready to state and prove the main result in this section:

Theorem 3.7.2 (Lanconelli and Proske (2004), explicit solution formula for SDEs). *Assume that (3.7.3) and (3.7.4) hold. Define*

$$
\Lambda(y) = \int_x^y \frac{1}{\sigma(u)}du \tag{3.7.8}
$$

Let $\phi : \mathbb{R} \to \mathbb{R}$ be measurable and such that

$$
\left(\phi \circ \Lambda^{-1}\right)\left(\hat{B}(t)\right) \in L^2(\hat{\mu}) \text{ for all } t \in [0,T] \tag{3.7.9}
$$

Let $X^x(t)$ be the strong solution of (3.7.1). Then

$$
\phi(X^x(t)) = \mathrm{E}_{\hat{\mu}}\left[\phi(\Lambda^{-1}(\hat{B}(t)))M_T^{\diamond}\right] \tag{3.7.10}
$$

where

$$
\begin{aligned}
M_T^{\diamond} = \exp^{\diamond}&\left[\int_0^T\left(W(s) + \frac{b(\Lambda^{-1}(\hat{B}(s)))}{\sigma(\Lambda^{-1}(\hat{B}(s)))} - \frac{1}{2}\sigma'(\Lambda^{-1}(\hat{B}(s)))\right)d\hat{B}(s)\right. \\
&\left.- \int_0^T\left(W(s) + \frac{b(\Lambda^{-1}(\hat{B}(s)))}{\sigma(\Lambda^{-1}(\hat{B}(s)))} - \frac{1}{2}\sigma'(\Lambda^{-1}(\hat{B}(s)))\right)^{\diamond 2}ds\right]. \tag{3.7.11}
\end{aligned}
$$

Proof Without loss of generality we can assume that the drift term is zero, i.e., that

$$dX^x(t) = \sigma(X^x(t))dB(t), \ X^x(0) = x \tag{3.7.12}$$

We first find the Hermite transform of $\phi(X^x(t))$: Choose $z \in (\mathbb{C}^N)_c$. Then by the relation between the $\mathcal{S}$-transform and the Hermite transform (Theorem 2.7.10) we have

$$\mathcal{H}(\phi(X^x(t)))(z) = E_\mu \left[\phi(X^x(t)) \exp^\diamond \left[\int_{\mathbb{R}} \{z_1\xi_1(s) + \cdots + z_k\xi_k(s)\}dB(s) \right] \right]$$

$$= E_\mu \left[\phi(X^x(t)) \exp^\diamond \left[\int_{\mathbb{R}} \mathcal{H}(W_s)(z)dB(s) \right] \right] \tag{3.7.13}$$

Hence by the Girsanov theorem (see Corollary 2.10.5)

$$\mathcal{H}(\phi(X^x(t))) = E_\mu[\phi(\tilde{X}^x(t))] \tag{3.7.14}$$

where

$$\tilde{X}^x(t,\omega) = X^x(t, \omega + \omega_0), \ \text{with} \ <\omega_0, \cdot> = (\mathcal{H}(W(s))(z), \cdot)_{L^2(\mathbb{R})} \tag{3.7.15}$$

Note that $\tilde{X}^x(t)$ solves the equation

$$d\tilde{X}^x(t) = h(t)\sigma(\tilde{X}^x(t))dt + \sigma(\tilde{X}^x(t))dB(t); \ \tilde{X}^x(0) = x \tag{3.7.16}$$

where $h(t) = \mathcal{H}(W(t))(z)$. Moreover, if we define

$$Z_t^x = \Lambda(\tilde{X}^x(t))$$

then by the Itô formula

$$dZ_t^x = \left(h(t) - \frac{1}{2} \left(\sigma' \circ \Lambda^{-1} \right) (Z_t^x) \right) dt + dB(t); \ Z_0^x = \Lambda(x) = 0 \tag{3.7.17}$$

Applying the classical Girsanov formula we obtain

$$\mathcal{H}(\phi(X^x(t))(z) = E_\mu[\phi(\tilde{X}^x(t))] = E_{\hat{\mu}}[(\phi \circ \Lambda^{-1})(Z_t^x)]$$

$$= E_{\hat{\mu}}[(\phi \circ \Lambda^{-1})(\hat{B}(t))M_T] \tag{3.7.18}$$

where

$$M_t = \exp \left[\int_0^T \left(h(s) - \frac{1}{2}(\sigma' \circ \Lambda^{-1})(\hat{B}(s)) \right) d\hat{B}(s) \right.$$

$$\left. - \int_0^T \left(h(s) - \frac{1}{2} \left(\sigma' \circ \Lambda^{-1} \right) (\hat{B}(s)) \right)^2 ds \right]. \tag{3.7.19}$$

By Lemma 7.3.1 we can apply the inverse Hermite transform to the last term in (3.7.18) and get

$$\mathcal{H}(\phi(X^x(t)))(z) = \mathcal{H}(\mathrm{E}_{\hat{\mu}}[(\phi \circ \Lambda^{-1})(\hat{B}(t))M_T^{\diamond}])(z) \qquad (3.7.20)$$

for $z \in \mathbb{K}_q(R)$ for some q, R. Then the conclusion follows by the characterization theorem for Hermite transforms (Theorem 2.6.11). □

Remark 3.7.3 In spite of its striking simplicity, the solution formula (3.7.10) is not easy to use to find standard, non-white-noise solution formulas, even in cases when the solution is easy to find using the Itô formula. See Exercise 3.11.

Exercises

3.1 Show that the processes $X_1(t), X_2(t)$ defined by (3.2.12) and (3.2.15), respectively, are not Markov processes. This illustrates that in general the solution of a Wick type stochastic differential equation does not have the Markov property.

3.2

a) Construct a solution $X(t)$ of equation (3.3.3) by proceeding as follows: Define $X_0(t) = X_0$ and, by induction, using Picard iteration,

$$X_n(t) = X_0 + \int_0^t F(s, X_{n-1}(s))ds.$$

Then $X_n(t)$ converges in $(\mathcal{S})_{-1,-k}^N$ to a solution $X(t)$ as $n \to \infty$.

b) Prove that equation (3.3.3) has only one solution by proceeding as follows: If $X_1(t), X_2(t)$ are two solutions, define $g(t) = \|X_1(t) - X_2(t)\|_{-1,-k}^2$ for $t \in [0, T]$. Then

$$g(t) \le A \int_0^t g(s)ds$$

for some constant A.

3.3 Deduce the inequalities in (3.5.19).

3.4 Use Wick calculus to solve the following Itô stochastic differential equations:

a) $dX(t) = rX(t)dt + \alpha X(t)dB(t), \ t > 0$
   $X_0 = x; \ x, r, \alpha$ constants;

b) $dX(t) = rX(t)dt + \alpha dB(t), \ t > 0$
   $X_0 = x; \ x, r, \alpha$ constants;

c) $dX(t) = rdt + \alpha X(t)dB(t), \ t > 0$
   $X_0 = x; \ x, r, \alpha$ constants;

d) $dX(t) = r(K - X(t))dt + \alpha(K - X(t))dB(t), \ t > 0$
   $X_0 = x; \ x, r, K, \alpha$ constants;

e) $dX(t) = (r + \rho X(t))dt + (\alpha + \beta X(t))dB(t), \ t > 0$
   $X_0 = x; \ x, r, \rho, \alpha, \beta$ constants.

f) If $X_0 = x$ is not constant, but an $\mathcal{F}_\infty$-measurable random variable
   such that $X_0(\omega) \in L^2(\mu)$, how does this affect the solutions of a) – e)
   above?

3.5 Solve the Skorohod stochastic differential equations

a) $dX(t) = rX(t)dt + \alpha X(t)\delta B(t); \ 0 < t < T$
   $X(T) = G(\omega) \in L^2(\mu), \ \mathcal{F}_\infty$-measurable;

b) $dX(t) = rX(t)dt + \alpha \delta B(t); \ 0 < t < T$
   $X(T) = G(\omega) \in L^2(\mu), \ \mathcal{F}_\infty$-measurable;

c) $dX(t) = rdt + \alpha X(t)\delta B(t); \ 0 < t < T$
   $X(T) = G(\omega) \in L^2(\mu), \ \mathcal{F}_\infty$-measurable;

d) $dX(t) = B(T)dt + X(t)\delta B(t); \ 0 < t < T$
   $X(0) = G(\omega) \in L^2(\mu), \ \mathcal{F}_\infty$-measurable.

3.6 Use Wick calculus to solve the following 2-dimensional system of
stochastic differential equations:

$$\begin{cases} \frac{dX_1(t)}{dt} = -X_2(t) + \alpha W_1(t) \\ \frac{dX_2(t)}{dt} = X_1(t) + \beta W_2(t); \ X_1(0), X_2(0) \quad \text{given}, \end{cases}$$

where $\mathbf{W}(t) = (W_1(t), W_2(t))$ is 2-dimensional, 1-parameter white noise.
This is a model for a vibrating string subject to a stochastic force.

3.7 In Grue and Øksendal (1997), the following second–order stochastic differential equation is studied as a model for the motion of a moored platform
in the sea exposed to random forces from the wind, waves and currents:

$$\ddot{x}(t) + [\alpha + \beta W(t)] \diamond \dot{x}(t) + \gamma x(t) = \theta W(t); \ t > 0$$
$$x(0), \ \dot{x}(0) \text{ given},$$

where $\alpha, \beta, \gamma, \theta$ are constants.

a) Transform this into a stochastic Volterra equation of the form

$$x(t) = J(t, \omega) + \int_0^t K(t, s, \omega) \diamond x(s) ds$$

for suitable stochastic distribution processes.

b) Verify that the conditions of Theorem 3.4.2 are satisfied in this case and hence conclude that the equation has a unique stochastic distribution solution $x(t) \in (\mathcal{S})_{-1}$.

3.8 Solve the second–order stochastic differential equation

$$\ddot{x}(t) + x(t) = W(t); \ t > 0$$
$$x(0), \dot{x}(0) \ \text{given},$$

by transforming it into a stochastic Volterra equation.

3.9 Use the method of Theorem 3.6.1 to solve the quasilinear SDE

$$dX(t) = f(X(t))dt + X(t)dB(t); \ t > 0$$
$$X(0) = x \in (0, 1) \ \text{(deterministic)},$$

where $f(x) = \min(x, 1); \ x \in \mathbb{R}$.

3.10 Use the method of Theorem 3.6.1 to solve the SDE

$$dX(t) = rX(t) + \alpha X(t)\delta B(t); \ t > 0$$
$$X(0) \in L^2(\mu), \ \mathcal{F}_\infty\text{-measurable},$$

and compare the result with the result in Exercise 3.4 a) and f).

3.11 Use Theorem 3.7.2 to show that the solution of the stochastic differential equation

$$dX^x(t) = \mu X^x(t)dt + \sigma X^x(t)dB(t); \ X^x(0) = x > 0$$

(where $\mu, \sigma > 0$ are constants) is the geometric Brownian motion

$$X^x(t) = x \exp\left[\left(\mu - \frac{1}{2}\sigma^2\right)t + \sigma B(t)\right]; \ t \geq 0$$

Chapter 4
Stochastic Partial Differential Equations Driven by Brownian White Noise

4.1 General Remarks

In this chapter we will apply the general theory developed in Chapter 2 to solve various stochastic partial differential equations (SPDEs) driven by Brownian white noise. In fact, as pointed out in Chapter 1, our main motivation for setting up this machinery was to enable us to solve some of the basic SPDEs that appear frequently in applications.

We can explain our general approach to SPDEs as follows:

Suppose that modeling considerations lead us to consider an SPDE expressed formally as

$$A(t, x, \partial_t, \nabla_x, U, \omega) = 0 \qquad (4.1.1)$$

where A is some given function, $U = U(t, x, \omega)$ is the unknown (generalized) stochastic process, and where the operators $\partial_t = \partial/\partial t, \nabla_x = (\partial/\partial x_1, \ldots, \partial/\partial x_d)$ when $x = (x_1, \ldots, x_d) \in \mathbb{R}^d$.

First we interpret all products as Wick products and all functions as their Wick versions, as explained in Definition 2.6.14. We indicate this as

$$A^{\diamond}(t, x, \partial_t, \nabla_x, U, \omega) = 0. \qquad (4.1.2)$$

Secondly, we take the Hermite transform of (4.1.2). This turns Wick products into ordinary products (between (possibly) complex numbers) and the equation takes the form

$$\widetilde{A}(t, x, \partial_t, \nabla_x, \widetilde{U}, z_1, z_2, \ldots) = 0, \qquad (4.1.3)$$

where $\widetilde{U} = \mathcal{H}U$ is the Hermite transform of U and z_1, z_2 are complex numbers. Suppose we can find a solution $u = u(t, x, z)$ of the equation

$$\widetilde{A}(t, x, \partial_t, \nabla_x, u, z) = 0, \qquad (4.1.4)$$

for each $z = (z_1, z_2, \ldots) \in \mathbb{K}_q(R)$ for some q, R (see Definition 2.6.4).

H. Holden et al., *Stochastic Partial Differential Equations*, 2nd ed., Universitext, 159
DOI 10.1007/978-0-387-89488-1_4, © Springer Science+Business Media, LLC 2010

Then, under certain conditions, we can take the inverse Hermite transform $U = \mathcal{H}^{-1}u \in (\mathcal{S})_{-1}$ and thereby obtain a solution U of the original (Wick) equation (4.1.2). See Theorem 4.1.1 below for details. This method has already been applied in Chapter 3. See, e.g., the proof of Theorem 3.2.1.

The first step of this procedure, to interpret all products as Wick products, reflects a certain choice regarding the exact mathematical model for the equation. As pointed out in Chapter 1, the solution U of (4.1.1) will in many cases only exist as a *generalized* stochastic process, and this makes it difficult to interpret the products as ordinary, pointwise products. Wick products, however, have the advantage of being well-defined (and well-behaved) on the space $(\mathcal{S})_{-1}$.

Moreover, it coincides with the ordinary product if one of the factors is deterministic, and it represents the natural extension of the principle of interpreting differential equations with white noise as Itô/Skorohod stochastic differential equations. See (1.1.9) and the other related comments in Chapter 1. However, regardless of all such good theoretical arguments for the use of the Wick product, the ultimate test for such a model is the comparison between the mathematical solution and the observed solution of the physical phenomenon we are modeling. See Section 3.5 for a 1-dimensional example.

The Hermite transform replaces a real–valued function depending on ω (or, more generally, an element of $(\mathcal{S})_{-1}$) by a complex–valued function depending on a complex parameter $z = (z_1, z_2, \ldots) \in (\mathbb{C}^{\mathbb{N}})_c$. So to solve (4.1.4) we have to solve a *deterministic* PDE with complex coefficients depending on the complex parameters $z_1, z_2, \ldots$. If we succeed in doing this, we proceed by taking inverse Hermite transforms to obtain a solution of the original equation. Sufficient conditions for this procedure to work are given in the next theorem.

Theorem 4.1.1. *Suppose $u(t, x, z)$ is a solution (in the usual strong, pointwise sense) of the equation*

$$\widetilde{A}(t, x, \partial_t, \nabla_x, u, z) = 0 \qquad (4.1.5)$$

for (t, x) in some bounded open set $G \subset \mathbb{R} \times \mathbb{R}^d$, and for all $z \in \mathbb{K}_q(R)$, for some q, R. Moreover, suppose that

$u(t, x, z)$ and all its partial derivatives, which are involved in (4.1.4), are(uniformly) bounded for $(t, x, z) \in G \times \mathbb{K}_q(R)$, continuous with respect to $(t, x) \in G$ for each $z \in \mathbb{K}_q(R)$ and analytic with respect to $z \in \mathbb{K}_q(R)$, for all $(t, x) \in G$. $\qquad (4.1.6)$

Then there exists $U(t, x) \in (\mathcal{S})_{-1}$ such that $u(t, x, z) = (\mathcal{H}U(t, x))(z)$ for all $(t, x, z) \in G \times \mathbb{K}_q(R)$ and $U(t, x)$ solves (in the strong sense in $(\mathcal{S})_{-1}$) the equation

$$A^{\diamond}(t, x, \partial_t, \nabla_x, U, \omega) = 0 \quad in \quad (\mathcal{S})_{-1}. \qquad (4.1.7)$$

Proof This result is a direct extension of Lemma 2.8.4 to the case involving higher order derivatives. It can be proved by applying the argument of Lemma 2.8.4 repeatedly. We omit the details. See Exercise 4.1.   □

Remark Note that it is enough to check condition (4.1.6) for the *highest-order* derivatives of each type, since from this the condition automatically holds for all lower-order derivatives, by the mean value property.

4.2 The Stochastic Poisson Equation

Let us illustrate the method described above on the following equation, called the *stochastic Poisson equation*:

$$\begin{cases} \Delta U(x) = -W(x); & x \in D \\ U(x) = 0; & x \in \partial D, \end{cases} \tag{4.2.1}$$

where $\Delta = \sum_{k=1}^{d} \partial^2 / \partial x_k^2$ is the Laplace operator in $\mathbb{R}^d$, $D \subset \mathbb{R}^d$ is a given bounded domain with regular boundary (see, e.g., Øksendal (1995), Chapter 9) and where $W(x) = \sum_{j=1}^{\infty} \eta_k(x) H_{\varepsilon_k}(\omega)$ is d-parameter white noise. As mentioned in Chapter 1, this equation models, for example, the temperature $U(x)$ in D when the boundary temperature is kept equal to 0 and there is a white noise heat source in D.

Taking the Hermite transform of (4.2.1), we get the equation

$$\begin{cases} \Delta u(x, z) = -\widetilde{W}(x, z); & x \in D \\ u(x, z) = 0; & x \in \partial D \end{cases} \tag{4.2.2}$$

for our candidate u for $\widetilde{U}$, where the Hermite transform $\widetilde{W}(x, z) = \sum_{j=1}^{\infty} \eta_j(x) z_j$ when $z = (z_1, z_2, \ldots) \in (\mathbb{C}^{\mathbb{N}})_c$ (see Example 2.6.2). By considering the real and imaginary parts of this equation separately, we see that the usual solution formula holds:

$$u(x, z) = \int_{\mathbb{R}^d} G(x, y) \widetilde{W}(y, z) dy, \tag{4.2.3}$$

where $G(x, y)$ is the classical Green function of D (so $G = 0$ outside D). (See, e.g., Port and Stone (1978), or Øksendal (1995), Chapter 9). Note that $u(x, z)$ exists for all $x \in (\mathbb{C}^{\mathbb{N}})_c, x \in D$, since the integral on the right of (4.2.3) converges for all such z, x. (For this we only need that $G(x, y) \in L^1(dy)$ for each x.) Moreover, for $z \in (\mathbb{C}^{\mathbb{N}})_c$, we have

$$|u(x,z)| = \left| \int G(x,y) \sum_j \eta_j(y) z_j dy \right| = \left| \sum_j z_j \int G(x,y)\eta_j(y)dy \right|$$

$$\leq \sup_{j,y} |\eta_j(y)| \int G(x,y)dy \sum_j |z_j| \leq C \sum_j |z_j| = C \sum_j |z^{\varepsilon_j}|$$

$$\leq C \left(\sum_j |z^{\varepsilon_j}|^2 (2\mathbb{N})^{2\varepsilon_j} \right)^{\frac{1}{2}} \left(\sum_j (2\mathbb{N})^{-2\varepsilon_j} \right)^{\frac{1}{2}}$$

$$\leq C R \left(\sum_j (2j)^{-2} \right)^{\frac{1}{2}}$$

$$< \infty, \tag{4.2.4}$$

if $z \in \mathbb{K}_2(R)$. Since $u(x,z)$ depends analytically on z, it follows from the Characterization Theorem (Theorem 2.6.11) that there exists $U(x) \in (\mathcal{S})_{-1}$ such that $\widetilde{U}(x,z) = u(x,z)$. Moreover, from the general theory of elliptic (deterministic) PDEs (see, e.g., Bers et al. (1964)), we know that, for each open $V \subset\subset D$ and $z \in (\mathbb{C}^{\mathbb{N}})_c$ there exists C such that

$$\|u(\cdot,z)\|_{C^{2+\alpha}(V)} \leq C(\|\Delta u(\cdot,z)\|_{C^\alpha(D)} + \|u(\cdot,z)\|_{C(D)}). \tag{4.2.5}$$

In particular, $\partial^2 u/\partial x_i^2(x,z)$ is bounded for $(x,z) \in \mathbb{K}_2(R)$ since both $\Delta u = -\widetilde{W}$ and u are. Therefore, by Theorem 4.1.1, $U(x)$ solves (4.2.1).

We recognize directly from (4.2.3) that u is the Hermite transform of

$$U(x) = \int_{\mathbb{R}^d} G(x,y)W(y)dy = \sum_{j=1}^\infty \int_{\mathbb{R}^d} G(x,y)\eta_j(y)dy H_{\varepsilon_j}(\omega),$$

which converges in $(\mathcal{S})^*$ because (see (2.3.26))

$$\sum_{j=1}^\infty \left(\int_{\mathbb{R}^d} G(x,y)\eta_j(y)dy \right)^2 (2j)^{-q} \leq C^2 \sum_{j=1}^\infty (2j)^{-q} < \infty \quad \text{for all} \quad q > 1.$$

Theorem 4.2.1. *The unique stochastic distribution process $U(x) \in (\mathcal{S})_{-1}$ solving (4.2.1) is given by*

$$U(x) = \int_{\mathbb{R}^d} G(x,y)W(y)dy$$

$$= \sum_{j=1}^\infty \int_{\mathbb{R}^d} G(x,y)\eta_j(y)dy H_{\varepsilon_j}(\omega). \tag{4.2.6}$$

We have $U(x) \in (\mathcal{S})^*$ for all $x \in \bar{D}$.

As mentioned in Chapter 1, equation (4.2.1) was studied in Walsh (1986). He showed that there is a unique *distribution valued* stochastic process $Y(x, \omega)$ solving (4.2.1). This means that there exists a Sobolev space $H^{-n}(D)$ such that $Y(\cdot, \omega) \in H^{-n}(D)$ for almost all ω and

$$\Delta Y(\cdot, \omega) = -W(\cdot, \omega) \quad \text{in} \quad H^{-n}(D), \text{ a.s.}$$

in the sense that

$$\langle \Delta Y(\cdot, \omega), \phi \rangle = -\langle W(\cdot, \omega), \phi \rangle,$$

i.e.,

$$\langle Y(\cdot, \omega), \Delta \phi \rangle = -\langle W(\cdot, \omega), \phi \rangle \quad \text{a.s., for all} \quad \phi \in H^n(D).$$

The Walsh solution is given explicitly by

$$\langle Y, \phi \rangle = \int_{\mathbb{R}^d} \int_{\mathbb{R}^d} G(x, y)\phi(x)dx dB(y); \quad \phi = \phi(x) \in H^n(\mathbb{R}^d). \tag{4.2.7}$$

For comparison, our solution $U(x)$ in (4.2.6) can be described by its action on its test functions $f \in (\mathcal{S})$:

$$\langle U(x), f \rangle = \int_{\mathbb{R}^d} G(x, y)\langle W(y), f \rangle dy; \quad f = f(\omega) \in (\mathcal{S}). \tag{4.2.8}$$

In short, the Walsh solution $Y(x, \omega)$ takes x-averages for almost all ω, while our solution $U(x, \omega)$ takes ω-averages for all x.

4.2.1 The Functional Process Approach

It is instructive to consider this equation from a functional process point of view (see Section 2.9). So we fix $\psi \in \mathcal{S}(\mathbb{R}^d)$ and consider the equation

$$\begin{cases} \Delta U(\psi, x) = -W_\psi(x); & x \in D \\ U(\psi, x) = 0; & x \in \partial D, \end{cases} \tag{4.2.9}$$

where $W_\psi(x) = w(\psi_x, \omega) = \int_{\mathbb{R}^d} \psi(\xi - x)dB(\xi)$ is ψ_x-smoothed white noise and $\psi_x(y) = \psi(y - x)$; $x, y \in \mathbb{R}^d$. This equation can be solved for each ω to give

$$U(\psi, x) = \int_{\mathbb{R}^d} G(x, y)W_\psi(y)dy = \int_{\mathbb{R}^d} \int_{\mathbb{R}^d} G(x, y)\psi(\xi - y)dy dB(\xi). \tag{4.2.10}$$

We have $U(\psi, x) \in L^p(\mu)$ for all $p < \infty$.

If ψ, or rather ψdy, tends to δ_0 in the weak star topology in the space of measures on $\mathbb{R}^d$, then $U(\psi, x) \to U(x)$ (given by (4.2.5)) in $(\mathcal{S})^*$. This is because, using the notation of Definition 2.7.1,

$$\|U(\psi, x) - U(x)\|_{0,r} \leq \int_{\mathbb{R}^d} G(x,y)\|W_\psi(y) - W(y)\|_{0,r} dy$$

$$\leq \int_{\mathbb{R}^d} G(x,y) \left(\sum_j |(\psi_y, \eta_j) - \eta_j(y)|^2 (2j)^r \right)^{\frac{1}{2}} dy \to 0 \quad \text{for} \quad r < -1.$$

See Lindstrøm et al. (1995), for a discussion of similar properties of the more general stochastic Poisson equation

$$\begin{cases} \mathrm{div}(a(x)\nabla U(x)) = -W(x); & x \in D \\ U(x) = 0; & x \in \partial D \end{cases}$$

where $a(x) = [a_{ij}(x)]$ is a $d \times d$ symmetric matrix with bounded measurable components $a_{ij}(x)$. The solutions are in this case interpreted in the Walsh sense.

This type of stability question, as well as the basic idea behind functional processes, are inspired by Colombeau's theory of distributions (see Colombeau (1990)), which makes it possible to define the product of certain distributions.

The Colombeau theory was adapted to stochastic analysis in Albeverio et al. (1996). There existence and uniqueness are proved for a Colombeau distribution solution of a certain class of nonlinear stochastic wave equations of space dimension 2. See also Russo (1994), Oberguggenberger (1992, 1995), and the references therein.

4.3 The Stochastic Transport Equation

4.3.1 Pollution in a Turbulent Medium

The transport of a substance that is dispersing in a moving medium can be modeled by an SPDE of the form

$$\begin{cases} \frac{\partial U}{\partial t} = \frac{1}{2}\sigma^2 \Delta U + V(t,x) \cdot \nabla U + K(t,x) \cdot U + g(t,x); & t > 0, x \in \mathbb{R}^d \\ U(0,x) = f(x); & x \in \mathbb{R}^d, \end{cases}$$

$$(4.3.1)$$

where $U(t,x)$ is the concentration of the substance at time t and at the point $x \in D$, $1/2\sigma^2 > 0$ (constant) is the dispersion coefficient, $V(t,x) \in \mathbb{R}^d$ is the velocity of the medium, $K(t, x) \in \mathbb{R}$ is the relative leakage rate and $g(t, x) \in \mathbb{R}$ is the source rate of the substance. The initial concentration is a given real function $f(x)$.

If one or several of these coefficients are assumed to be stochastic, we call equation (4.3.1) a *stochastic transport equation*. For example, the case when we have $K = g = 0$ and $V = \mathbf{W}(x)$ (d-dimensional white noise) models the transport of a substance in a turbulent medium. This case has been studied by several authors. When $d = 1$ and the product $W(x) \cdot \nabla U(t, x)$ is interpreted by means of a Stratonovich integration, the equation has been studied by Chow (1989), Nualart and Zakai (1989), and Potthoff (1992), and in the Hitsuda–Skorohod interpretation by Potthoff (1994). See also Deck and Potthoff (1998), for a more general approach. For arbitrary d and with $V(x) = \mathbf{W}_\phi(x) = (W_\phi^{(1)}(x), \ldots, W_\phi^{(d)}(x))$, d-dimensional ϕ-smoothed white noise ($\phi \in \mathcal{S}$), and the product $\mathbf{W}_\phi(x) \cdot \nabla U(t, x)$ interpreted as a Wick product $\mathbf{W}_\phi(x) \diamond \nabla U(t, x)$ (and still $K = g = 0$), an explicit solution was found in Gjerde et al. (1995). There the initial value $f(x)$ was allowed to be stochastic and anticipating and it was shown that the solution was actually a strong solution in $(\mathcal{S})^*$ for all t, x. Equation (4.3.1), with V, K deterministic, but $g(t, x)$ random, was studied by Kallianpur and Xiong (1994), as a model for pollution dispersion. Then Gjerde (1994) combined the two cases studied by Gjerde et al. (1995), and Kallianpur and Xiong (1994) and solved the following generalization:

$$\begin{cases} \frac{\partial U}{\partial t} = \frac{1}{2}\sigma^2 \Delta U + \mathbf{W}_\phi(x) \diamond \nabla U + K(t, x) \diamond U + g(t, x); & t > 0, x \in \mathbb{R}^d \\ U(0, x) = f(x); & x \in \mathbb{R}^d, \end{cases}$$

$$(4.3.2)$$

where σ is a constant and $K(t, x)$, $g(t, x)$ and $f(x)$ are given stochastic distribution processes.

Our presentation here is a synthesis of the methods in Gjessing (1994), Gjerde et al (1995), Holden et al. (1995a), and Holden et al. (1995b).

Theorem 4.3.1 Gjerde (1994) (The stochastic transport equation).
Assume that $K : \mathbb{R}^+ \times \mathbb{R}^d \to (\mathcal{S})_{-1}, g : \mathbb{R}^+ \times \mathbb{R}^d \to (\mathcal{S})_{-1}$ and $f : \mathbb{R}^d \to (\mathcal{S})_{-1}$ satisfy the following conditions:

There exist q and $R < \infty$ such that $|\widetilde{K}(t, x, z)| + |\widetilde{g}(t, x, z)| + |\widetilde{f}(x, z)|$ is uniformly bounded for $(t, x, z) \in \mathbb{R}^+ \times \mathbb{R}^d \times \mathbb{K}_q(R)$. $\qquad(4.3.3)$

There exists $\mathbb{K}_q(R)$ such that for each $z \in \mathbb{K}_q(R)$ we can find $\gamma \in (0, 1)$ such that $\widetilde{K}(t, x, z) \in C^{1,\gamma}(\mathbb{R}^+ \times \mathbb{R}^d)$, $\widetilde{g}(t, x, z) \in C^{1,\gamma}(\mathbb{R}^+ \times \mathbb{R}^d)$ and $\widetilde{f}(x, z) \in C^{2+\gamma}(\mathbb{R}^d)$. $\qquad(4.3.4)$

Then there exists a unique stochastic distribution process $U(t,x)$ *solving*
(4.3.2), namely

$$
U(t,x) = \hat{E}^x \left[\left(f(\sigma b_t) \diamond \exp^\diamond \left[\int_0^t K(t-r, \sigma b_r) dr \right] \right. \right.
$$
$$
\left. \left. + \int_0^t g(t-s, \sigma b_s) \diamond \exp^\diamond \left[\int_0^s K(s-r, \sigma b_r) dr \right] ds \right) \diamond M_t^\diamond \right], \quad (4.3.5)
$$

where

$$
M_t^\diamond = \exp^\diamond \left[-\sum_{k=1}^d \sigma^{-1} \int_0^t W_\phi^{(k)}(\sigma b_s) db_s^{(k)} - \frac{1}{2} \sum_{k=1}^d \sigma^{-2} \int_0^t (W_\phi^{(k)}(\sigma b_s))^{\diamond 2} ds \right].
$$
$$
(4.3.6)
$$

Here $(b_t)_{t\geq 0} = (b_t^{(1)}, \ldots, b_t^{(d)})_{t\geq 0}$ *is an auxiliary Brownian motion in* $\mathbb{R}^d$
(independent of $\{B(t)\}_{t\geq 0}$*) with probability law* $\hat{P}^x$ *when starting at* $x \in \mathbb{R}^d$
at time $t = 0$*, and* $\hat{E}^x$ *denotes the expectation with respect to* $\hat{P}^x$*. Thus the*
integrals in the Wick exponent are $(\mathcal{S})_{-1}$*-valued Itô integrals and*
$(\mathcal{S})_{-1}$*-valued Lebesgue integrals (Bochner integrals), respectively.*

Proof Taking the Hermite transform of (4.3.2), we get the equation

$$
\begin{cases}
\frac{\partial u}{\partial t} = \frac{1}{2}\sigma^2 \Delta u + \widetilde{\mathbf{W}}_\phi(x) \cdot \nabla u + \widetilde{K}(t,x) \cdot u + \widetilde{g}(t,x); & t \geq 0, \ x \in \mathbb{R}^d \\
u(0,x) = \widetilde{f}(x); & x \in \mathbb{R}^d
\end{cases}
$$
$$
(4.3.7)
$$

for the Hermite transform $u = u(t,x,z) = (\mathcal{H}Y(t,x))(z) = \widetilde{Y}(t,x,z)$, where
$z = (z_1, z_2, \ldots) \in (\mathbb{C}^\mathbb{N})_c$.

Let us first assume that $z = \lambda = (\lambda_1, \lambda_2, \ldots) \in (\mathbb{R}^\mathbb{N})_c$ and then define the
operator

$$
A^\lambda = \sum_{k=1}^d \frac{1}{2}\sigma^2 \frac{\partial^2}{\partial x_k^2} + \sum_{k=1}^d \widetilde{W}_\phi^{(k)}(x,\lambda) \frac{\partial}{\partial x_k}, \quad (4.3.8)
$$

where $W_\phi^{(k)}(x)$ is component number k of $\mathbf{W}_\phi(x)$, so that

$$
\widetilde{W}_\phi^{(k)}(x,\lambda) = \sum_{j=1}^\infty (\phi_x^{(k)}, \eta_j) \lambda_{(j-1)k+\alpha}; \quad 1 \leq k \leq d \quad (4.3.9)
$$

(see (2.6.10)), where $\phi_x^{(k)}$ is component number k of $\phi_x \in \mathcal{S}$. Then equation
(4.3.7) can be written

$$
\begin{cases}
-\frac{\partial u}{\partial t} + A^\lambda u + \widetilde{K}u = -\widetilde{g}; & t > 0, x \in \mathbb{R}^d \\
u(0,x) = \widetilde{f}(x); & x \in \mathbb{R}^d.
\end{cases}
$$
$$
(4.3.10)
$$

Let $X_t^\lambda = X_t^{\lambda,x}$ be the unique, strong solution of the Itô stochastic differential equation

$$dX_t^\lambda = \widetilde{\mathbf{W}}_\phi(X_t^\lambda, \lambda)dt + \sigma db_t; \; t \geq 0, \; X_0^\lambda = x. \qquad (4.3.11)$$

Note that the coefficient $\widetilde{\mathbf{W}}_\phi(x, \lambda)$ is Lipschitz continuous in x and has at most linear growth, so by the general theory (see Appendix B) a unique, strong, global solution X_t^λ exists. Note also that X_t^λ has generator A^λ. Therefore, by the Feynman–Kac formula (see, e.g., Karatzas and Shreve (1991), Friedman (1976)), the unique solution u of (4.3.10) is given by

$$u(t, x, \lambda) = E_Q^x \left[\widetilde{f}(X_t, \lambda) \exp\left[\int_0^t \widetilde{K}(t - r, X_r^\lambda, \lambda)dr \right] \right.$$
$$\left. + \int_0^t \widetilde{g}(t - s, X_s^\lambda, \lambda) \exp\left[\int_0^s \widetilde{K}(s - r), X_r^\lambda, \lambda)dr \right] ds \right], \qquad (4.3.12)$$

where E_Q^x denotes the expectation with respect to the law Q^x of $\{X_t^\lambda\}_{t \geq 0}$ when we have $X_0^\lambda = x$. Using the Girsanov transformation (see Appendix B) this can be formulated in terms of the expectation $\hat{E}^x$ as follows:

$$u(t, x, \lambda) = \hat{E}^x \left[\left(\widetilde{f}(\sigma b_t, \lambda) \exp\left[\int_0^t \widetilde{K}(t - r, \sigma b_r, \lambda)dr \right] \right. \right.$$
$$\left. \left. + \int_0^t \widetilde{g}(t - s, \sigma b_s, \lambda) \exp\left[\int_0^s \widetilde{K}(s - r, \sigma b_r, \lambda)dr \right] ds \right) M_t(\lambda) \right],$$
$$(4.3.13)$$

where

$$M_t(\lambda) = \exp\left[\sum_{k=1}^d \sigma^{-1} \int_0^t \widetilde{W}_\phi^{(k)}(\sigma b_s, \lambda)db_s^{(k)} \right.$$
$$\left. - \frac{1}{2} \sum_{k=1}^d \sigma^{-2} \int_0^t (\widetilde{W}_\phi^{(k)}(\sigma b_s, \lambda))^2 ds \right]. \qquad (4.3.14)$$

Clearly this function $\lambda \to u(t, x, \lambda); \lambda \in (\mathbb{R}^N)_c$, extends analytically to a function $z \to u(t, x, z); z \in (\mathbb{C}^N)_c$, obtained by substituting z for λ in (4.3.13)–(4.3.14). To prove that this analytic function $u(t, x, z)$ is the Hermite transform of some element $U(t, x) \in (\mathcal{S})_{-1}$, we must verify that $u(t, x, z)$ is bounded for $z \in \mathbb{K}_q(R)$ for some q, R. To this end it suffices, because of our assumptions, to prove that we have

$$\hat{E}^x[|M_t(z)|] \quad \text{is bounded for} \quad z \in \mathbb{K}_q(R).$$

Choose $q = 2$ and $z = \lambda + iy = (\lambda_1 + iy_1, \lambda_2 + iy_2, \ldots) \in \mathbb{K}_2(R)$. Then, since M_t is an exponential martingale, we get

$$\hat{E}^x[\|M_t(z)\|]$$

$$= \hat{E}^x\left[\exp\left[\sum_{k=1}^d \sigma^{-1} \int_0^t \mathrm{Re}[\widetilde{W}_\phi^{(k)}(\sigma b_s, z)]db_s^{(k)}\right.\right.$$

$$\left.\left. -\frac{1}{2}\sum_{k=1}^d \sigma^{-2} \int_0^t \mathrm{Re}[(\widetilde{W}_\phi^{(k)}(\sigma b_s, z))^2]ds\right]\right]$$

$$= \hat{E}^x\left[\exp\left[\sum_{k=1}^d \sigma^{-1} \int_0^t \widetilde{W}_\phi^{(k)}(\sigma b_s, \lambda)db_s^{(k)} - \frac{1}{2}\sum_{k=1}^d \sigma^{-2} \int_0^t (\widetilde{W}_\phi^{(k)}(\sigma b_s, \lambda))^2 ds\right.\right.$$

$$\left.\left. +\frac{1}{2}\sum_{k=1}^d \sigma^{-2} \int_0^t (\widetilde{W}_\phi^{(k)}(\sigma b_s, y))^2 ds\right]\right]$$

$$\leq \sup_{\beta \in \mathbb{R}^d} \exp\left[\frac{1}{2}\sum_{k=1}^d \sigma^{-2} \int_0^t (\widetilde{W}_\phi^{(k)}(\beta, y))^2 dy\right]$$

$$\leq \exp\left[\frac{1}{2}d\sigma^{-2}t\|\phi\|\left(\sum_{j=1}^\infty |y_i|\right)^2\right] \leq \exp\left[C\sum_{j=1}^\infty y_j^2(2\mathbb{N})^{2\epsilon^{(j)}} \cdot \sum_{j=1}^\infty (2\mathbb{N})^{-2\epsilon^{(j)}}\right]$$

$$\leq \exp\left[C\sum_{\alpha \neq 0} |y^\alpha|^2(2\mathbb{N})^{2\alpha} \cdot \sum_{j=1}^\infty (2j)^{-2}\right] < \infty.$$

Hence there exists $U(t,x) \in (\mathcal{S})_{-1}$ such that $\widetilde{U}(t,x) = u(t,x)$. By comparing the expansions for $U(t,x) = \sum_\alpha a_\alpha(t,x)H_\alpha$ and $u(t,x) = \sum_\alpha b_\alpha(t,x)z^\alpha$, we see that $U(t,x)$ can be expressed by (4.3.5)–(4.3.6).

Verifying that $U(t,x)$ solves (4.3.2) remains. Define the partial differential operator L by

$$Lu(t,x) = \frac{\partial u}{\partial t} - \frac{1}{2}\sigma^2 \Delta u - \widetilde{\mathbf{W}}_\phi \cdot \nabla u - \widetilde{K}(t,x)u. \tag{4.3.15}$$

Then from the general theory of (deterministic) parabolic differential operators we have (see Egorov and Shubin (1992), Theorem 2.78, and the references therein) that for every open set $G = (0,T) \times D \subset\subset \mathbb{R}^+ \times \mathbb{R}^d$ there exists $C < \infty$ such that

$$\|u\|_{C^{1,2+\gamma}(G)} \leq C(\|Lu\|_{C^{1,\gamma}(G)} + \|\widetilde{f}\|_{C^{2+\gamma}(\partial D)}). \tag{4.3.16}$$

Combining this with our estimate for the function $u(t,x,z)$ and our assumption about $Lu(t,x,z) = \widetilde{g}(t,x,z)$, we see that the conditions of

Theorem 4.1.1 are satisfied for $u(t, x, z)$, and we conclude that $U(t, x)$ solves (4.3.2), as claimed. $\square$

Theorem 4.3.1 has several important special cases, some of which were already mentioned at the beginning of this section. We state two more:

4.3.2 The Heat Equation with a Stochastic Potential

If we choose $V = g = 0$ in (4.3.1), we get the following *stochastic heat equation*

$$\begin{cases} \frac{\partial U}{\partial t}(t, x) = \frac{1}{2}\sigma^2 \Delta U(t, x) + K(t, x) \diamond U(t, x); & t \geq 0, x \in \mathbb{R}^d \\ U(0, x) = f(x); & x \in \mathbb{R}^d. \end{cases} \tag{4.3.17}$$

This equation was studied in Nualart and Zakai (1989), in the case when $K(t, x)$ is white noise. They prove the existence of a solution of a type called *generalized Wiener functionals*. In Holden et al. (1995b), this equation was solved in the $(\mathcal{S})_{-1}$ setting presented here.

Corollary 4.3.2 Holden et al. (1995b). *Suppose that $K(t, x)$ and $f(x)$ are stochastic distribution processes satisfying the conditions of Theorem 4.3.1. Then the unique $(\mathcal{S})_{-1}$ solution $U(t, x)$ of (4.3.17) is given by*

$$U(t, x) = \hat{E}^x \left[f(\sigma b_t) \diamond \exp^\diamond \left[\int_0^t K(t - s, \sigma b_s) ds \right] \right]. \tag{4.3.18}$$

4.4 The Stochastic Schrödinger Equation

An equation closely related to equation (4.3.2) is the *stationary Schrödinger equation with a stochastic potential*

$$\begin{cases} \frac{1}{2}\Delta U(x) + V(x) \diamond U(x) = -f(x); & x \in D \\ U(x) = 0; & x \in \partial D. \end{cases} \tag{4.4.1}$$

Here D is a bounded domain in $\mathbb{R}^d$ and $V(x)$ and $f(x)$ are given stochastic distribution processes. This equation was studied in Holden et al. (1993a), in the case when the potential $V(x)$ is proportional to the Wick exponential of smoothed white noise; or more precisely

$$V(x) = \rho \exp^\diamond[W_\phi(x)], \quad \phi \in \mathcal{S}(\mathbb{R}), \tag{4.4.2}$$

where $\rho \in \mathbb{R}$ is a constant. If $\rho > 0$, this is called the *attractive case*.

Let λ_0 be the smallest eigenvalue for the operator $-1/2\Delta$ in D, i.e., $\lambda_0 > 0$ is the smallest λ for which the boundary value problem

$$\begin{cases} -\frac{1}{2}\Delta u(x) = \lambda u(x); & x \in D \\ u(x) = 0; & x \in \partial_R D \text{ (the regular boundary of } D) \end{cases} \tag{4.4.3}$$

has a bounded solution $u \in C^2(D)$. (As usual, the boundary condition $u(x) = 0$ for $x \in \partial D$ is a shorthand notation for $\lim_{\substack{y \to x \\ y \in D}} u(y) = 0$ for $x \in \partial D$.) As in the previous section, we let $\{b_t\}_{t \geq 0}$, denote an auxiliary Brownian motion in $\mathbb{R}^d$ (independent of $\{B(t)\}_{t \geq 0}$), and $\hat{E}^x$ denotes the expectation with respect to the law $\hat{P}^x$ of b_t starting at x. Define the first exit time τ_D for b_t from D by

$$\tau_D = \inf\{t > 0; b_t \notin D\}.$$

The following result will be useful: $\lambda_0 > 0$ is related to τ_D

$$\lambda_0 = \sup\{\rho \in \mathbb{R}; \hat{E}^x[\exp[\rho \tau_D]] < \infty\}, \quad \text{for all} \quad x \in D. \tag{4.4.4}$$

(See, e.g., Durrett (1984), Chapter 8B.)

Theorem 4.4.1. *Suppose $f(x)$ is a stochastic distribution process such that*

$$\tilde{f}(x, z) \quad \text{is bounded for} \quad (x, z) \in D \times \mathbb{K}_{q_1}(R_1), \quad \text{for some} \quad q_1, R_1. \tag{4.4.5}$$

Let D be a bounded domain in $\mathbb{R}^d$ with all its points regular for the classical Dirichlet problem in D. Let $\rho < \lambda_0$ be a constant. Then there is a unique $(\mathcal{S})_{-1}$ solution $U(x)$ of the stochastic Schrödinger equation

$$\begin{cases} \frac{1}{2}\Delta U(x) + \rho \exp^{\diamond}[W(x)] \diamond U(x) = -f(x); & x \in D \\ U(x) = 0; & x \in \partial D, \end{cases} \tag{4.4.6}$$

and it is given by

$$U(x) = \hat{E}^x \left[\int_0^{\tau_D} \exp^{\diamond}\left[\rho \int_0^t \exp^{\diamond}(W(b_s))ds \right] \diamond f(b_t)dt \right]. \tag{4.4.7}$$

Proof This result can be proved by modifying the proof of Theorem 4.3.1. For completeness we give the details.

By taking Hermite transforms we get the equation

$$\begin{cases} \frac{1}{2}\Delta u(x, z) + \rho \exp[\widetilde{W}(x)(z)] \cdot u(x, z) = -\tilde{f}(x, z); & x \in D \\ u(x, z) = 0; & x \in \partial D. \end{cases} \tag{4.4.8}$$

Choose $z \in (\mathbb{C}^N)_c$. Then by a complex version of the Feynman–Kac formula (see, e.g., Karatzas and Shreve (1991), Friedman (1976), in the real case) we can express the unique solution $u(x, z)$ of (4.4.8) as

$$u(x, z) = \hat{E}^x \left[\int_0^{\tau_D} \tilde{f}(b_t, z) \exp \left[\rho \int_0^t \exp[\widetilde{W}(b_s, z)] ds \right] dt \right], \qquad (4.4.9)$$

provided the expression converges. Note that, by (C.15),

$$
\begin{aligned}
|\widetilde{W}_\phi(b_s, z)|^2 &= \left| \sum_{j=1}^\infty \eta_j(b_s) z_j \right|^2 \leq \sup_{j,x} |\eta_j(x)|^2 \left(\sum_{j=1}^\infty |z_j| \right)^2 \\
&\leq \sup_{j,x} |\eta_j(x)|^2 \sum_{j=1}^\infty |z_j|^2 (2\mathbb{N})^{q\epsilon^{(j)}} \sum_{j=1}^\infty (2\mathbb{N})^{-q\epsilon^{(j)}} \\
&\leq \sup_{j,x} |\eta_j(x)|^2 R^2 \sum_{j=1}^\infty (2j)^{-q} =: C(q, R)^2 < \infty
\end{aligned}
$$

if $z \in \mathbb{K}_q(R)$ for $q > 1$.

Therefore, by our assumption on $\tilde{f}$, for $z \in \mathbb{K}_q(R)$ we get,

$$
\begin{aligned}
|u(x, z)| &\leq M \hat{E}^x \left[\int_0^{\tau_D} \exp \left[\rho \int_0^t \exp[C(q, R)] ds \right] dt \right] \\
&\leq M \hat{E}^x \left[\int_0^{\tau_D} \exp \left[\rho \exp[C(q, R)] t \right] dt \right] \\
&\leq \frac{M}{\rho \exp[C(q, R)]} \hat{E}^x [\exp[\rho \exp[C(q, R)] \tau]],
\end{aligned}
$$

where $M = \sup\{|\tilde{f}(x, z)|; (x, z) \in D \times \mathbb{K}_{q_1}(R_1)\}$.

Now choose q_2, R_2 and $\epsilon > 0$ such that

$$\rho e^{C(q_2, R_2)} < (1 - \epsilon)\lambda_0.$$

Then for $q \geq \max(q_1, q_2)$ and $R \leq \min(R_1, R_2)$ we have

$$|u(x, z)| \leq \frac{M}{\rho \exp[C(q, R)]} \hat{E}^x \left[\exp \left[(1 - \epsilon)\lambda_0 \tau \right] \right] < \infty$$

for all $(x, z) \in D \times \mathbb{K}_q(R)$.

Since $u(x, z)$ is clearly analytic in $z \in (\mathbb{C}^N)_c \cap \mathbb{K}_q(R)$, we conclude that there exists $U(x) \in (\mathcal{S})_{-1}$ such that $\mathcal{H}U = u$. Moreover, since $L = 1/2\Delta$ is uniformly elliptic, it follows by the estimate (4.2.5) that the double derivatives

$\partial^2 u / \partial x_i^2(x, z)$; $1 \leq i \leq d$, are uniformly bounded for $(x, z) \in V \times \mathbb{K}_q(R)$ for each open set $V \subset\subset D$. Hence by Theorem 4.1.1, $U(x)$ does indeed solve (4.4.6). Moreover, we verify directly that the Hermite transform of the expression in (4.4.7) is the expression in (4.4.9). □

4.4.1 $L^1(\mu)$-Properties of the Solution

Working in the space $(\mathcal{S})_{-1}$ can be technically convenient, because of the useful properties of the Hermite transform and the characterization theorem. However, we pay a price: The space $(\mathcal{S})_{-1}$ is large and relatively abstract. Therefore it will always be of interest to identify the solutions as members of smaller or more concrete spaces, such as $(\mathcal{S})^*$ or, preferably, $L^p(\mu)$ for some $p > 1$. We have already seen examples of equations whose solutions belong to $(\mathcal{S})_{-1}$ but not to $(\mathcal{S})^*$ and hence not to $L^p(\mu)$ for any $p > 1$. (See, e.g., Section 3.2). Nevertheless, it turns out that the solution sometimes is also in $L^1(\mu)$ (like in Section 3.2). This useful feature is more often achieved when we apply the functional process approach, i.e., smoothing the white noise with a test function $\phi \in \mathcal{S}(\mathbb{R})$ as we did above. We now prove that, under certain conditions, our solution in Theorem 4.4.1 is actually in $L^1(\mu)$, provided that we interpret the equation in the weak (distributional) sense with respect to x.

Theorem 4.4.2 Holden et al. (1993a). *Assume as before that D is a bounded domain in $\mathbb{R}^d$ with $\partial D = \partial_R D$. Moreover, assume that*

$$f(x) \text{ is deterministic and bounded in } \bar{D} \text{ and} \qquad (4.4.10)$$

$$\rho < \lambda_0 \text{ (defined by (4.4.4)).} \qquad (4.4.11)$$

For $x \in \bar{D}$ and $\phi \in \mathcal{S}(\mathbb{R}^d)$, define

$$U(x) = U(\phi, x, \omega) = \hat{E}^x \left[\int_0^{\tau_D} \exp^\diamond \left[\rho \int_0^t \exp^\diamond [W_\phi(b_s)] ds \right] f(b_t) dt \right].$$

Then $U(x) \in L^1(\mu)$, $x \to U(x) \in L^1(\mu)$ is continuous for $x \in \bar{D}$, and $U(x)$ satisfies the stochastic Schrödinger equation

$$\begin{cases} \frac{1}{2} \Delta U(x) + \rho \exp^\diamond [W_\phi(x)] \diamond U(x) = -f(x); & x \in D \\ U(x) = 0; & x \in \partial D \end{cases} \qquad (4.4.12)$$

in the weak distributional sense with respect to $x \in D$, i.e., that there exists $\Omega_\phi \subset \mathcal{S}'(\mathbb{R}^d)$ with $\mu(\Omega_\phi) = 1$ such that

$$\frac{1}{2}(U, \Delta \psi) + \rho(\exp^\diamond [W_\phi] \diamond U, \psi) = -(f, \psi) \qquad (4.4.13)$$

for all $\omega \in \Omega_\phi$ *and for all* $\psi \in C_0^\infty(D)$ *(where* $(\cdot, \cdot)$ *is the inner product on* $L^2(\mathbb{R}^d)$*).*

Remark The Wick product in (4.4.13) is now interpreted in the $L^1(\mu)$ sense. See Definition 2.9.4.

Proof Expanding the first Wick exponential according to its Taylor series, we find

$$
U(x) = \hat{E}^x \left[\int\limits_0^{\tau_D} \sum_{k=0}^\infty \frac{\left(\rho \int\limits_0^t \exp^\diamond [W_\phi(b_s)] \right)^{\diamond k}}{k!} f(b_t) dt \right]
$$

$$
= \sum_{k=0}^\infty \frac{\rho^k}{k!} \hat{E}^x \left[\int\limits_0^{\tau_D} \left(\int\limits_0^t \exp^\diamond [W_\phi(b_s)] ds \right)^{\diamond k} f(b_t) dt \right] = \sum_{k=0}^\infty \frac{\rho^k}{k!} V_k(x)
$$

$$(4.4.14)$$

with

$$
V_k(x) = V_k(\phi, x, \omega) = \hat{E}^x \left[\int\limits_0^{\tau_D} \left(\int\limits_0^t \exp^\diamond (W_\phi(b_s)) ds \right)^{\diamond k} f(b_t) dt \right]; \quad (4.4.15)
$$

$k = 0, 1, 2, \ldots$. The key identity we will establish is

$$
\frac{1}{2} \Delta V_k(x) = \begin{cases} -k \exp^\diamond W_\phi(x) \diamond V_{k-1}(x); & k \in \mathbb{N} \\ -f(x); & k = 0. \end{cases}
$$

$$(4.4.16)$$

for all $x \in D$.

To this end we perform a Hermite transform of $V_k(x)$ to obtain

$$
\widetilde{V}_k(x) = \widetilde{V}_k(x, z) = \hat{E}^x \left[\int\limits_0^{\tau_D} \left(\int\limits_0^t \widetilde{G}(b_s) ds \right)^k f(b_t) dt \right],
$$

$$(4.4.17)$$

where $\widetilde{G}(y) = \exp[\widetilde{W}_\phi(y)]$. For $x \in D$ let $N_j = N_j(x); \ j = 1, 2, \ldots$, be a sequence of open sets such that $\bar{N}_j \subset D$ and $\bigcap_{j=1}^\infty N_j = \{x\}$. Define

$$
\sigma_j = \inf\{t > 0; \ b_t \notin N_j\}; \ j = 1, 2, \ldots. \qquad (4.4.18)
$$

Let $\mathcal{A}$ denote Dynkin's characteristic operator, so that

$$
\mathcal{A}\widetilde{V}_k(x) = \lim_{j \to \infty} \frac{1}{\hat{E}^x[\sigma_j]} (\hat{E}^x[\widetilde{V}_k(b_{\sigma_j})] - \widetilde{V}_k(x)). \qquad (4.4.19)
$$

Write $\sigma = \sigma_j$ and consider

$$J := \hat{E}^x[\tilde{V}_k(b_\sigma)] = \hat{E}^x\left[\hat{E}^{b_\sigma}\left[\int_0^\tau\left(\int_0^t\tilde{G}(b_s)ds\right)^k f(b_t)dt\right]\right]$$

$$= \hat{E}^x\left[\hat{E}^x\left[\theta_\sigma\left(\int_0^\tau\left(\int_0^t\tilde{G}(b_s)ds\right)^k f(b_t)dt\right)\middle|\mathcal{F}_\sigma\right]\right]$$

$$= \hat{E}^x\left[\int_0^{T^{(\sigma)}}\left(\int_0^t\tilde{G}(b_{s+\sigma})ds\right)^k f(b_{t+\sigma})dt\right], \qquad (4.4.20)$$

where θ_σ is the shift operator $(\theta_\sigma(b_t) = b_{t+\sigma}), T^{(\sigma)} = \inf\{t > 0; \ b_{t+\sigma} \notin D\}$, and we have used the strong Markov property of $\{b_t\}_{t\geq 0}$ (see, e.g., Øksendal (1995), Chapter 7). Since $\sigma < \tau$, we have $\sigma + T^{(\sigma)} = \inf\{s > \sigma; \ b_s \notin D\} = \tau$ and therefore

$$J = \hat{E}^x\left[\int_0^{T^{(\sigma)}}\left(\int_\sigma^{t+\sigma}\tilde{G}(b_r)dr\right)^k f(b_{t+\sigma})dt\right]$$

$$= \hat{E}^x\left[\int_\sigma^\tau\left(\int_\sigma^s\tilde{G}(b_r)dr\right)^k f(b_s)ds\right]$$

$$= \hat{E}^x\left[\int_0^\tau\left(\int_\sigma^s\tilde{G}(b_r)dr\right)^k f(b_s)ds\right] - \hat{E}^x\left[\int_0^\sigma\left(\int_\sigma^s\tilde{G}(b_r)dr\right)^k f(b_s)ds\right].$$

$$(4.4.21)$$

Now

$$\frac{1}{\hat{E}^x[\sigma_j]}\cdot\left|\hat{E}^x\left[\int_0^{\sigma_j}\left(\int_{\sigma_j}^s\tilde{G}(b_r)dr\right)^k f(b_s)ds\right]\right|$$

$$\leq \frac{1}{\hat{E}^x[\sigma_j]}\cdot\hat{E}^x\left[\int_0^{\sigma_j}(Ms)^k M ds\right] \to 0 \quad \text{as} \quad j \to \infty, \qquad (4.4.22)$$

where $M = \sup\{|\tilde{G}(y)| + |f(y)|; y \in D\}$. Therefore, writing $\tilde{H}(s) = \int_0^s\tilde{G}(b_r)dr$ and assuming $k \geq 1$, we get

$$\mathcal{A}\tilde{V}_k(x)$$

$$= \lim_{j\to\infty}\frac{1}{\hat{E}^x[\sigma_j]}\hat{E}^x\left[\int_0^\tau\left(\left(\int_\sigma^s\tilde{G}(b_r)dr\right)^k - \left(\int_0^s\tilde{G}(b_r)dr\right)^k\right)f(b_s)ds\right]$$

$$= \lim_{j \to \infty} \frac{1}{\hat{E}^x[\sigma_j]} \hat{E}^x \left[\int_0^\tau \left((\tilde{H}(s) - \tilde{H}(\sigma_j))^k - \tilde{H}(s)^k \right) f(b_s) ds \right]$$

$$= \lim_{j \to \infty} -\frac{1}{\hat{E}^x[\sigma_j]} \hat{E}^x \left[\int_0^\tau k \, (\check{H}_j(s))^{k-1} f(b_s) ds \, \tilde{H}(\sigma_j) \right] \tag{4.4.23}$$

by the mean value theorem, where $\check{H}_j(s)$ lies on the line segment between the points $\tilde{H}(s)$ and $\tilde{H}(s) - \tilde{H}(\sigma_j)$. Since $\check{H}_j(s) \to \tilde{H}(s)$ pointwise boundedly as $j \to \infty$ and (see Exercise 4.2)

$$\frac{\hat{E}^x[\tilde{H}(\sigma_j)]}{\hat{E}^x[\sigma_j]} \to \tilde{G}(x) \quad \text{as} \quad j \to \infty,$$

we see from (4.4.23) and (4.4.19) that

$$\mathcal{A}\tilde{V}_k(x) = -k\tilde{G}(x)\tilde{V}_{k-1}(x) \quad \text{for} \quad k \geq 1. \tag{4.4.24}$$

Similarly, but much more easily, we see from (4.4.23) that

$$\mathcal{A}\tilde{V}_0 = -f(x). \tag{4.4.25}$$

In general we know that the solution of the generalized boundary value problem

$$\begin{cases} \mathcal{A}u(x) = -g(x); & x \in D \\ u(x) = 0; & x \in \partial D \end{cases} \tag{4.4.26}$$

(where g is a given bounded, continuous function) is unique and given by (see, e.g., Dynkin (1965), or Freidlin (1985))

$$u(x) = \hat{E}^x \left[\int_0^{\tau_D} g(b_s) ds \right]; \; x \in \bar{D}. \tag{4.4.27}$$

On the other hand, from the theory of (deterministic) elliptic boundary value problems, we know that the equation

$$\begin{cases} \frac{1}{2}\Delta v(x) = -g(x); & x \in D \\ v(x) = 0; & x \in \partial D \end{cases} \tag{4.4.28}$$

has a unique solution $v \in C^2(D)$. Since $\mathcal{A}$ coincides with $1/2\Delta$ on C^2, the two solutions must be the same, and u must be C^2. If we apply this to (4.4.25), we get that $\tilde{V}_0 \in C^2(D)$ and then by induction, using (4.4.24), we get that

$\tilde{V}_k \in C^2(D)$ for all $k \in \mathbb{N}_0$. Hence we have proved that, in the strong sense,

$$\frac{1}{2}\Delta\tilde{V}_k(x) = \begin{cases} -k\exp[\widetilde{W}_\phi(x)]\tilde{V}_{k-1}(x); & k \in \mathbb{N} \\ -f(x); & k = 0 \end{cases} \tag{4.4.29}$$

Moreover, from this we see, again by induction, that $\Delta\tilde{V}_k(x)$ is bounded for $(x,z) \in D \times \mathbb{K}_q(R)$ and continuous with respect to x for each $z \in \mathbb{K}_q(R)$. Hence we can take inverse Hermite transforms and conclude (by Theorem 4.1.1) that (4.4.16) holds in $(\mathcal{S})_{-1}$.

Define

$$U_n(x) = \sum_{k=0}^{n} \frac{\rho^k}{k!} V_k(x); \quad n = 0, 1, 2, \ldots. \tag{4.4.30}$$

Then from (4.4.16) we get

$$\begin{aligned} \frac{1}{2}\Delta U_n(x) &= \frac{1}{2}\sum_{k=1}^{n} \frac{\rho^k}{k!}\Delta V_k(x) - f(x) \\ &= -\rho\sum_{k=1}^{n} \frac{\rho^{k-1}}{(k-1)!} V_{k-1}(x) \diamond \exp^\diamond(W_\phi(x)) - f(x) \\ &= -\rho U_{n-1}(x) \diamond \exp^\diamond[W_\phi(x)] - f(x). \end{aligned} \tag{4.4.31}$$

We will now prove the following three statements:

$$U_n(x) \in L^2(\mu) \quad \text{for all} \quad x \in D, n \in \mathbb{N}; \tag{4.4.32}$$

$$\sup_{x \in D} \|U_n(x) \diamond \exp^\diamond W_\phi(x) - U_m(x) \diamond \exp^\diamond W_\phi(x)\|_{L^1(\mu)} \to 0 \tag{4.4.33}$$

as $m, n \to \infty$; and

$$\sup_{x \in D} \|U_n(x) - U(x)\|_{L^1(\mu)} \to 0 \quad \text{as} \quad n \to \infty. \tag{4.4.34}$$

To prove (4.4.32), consider

$$V_k^2(x) \le C^2 \left(\hat{E}^x \left[\int_0^{\tau_D} \left(\int_0^t \exp^\diamond[W_\phi(b_s)]ds \right)^{\diamond k} dt \right] \right)^2$$

$$= C^2 \left(\hat{E}^x \left[\int_0^{\tau_D} \left(\int_0^t \cdots \int_0^t \exp^\diamond\left[\sum_{i=1}^{k} W_\phi(b_{s_i}) \right] ds_1 \cdots ds_k \right) dt \right] \right)^2$$

$$= C^2 \left(\hat{E}^x \left[\int_0^{\tau_D} \left(\int_0^t \cdots \int_0^t \exp^\diamond\left[w\left(\sum_{i=1}^{k} \phi_{b_{s_i}} \right) \right] ds_1 \cdots ds_k \right) dt \right] \right)^2$$

$$\le C^2 \hat{E}^x \left[\tau_D \int_0^{\tau_D} \left(\int_0^t \cdots \int_0^t \exp^\diamond\left[w\left(\sum_{i=1}^{k} \phi_{b_{s_i}} \right) \right] ds_1 \cdots ds_k \right)^2 dt \right]$$

$$\leq C^2 \hat{E}^x \left[\tau_D \int_0^{\tau_D} t^k \int_0^t \cdots \int_0^t \left(\exp^\diamond \left[w\left(\sum_{i=1}^k \phi_{b_{s_i}} \right) \right] \right)^2 ds_1 \cdots ds_k dt \right]$$

$$= C^2 \hat{E}^x \left[\tau_D \int_0^{\tau_D} t^k \int_0^t \cdots \int_0^t \left(\exp \left[w\left(\sum_{i=1}^k \phi_{b_{s_i}} \right) - \frac{1}{2} \left\| \sum_{i=1}^k \phi_{b_{s_i}} \right\|_2^2 \right] \right)^2 ds_1 \cdots ds_k dt \right]$$

$$= C^2 \hat{E}^x \left[\tau_D \int_0^{\tau_D} t^k \int_0^t \cdots \int_0^t \exp \left[w\left(2\sum_{i=1}^k \phi_{b_{s_i}} \right) - \left\| \sum_{i=1}^k \phi_{b_{s_i}} \right\|_2^2 \right] ds_1 \cdots ds_k dt \right]$$

$$= C^2 \hat{E}^x \left[\tau_D \int_0^{\tau_D} t^k \int_0^t \cdots \int_0^t \exp^\diamond \left[w\left(\sum_{i=1}^k 2\phi_{b_{s_i}} \right) + \left\| \sum_{i=1}^k \phi_{b_{s_i}} \right\|_2^2 \right] ds_1 \cdots ds_k dt \right].$$

This gives

$$E_\mu[V_k^2(x)] \leq C^2 \hat{E}^x \left[\tau_D \int_0^{\tau_D} t^k \int_0^t \cdots \int_0^t \exp \left[\left\| \sum_{i=1}^k \phi_{b_{s_i}} \right\|_2^2 \right] ds_1 \cdots ds_k dt \right]$$

$$\leq C^2 \hat{E}^x \left[\tau_D \int_0^{\tau_D} t^k \int_0^t \cdots \int_0^t \exp[k^2 \|\phi\|_2^2] ds_1 \cdots ds_k dt \right]$$

$$= C^2 \hat{E}^x \left[\tau_D \int_0^{\tau_D} t^{2k} \exp[k^2 \|\phi\|_2^2] dt \right]$$

$$= C^2 \exp\left[k^2 \|\phi\|_2^2 \right] \frac{1}{2k+1} \hat{E}^x[\tau_D^{2k+2}] < \infty$$

by (4.4.4). This proves (4.4.32).

Next, to prove (4.4.33) choose $m < n$ and consider

$$H_{n,m}(x) := U_n(x) \diamond \exp^\diamond W_\phi(x) - U_m(x) \diamond \exp^\diamond W_\phi(x)$$

$$= \sum_{k=m+1}^n \frac{\rho^k}{k!} \hat{E}^x \left[\int_0^{\tau_D} \left(\int_0^t \exp^\diamond[W_\phi(b_s)] ds \right)^{\diamond k} f(b_t) dt \right] \diamond \exp^\diamond[W_\phi(x)]$$

$$= \sum_{k=m+1}^n \frac{\rho^k}{k!} V_k(x) \diamond \exp^\diamond[W_\phi(x)].$$

We see that, by an argument similar to the one above,

$$E_\mu[|H_{n,m}|] \leq \sum_{k=m+1}^n \frac{\rho^k}{k!} E_\mu[|V_k(x)|]$$

$$\leq \sum_{k=m+1}^{n} \frac{\rho^k}{k!} C E_\mu \left[\hat{E}^x \left[\int_0^{\tau_D} \int_0^t \cdots \int_0^t \exp^\diamond \left[\sum_{i=1}^k W_\phi(b_{s_i}) \right] ds_1 \cdots ds_k dt \right] \right]$$

$$= \sum_{k=m+1}^{n} \frac{C\rho^k}{k!} \hat{E}^x \left[\int_0^{\tau_D} t^k dt \right] = C \sum_{k=m+1}^{n} \frac{\rho^k}{(k+1)!} \hat{E}^x [\tau_D^{k+1}]$$

$$\leq \frac{C}{\rho} \hat{E}^x \left[\sum_{k=m+2}^{\infty} \frac{\rho^k}{k!} \tau_D^k \right] \to 0 \quad \text{as} \quad m \to \infty,$$

uniformly for $x \in \bar{D}$, since $\rho < \lambda_0$. Finally, we note that (4.4.34) follows by the same proof as for (4.4.33).

We can now complete the proof of Theorem 4.4.2: Choose $\psi \in C_0^\infty(D)$. Then by (4.4.31) we have

$$\left(\frac{1}{2} U_n, \Delta \psi \right) = -\rho(U_{n-1} \diamond \exp^\diamond W_\phi, \psi) - (f, \psi) \quad \text{for} \quad n \in \mathbb{N}.$$

Letting $n \to \infty$ we use (4.4.32)–(4.4.34) and the definition of Wick product in $L^1(\mu)$ to obtain that (4.4.13) holds. Moreover, since $x \to U_n(x) \in L^2(\mu)$ is continuous on $\bar{D}$ for each n we also have that $x \to U_n(x)$ is continuous as a map into $L^1(\mu)$. Moreover, by (4.4.34), $U_n(x) \to U(x)$ uniformly for $x \in \bar{D}$ and hence $x \to U(x)$ is continuous as a map from $\bar{D}$ into $L^1(\mu)$. Since it is obvious from the definition of $U(x)$ that $U(x) = 0$ for $x \in \partial D = \partial_R D$, the proof is complete. $\square$

4.5 The Viscous Burgers Equation with a Stochastic Source

The (1-dimensional) viscous Burgers equation with a source f has the form (λ, ν positive constants)

$$\begin{cases} \frac{\partial u}{\partial t} + \lambda u \frac{\partial u}{\partial x} = \nu \frac{\partial^2 u}{\partial x^2} + f; & t > 0, x \in \mathbb{R} \\ u(0, x) = g(x); & x \in \mathbb{R}. \end{cases} \tag{4.5.1}$$

This equation has been used as a prototype model of various nonlinear phenomena. It was introduced in Forsyth (1906), p. 101, where he also presented what is known as the Cole–Hopf transformation in order to solve it. The equation was extensively analyzed by Burgers (1940, 1974), who in fact studied it (heuristically) with white noise initial data and no source, i.e., $f = 0$.

Burgers equation is a simplified version of the Navier–Stokes equation where $R = \nu^{-1}$ corresponds to the Reynolds number. Applications vary from the formation and the structure of the universe (see Albeverio et al. (1996) and Shandarin and Zeldovich (1989)) to the growth of interfaces. For a discussion of some applications of (4.5.1) we refer to Gurbatov et al. (1991). For other recent discussions of Burgers equation related to the presentation here, we refer to Da Prato et al. (1994), Truman and Zhao (1996) and Bertini et al. (1994). For a different approach to SPDEs of the type $\partial u/\partial t + \partial f(u)/\partial x = g$, $u(x, 0) = u_0(x)$ with either random noise g or random initial data u_0, we refer to Holden and Risebro (1991, 1997).

We will, however, mention one application in some detail here, namely the Kardar–Parisi–Zhang (KPZ) model of growth of interfaces of solids (see Kardar et al. (1986) and Medina et al. (1989).)

Let $h(t, x)$ denote the location of an interface measured from some given reference plane. We assume that there are two opposing forces that act on the interface. One force is a surface tension contribution given by $\nu \Delta h$ and another force is a nonlinear function σ of ∇h that tends to support and promote higher irregularity of the interface. Thus we may write

$$\frac{\partial h}{\partial t} = \nu \Delta h + \sigma(\nabla h) + N \quad \text{(KPZ)}, \tag{4.5.2}$$

where N is some external source term. Now assume that

$$\sigma(x) = \frac{1}{2} \lambda x^2; \ x \in \mathbb{R}^d, \tag{4.5.3}$$

and introduce

$$u = -\nabla h, \ f = -\nabla N. \tag{4.5.4}$$

Taking the gradient of both sides of (4.5.2), we get

$$-\frac{\partial}{\partial t} \nabla h + \nabla(\sigma(\nabla h)) = -\nu \nabla(\Delta h) - \nabla N$$

or

$$\frac{\partial u}{\partial t} + \frac{1}{2} \lambda \nabla(u^2) = \nu \Delta u + f. \tag{4.5.5}$$

With $u = (u_1, \ldots, u_d), x = (x_1, \ldots, x_d), f = (f_1, \ldots, f_d)$, this can be written

$$\frac{\partial u_k}{\partial t} + \lambda \sum_{j=1}^{d} u_j \frac{\partial u_k}{\partial x_j} = \nu \Delta u_k + f_k; \ 1 \leq k \leq d. \tag{4.5.6}$$

Sometimes this is also written

$$\frac{\partial u}{\partial t} + \lambda(u, \nabla)u = \nu \Delta u + f,$$

where $(u, \nabla) = \sum_{j=1}^{d} u_j \partial/\partial x_j$. The equation (4.5.6) is a multidimensional system generalization of the classical Burgers equation. We will study a stochastic version of this equation in this section.

Formally, the KPZ equation (4.5.2)–(4.5.3) can be linearized by what is traditionally called the *Cole–Hopf transformation*.

If we introduce

$$\phi = \exp\left[\frac{1}{2\nu}h\right], \tag{4.5.7}$$

then the KPZ equation becomes a heat equation:

$$\frac{\partial \phi}{\partial t} = \nu\Delta\phi + \frac{\lambda}{2\nu}N\phi. \tag{4.5.8}$$

By solving this and transforming back,

$$h = \frac{2\nu}{\lambda}\ln\phi, \tag{4.5.9}$$

we get a solution of (4.5.2)–(4.5.3).

If the source components $f_1, \ldots, f_d$ are functionals of white noise, then it is not clear how to interpret the products $u_j\partial u_j/\partial x_k$ in (4.5.6), nor is it clear how to interpret the Cole–Hopf transformation and its inverse. However, we will show that if the products are interpreted as Wick products $u_j \diamond \partial u_j/\partial x_k$ and the equation is regarded as an equation in $(\mathcal{S})_{-1}$, then a Wick version of the Cole–Hopf solution method can be carried out and gives us a unique solution of the stochastic Burgers equation. We now formulate this rigorously.

Lemma 4.5.1 Holden et al. (1994), Holden et al. (1995b) (The Wick Cole–Hopf transformation). *Let us assume that $N = N(t,x)$ and $G(x) = (G_1(x), \ldots, G_d(x))$ be $(\mathcal{S})_{-1}$ and $(\mathcal{S})_{-1}^d$ processes, respectively. Assume that $N(t,x) \in C^{0,1}$, and define*

$$F = -\nabla_x N. \tag{4.5.10}$$

Assume that there exists an $(\mathcal{S})_{-1}$-valued $C^{1,3}$-process $Z(t,x)$ such that the process

$$U = -\nabla_x Z \tag{4.5.11}$$

solves the multidimensional stochastic Burgers equation

$$\begin{cases} \frac{\partial U_k}{\partial t} + \lambda\sum_{j=1}^d U_j \diamond \frac{\partial U_k}{\partial x_j} = \nu\Delta U_k + F_k; \quad t > 0, x \in \mathbb{R}^d \\ U_k(0,x) = G_k(x); \quad 1 \le k \le d \end{cases} \tag{4.5.12}$$

Then the Wick Cole–Hopf transform Y of U defined by

$$Y := \exp^\diamond\left[\frac{\lambda}{2\nu}Z\right] \tag{4.5.13}$$

solves the stochastic heat equation

$$\begin{cases} \frac{\partial Y}{\partial t} = \nu \Delta Y + \frac{\lambda}{2\nu} Y \diamond [N + C]; & t > 0, x \in \mathbb{R}^d \\ Y(0, x) = \exp^{\diamond} \left[\frac{\lambda}{2\nu} Z(0, x) \right] \end{cases} \tag{4.5.14}$$

for some t-continuous $(\mathcal{S})_{-1}$-process $C(t)$ (independent of x).

Proof Substituting (4.5.10) and (4.5.11) in (4.5.12) gives

$$-\frac{\partial}{\partial x_k}\left(\frac{\partial Z}{\partial t}\right) + \lambda \sum_j \frac{\partial Z}{\partial x_j} \diamond \frac{\partial}{\partial x_j}\left(\frac{\partial Z}{\partial x_k}\right) = -\nu \sum_j \frac{\partial^2}{\partial x_j^2}\left(\frac{\partial Z}{\partial x_k}\right) - \frac{\partial N}{\partial x_k}$$

or

$$\frac{\partial Z}{\partial t} = \frac{\lambda}{2}\sum_j \left(\frac{\partial Z}{\partial x_j}\right)^{\diamond 2} + \nu \Delta Z + N + C, \tag{4.5.15}$$

where $C = C(t)$ is a t-continuous, x-independent $(\mathcal{S})_{-1}$-process. Basic Wick calculus rules give that

$$\frac{\partial Y}{\partial t} = \frac{\lambda}{2\nu} Y \diamond \frac{\partial Z}{\partial t} \tag{4.5.16}$$

and

$$\frac{\partial Y}{\partial x_j} = \frac{\lambda}{2\nu} Y \diamond \frac{\partial Z}{\partial x_j}. \tag{4.5.17}$$

Hence

$$\begin{aligned} \Delta Y - \sum_j \frac{\partial}{\partial x_j}\left(\frac{\partial Y}{\partial x_j}\right) &= \sum_j \frac{\partial}{\partial x_j}\left(\frac{\lambda}{2\nu} Y \diamond \frac{\partial Z}{\partial x_j}\right) \\ &= \sum_j \left(\frac{\lambda}{2\nu}\right)^2 Y \diamond \left(\frac{\partial Z}{\partial x_j}\right)^{\diamond 2} + \sum_j \frac{\lambda}{2\nu} Y \diamond \frac{\partial^2 Z}{\partial x_j^2} \\ &= \frac{\lambda}{2\nu} Y \diamond \left(\frac{\lambda}{2\nu}\sum_j \left(\frac{\partial Z}{\partial x_j}\right)^{\diamond 2} + \Delta Z\right). \end{aligned} \tag{4.5.18}$$

Now apply (4.5.16), (4.5.15) and (4.5.18) to get

$$\begin{aligned} \frac{\partial Y}{\partial t} &= \frac{\lambda}{2\nu} Y \diamond \left(\frac{\lambda}{2}\sum_j \left(\frac{\partial Z}{\partial x_j}\right)^{\diamond 2} + \nu \Delta Z + N + C\right) \\ &= \frac{\lambda}{2\nu} \diamond \left(\frac{\lambda}{2}\sum_j \left(\frac{\partial Z}{\partial x_j}\right)^{\diamond 2} + \nu \Delta Z\right) + \frac{\lambda}{2\nu} Y \diamond (N + C) \\ &= \nu \Delta Y + \frac{\lambda}{2\nu} Y \diamond [N + C], \end{aligned}$$

as claimed. $\square$

Next we consider the deduced stochastic heat equation (4.5.14). By Corollary 4.3.2 we have

Lemma 4.5.2. *Suppose that $K(t,x)$ and $M(x)$ are $(\mathcal{S})_{-1}$-processes such that*

there exists $\mathbb{K}_q(R)$ such that both $\widetilde{K}(t,x,z)$ and $\widetilde{M}(x,z)$ are uniformly bounded for $(t,x,z) \in \mathbb{R}^+ \times \mathbb{R}^d \times \mathbb{K}_q(R)$, $\qquad$ (4.5.19)

and

$\widetilde{K}(t,x,z)$ and $\widetilde{M}(x,z)$ are locally Hölder continuous in x, uniformly in t, for each $z \in \mathbb{K}_q(R)$. $\qquad$ (4.5.20)

Then there is a unique $(\mathcal{S})_{-1}$-valued $C^{1,2}$ solution $Y(t,x)$ of the stochastic heat equation

$$
\begin{cases}
\frac{\partial Y}{\partial t} = \nu \Delta Y + K \diamond Y; & t > 0, x \in \mathbb{R}^d \\
Y(0,x) = M(x); & x \in \mathbb{R}^d,
\end{cases}
\qquad (4.5.21)
$$

namely

$$
Y(t,x) = \hat{E}^x \left[M(\sqrt{2\nu} b_t) \diamond \exp^\diamond \left[\int_0^t K(t-s, \sqrt{2\nu} b_s) ds \right] \right], \qquad (4.5.22)
$$

where, as before, b_t is a standard Brownian motion in $\mathbb{R}^d$, and $\hat{E}^x$ denotes expectation with respect to the law $\hat{P}^x$ of b_t starting at x.

Finally we show how to get from a solution of the stochastic heat equation (4.5.21) back to a solution of the stochastic Burgers equation (4.5.12):

Lemma 4.5.3 Holden et al. (1994), Holden et al. (1995b) The inverse Wick Cole–Hopf transformation. *Let $Y(t,x)$ be the $(\mathcal{S})_{-1}$-process given by (4.5.22) that solves the stochastic heat equation (4.5.21), where $K(t,x)$ and $M(x)$ are continuously differentiable $(\mathcal{S})_{-1}$-processes satisfying (4.5.19) and (4.5.20). Moreover, assume that*

$$
E_\mu[M(x)] > 0 \quad for \quad x \in \mathbb{R}^d, \qquad (4.5.23)
$$

where E_μ denotes generalized expectation (Definition 2.6.13). Then

$$
U(t,x) := -\frac{2\nu}{\lambda} \nabla_x (\log^\diamond Y(t,x)) \qquad (4.5.24)
$$

belongs to $(\mathcal{S})_{-1}^d$ for all $t \geq 0, x \in \mathbb{R}^d$ and $U(t,x) = (U_1(t,x), \dots, U_d(t,x))$ solves the stochastic Burgers equation

$$\begin{cases} \frac{\partial U_k}{\partial t} + \lambda \sum_{j=1}^{d} U_j \diamond \frac{\partial U_k}{\partial x_j} = \nu \Delta U_k + F_k; & t > 0, x \in \mathbb{R}^d \\ U_k(0, x) = G_k(x); & x \in \mathbb{R}^d; 1 \le k \le n, \end{cases} \tag{4.5.25}$$

where

$$F_k(t, x) = -\frac{2\nu}{\lambda} \frac{\partial K}{\partial x_k}(t, x) \tag{4.5.26}$$

and

$$G_k(x) = -\frac{2\nu}{\lambda} M(x)^{\diamond(-1)} \diamond \frac{\partial M}{\partial x_k}(x); \ 1 \le k \le n. \tag{4.5.27}$$

Proof From (4.5.22) and (2.6.54) we see that

$$E_\mu[Y(t, x)] = \hat{E}^x \left[E_\mu[M(\sqrt{2\nu}b_t)] \cdot \exp \left[\int_0^t E_\mu[K(t - s, \sqrt{2\nu}b_s)]ds \right] \right] > 0.$$

Therefore the Wick log of $Y(t, x), \log^\diamond Y(t, x)$, exists in $(\mathcal{S})_{-1}$ (see (2.6.51)). Hence we can reverse the argument in the proof of Lemma 4.5.1.
Set

$$Z(t, x) := \frac{2\nu}{\lambda} \log^\diamond Y(t, x) \tag{4.5.28}$$

and

$$U(t, x) = -\nabla_x Z(t, x). \tag{4.5.29}$$

Then

$$Y(t, x) = \exp^\diamond \left[\frac{\lambda}{2\nu} Z(t, x) \right], \tag{4.5.30}$$

so by (4.5.16) and (4.5.18) we get

$$\frac{\partial U_k}{\partial t} + \lambda \sum_j U_j \diamond \frac{\partial U_k}{\partial x_j} - \nu \Delta U_k - F_k$$

$$= -\frac{\partial}{\partial t} \left(\frac{\partial Z}{\partial x_k} \right) + \lambda \sum_j \frac{\partial Z}{\partial x_j} \diamond \frac{\partial}{\partial x_j} \left(\frac{\partial Z}{\partial x_k} \right) + \nu \sum_j \frac{\partial^2}{\partial x_j^2} \left(\frac{\partial Z}{\partial x_k} \right) + \frac{2\nu}{\lambda} \frac{\partial K}{\partial x_k}$$

$$= \frac{\partial}{\partial x_k} \left(-\frac{\partial Z}{\partial t} + \frac{\lambda}{2} \sum_j \left(\frac{\partial Z}{\partial x_j} \right)^{\diamond 2} + \nu \Delta Z + \frac{2\nu}{\lambda} K \right)$$

$$= \frac{\partial}{\partial x_k} \left(Y^{\diamond(-1)} \diamond \left(-\frac{2\nu}{\lambda} \frac{\partial Y}{\partial t} + \frac{2\nu^2}{\lambda} \Delta Y + \frac{2\nu}{\lambda} K \diamond Y \right) \right)$$

$$= \frac{2\nu}{\lambda} \cdot \frac{\partial}{\partial x_k} \left(Y^{\diamond(-1)} \diamond \left(-\frac{\partial Y}{\partial t} + \nu \Delta Y + K \diamond Y \right) \right) = 0,$$

which shows that U_k satisfies the first part of (4.5.25).

To prove the second part, observe that

$$U_k(0,x) = -\frac{\partial Z}{\partial x_k}(0,x) = -\frac{2\nu}{\lambda} Y(0,x)^{\diamond(-1)} \diamond \frac{\partial Y}{\partial x_k}(0,x)$$

$$= -\frac{2\nu}{\lambda} M(x)^{\diamond(-1)} \diamond \frac{\partial M}{\partial x_k}(x); 1 \leq k \leq d,$$

as claimed.   □

To summarize, we have now shown how to get from a solution of the Burgers equation (4.5.12) to a solution of the heat equation (4.5.14), which is easily solved. Then we can reverse the process and obtain the solution of the Burgers equation. This gives us the following existence and uniqueness result:

Theorem 4.5.4 Holden et al. (1995b) (Solution of the stochastic Burgers equation). *Let $N(t,x), R(x)$ be $(\mathcal{S})_{-1}$-valued $C^{0,1}$ and C^1 processes, respectively, satisfying the following conditions:*

There exists $\mathbb{K}_q(r)$ such that $\widetilde{N}(t,x,z)$ and $\exp\left[-\frac{\lambda}{2\nu}\widetilde{R}(x,z)\right]$ are

uniformly bounded for $(t,x,z) \in \mathbb{R} \times \mathbb{R}^d \times \mathbb{K}_q(r)$ and   (4.5.31)

$\widetilde{N}(t,x,z)$ and $\exp\left[-\frac{\lambda}{2\nu}\widetilde{R}(x)\right]$ are locally Hölder continuous in x,

uniformly in t, for each $z \in \mathbb{K}_q(r)$.   (4.5.32)

Then there exists an $(\mathcal{S})_{-1}^d$-valued $C^{1,2}$ process $U(t,x) = (U_1(t,x), \cdot, U_d(t,x))$ that solves the stochastic Burgers equation

$$\begin{cases} \frac{\partial U_k}{\partial t} + \lambda \sum_{j=1}^d U_j \diamond \frac{\partial U_k}{\partial x_j} = \nu \Delta U_k - \frac{\partial N}{\partial x_k}; & t > 0, \ x \in \mathbb{R}^d \\ U_k(0,x) = -\frac{\partial R}{\partial x_k}(x); & x \in \mathbb{R}^d; 1 \leq k \leq d. \end{cases}$$ (4.5.33)

This solution is given by

$$U(t,x) := -\frac{2\nu}{\lambda} \nabla_x (\log^{\diamond} Y(t,x)),$$ (4.5.34)

where

$$Y(t,x) = \hat{E}^x\left[\exp^{\diamond}\left[-\frac{\lambda}{2\nu}R(\sqrt{2\nu}b_t)\right] \diamond \exp^{\diamond}\left[-\frac{\lambda}{2\nu}\int_0^t N(t-s, \sqrt{2\nu}b_s)ds\right]\right]$$

(4.5.35)

is the unique $(\mathcal{S})_{-1}$*-valued* $C^{1,2}$ *solution of the stochastic heat equation*

$$\begin{cases} \frac{\partial Y}{\partial t} = \nu \Delta Y + \frac{\lambda}{2\nu} N \diamond Y; & t > 0, x \in \mathbb{R}^d \\ Y(0, x) = \exp^\diamond \left[\frac{\lambda}{2\nu} R(x) \right]; & x \in \mathbb{R}^d. \end{cases} \tag{4.5.36}$$

Moreover, the process U *given by (4.5.34) is the unique solution of (4.5.33) of gradient form, i.e., the gradient with respect to* x *of some* $(\mathcal{S})_{-1}$*-valued* $C^{1,3}$ *process.*

Proof a) *Existence*: By Lemma 4.5.2 the process $Y(t, x)$ given by (4.5.35) solves (4.5.36). Hence by Lemma 4.5.3 the process $U(t, x)$ given by (4.5.34) solves the stochastic Burgers equation (4.5.25) with

$$F_k = -\frac{2\nu}{\lambda} \frac{\partial K}{\partial x_k} = -\frac{2\nu}{\lambda} \cdot \frac{\partial}{\partial x_k} \left(\frac{\lambda}{2\nu} N \right) = -\frac{\partial N}{\partial x_k}$$

and

$$\begin{aligned} G_k &= -\frac{2\nu}{\lambda} M^{\diamond(-1)} \diamond \frac{\partial M}{\partial x_k} \\ &= -\frac{2\nu}{\lambda} \exp^\diamond \left[-\frac{\lambda}{2\nu} R \right] \diamond \exp^\diamond \left[\frac{\lambda}{2\nu} R \right] \left(\frac{\lambda}{2\nu} \frac{\partial R}{\partial x_k} \right) \\ &= -\frac{\partial R}{\partial x_k}. \end{aligned}$$

b) *Uniqueness*: If $U(t, x) = -\nabla_x Z(t, x)$ solves (4.5.33), then by Lemma 4.5.1 the process

$$Y := \exp^\diamond \left[\frac{\lambda}{2\nu} Z \right]$$

solves the equation

$$\begin{cases} \frac{\partial Y}{\partial t} = \nu \Delta Y + \frac{\lambda}{2\nu} Y \diamond [N + C]; & t > 0, x \in \mathbb{R}^d \\ Y(0, x) = \exp^\diamond \left[\frac{\lambda}{2\nu} Z(0, x) \right] \end{cases} \tag{4.5.37}$$

for some t-continuous process $C(t)$ independent of x. Hence by Lemma 4.5.2 we have

$$Y(t, x) = \hat{E}^x \left[\exp^\diamond \left[\frac{\lambda}{2\nu} Z(0, \sqrt{2\nu} b_t) \right] \right.$$

$$\left. \diamond \exp^\diamond \left[\int_0^t N(t - s, \sqrt{2\nu} b_s) ds + \int_0^t C(s) ds \right] \right]$$

$$= Y^{(0)}(t, x) \diamond \exp^\diamond \left[\int_0^t C(s) ds \right],$$

where $Y^{(0)}(t,x)$ is the solution of (4.5.37) with $C = 0$. Hence

$$Z = \frac{2\nu}{\lambda} \log^\diamond Y = \frac{2\nu}{\lambda} \log^\diamond Y^{(0)} + \frac{2\nu}{\lambda} \int_0^t C(s)\,ds,$$

so that

$$U = -\nabla Z = -\frac{2\nu}{\lambda}\nabla(\log^\diamond Y^{(0)}),$$

which in turn implies that U is unique. □

4.6 The Stochastic Pressure Equation

We now return to one of the equations that we discussed in the introduction (Chapter 1). This equation was introduced as an example of a physical situation where rapidly fluctuating, apparently stochastic, parameter values lead naturally to an SPDE model:

$$\begin{cases} \operatorname{div}(K(x) \cdot \nabla p(x)) = -f(x); & x \in D \\ p(x) = 0; & x \in \partial D. \end{cases} \tag{4.6.1}$$

Here D is a given bounded domain in $\mathbb{R}^d$, and $f(x), K(x)$ are given functions. This corresponds to equations (1.1.2)–(1.1.3) in Chapter 1 for a fixed instant of time t (deleted from the notation). With this interpretation $p(x)$ is the (unknown) *pressure* of the fluid at the point x, $f(x)$ is the *source rate* of the fluid, and $K(x) \geq 0$ is the *permeability* of the rock at the point x. As argued in Chapter 1, it is natural to represent $K(x)$ by a stochastic quantity. Moreover, it is commonly assumed that probabilistically K has — at least approximately — the three properties (2.6.57)–(2.6.59). These are the properties of the *smoothed positive noise process* K_ϕ in (2.6.56), i.e.,

$$K_\phi(x, \omega) := \exp^\diamond W_\phi(x). \tag{4.6.2}$$

Thus we set $K = K_\phi$ for some fixed $\phi \in \mathcal{S}(\mathbb{R}^d)$. In view of (2.6.57), one can say that the diameter of the support of ϕ is the maximal distance within which there is correlation in permeability values. So, from a modeling point of view, this diameter should be on the same order of magnitude as the maximal size of the pores of the rock.

Alternatively, one could insist on the idealized, *singular positive noise process*

$$K(x, \cdot) := \exp^\diamond W(x, \cdot) \in (\mathcal{S})^*, \tag{4.6.3}$$

corresponding to the limiting case of $K_\phi(x)$ when $\phi \to \delta_0$. Indeed, this is the usual attitude in stochastic *ordinary* differential equations, where one prefers to deal with singular white noise rather than smoothed white noise, even in cases where the last alternative could be more natural from a modeling point of view.

In view of this, we will discuss both cases. However, in either case we will, as before, interpret the product in (4.6.1) as the Wick product. With K as in (4.6.3) it is not clear how to make the equation well-defined with the pointwise product, although both products would make sense (and give different results) in the smoothed case (see Section 3.5).

Since the proofs in the two cases are so similar, we give the details only in the smoothed case and merely state the corresponding solution in the singular case afterwards.

4.6.1 The Smoothed Positive Noise Case

Theorem 4.6.1 Holden et al. (1995). *Let D be a bounded C^2 domain in $\mathbb{R}^d$ and let $f(x)$ be an $(\mathcal{S})_{-1}$-valued function satisfying the condition*

there exists $\mathbb{K}_q(R)$ such that $\widetilde{f}(x, z)$ is uniformly bounded for $(x, z) \in D \times \mathbb{K}_q(R)$ and for each $z \in \mathbb{K}_q(R)$ there exists $\lambda \in (0, 1)$ such that $\widetilde{f}(x, z)$ is $\lambda - H\ddot{o}lder$ continuous with respect to $x \in D$.   (4.6.4)

Fix $\phi \in \mathcal{S}(\mathbb{R}^d)$. Then the smoothed stochastic pressure equation

$$\begin{cases} div(K_\phi(x) \diamond \nabla p(x)) = -f(x); & x \in D \\ p(x) = 0; & x \in \partial D \end{cases}$$ (4.6.5)

has a unique $(\mathcal{S})_{-1}$-valued solution $p(x) = p_\phi(x) \in C^2(D) \cap C(\bar{D})$ given by

$$p_\phi(x) = \frac{1}{2} \exp^\diamond \left[-\frac{1}{2} W_\phi(x) \right] \diamond \hat{E}^x \left[\int_0^{\tau_D} f(b_t) \diamond \exp^\diamond \left[-\frac{1}{2} W_\phi(b_t) \right. \right.$$

$$\left. \left. -\frac{1}{4} \int_0^t \left[\frac{1}{2} (\nabla W_\phi(y))^{\diamond 2} + \Delta W_\phi(y) \right]_{y=b_s} ds \right] dt \right],$$ (4.6.6)

where $(b_t(\hat{\omega}), \hat{P}^x)$ is a (1-parameter) standard Brownian motion in $\mathbb{R}^d$ (independent of $B_x(\omega)$), $\hat{E}^x$ denotes expectation with respect to $\hat{P}^x$ and

$$\tau_D = \tau_D(\hat{\omega}) = \inf\{t > 0; b_t(\hat{\omega}) \notin D\}.$$

Proof   Taking Hermite transforms of (4.6.5) we get the following equation in the unknown $u(x) = u(x, z) = \widetilde{p}(x, z)$, for z in some $\mathbb{K}_q(\mathcal{S})$,

$$\begin{cases} div(\widetilde{K}_\phi(x) \cdot \nabla u(x)) = -\widetilde{f}(x); & x \in D \\ u(x) = 0; & x \in \partial D \end{cases}$$ (4.6.7)

or

$$\begin{cases} L^{(z)}u(x,z) = -F(x,z); & x \in D \\ u(x) = 0; & x \in \partial D, \end{cases} \tag{4.6.8}$$

where

$$L^{(z)}u(x) = \frac{1}{2}\Delta u(x) + \frac{1}{2}\nabla\gamma(x) \cdot \nabla u(x) \tag{4.6.9}$$

with

$$\gamma(x) = \gamma_\phi(x) = \gamma_\phi(x,z) = \widetilde{W}_\phi(x,z) = \sum_{j=1}^{\infty}(\phi_x, \eta_j)z_j, \tag{4.6.10}$$

and

$$F(x) = F(x,z) = \frac{1}{2}\widetilde{f}(x,z) \cdot \exp\left[-\gamma(x,z)\right]. \tag{4.6.11}$$

First assume that $z = (z_1, z_2, \dots) = (\xi_1, \xi_2, \dots)$ with $\xi_k \in \mathbb{R}$ for all k. Since the operator $L^{(\xi)}$ is uniformly elliptic in D we know by our assumption on f that the boundary value problem (4.6.8) has a unique $C^{2+\lambda}(D)$ solution $u(x) = u(x,\xi)$ for each ξ, where $\lambda = \lambda(\xi) > 0$ may depend on ξ. Moreover, we can express this solution probabilistically as follows:

Let $(x_t = x_t^{(\xi)}(\hat{\omega}), \widetilde{P}^x)$ be the solution of the (ordinary) Itô stochastic differential equation

$$dx_t = \frac{1}{2}\nabla\gamma(x_t, \xi)dt + db_t; \quad x_0 = x \tag{4.6.12}$$

where $(b_t(\hat{\omega}), \hat{P}^x)$ is the d-dimensional Brownian motion we described above. Then the generator of $x_t^{(\xi)}$ is $L^{(\xi)}$, so by Dynkin's formula (see, e.g., Theorem 7.12 in Øksendal (1995)), we have that, for $x \in U \subset\subset D$,

$$\widetilde{E}^x[u(x_{\widetilde{\tau}_U}, \xi)] = u(x, \xi) + \widetilde{E}^x\left[\int_0^{\widetilde{\tau}_U} L^{(\xi)}u(x_t, \xi)dt\right], \tag{4.6.13}$$

where $\widetilde{E}^x$ denotes expectation with respect to $\widetilde{P}^x$ and

$$\widetilde{\tau}_U = \widetilde{\tau}_U(\hat{\omega}) = \inf\{t > 0; x_t(\hat{\omega}) \notin U\}$$

is the first exit time from U for x_t. By the Girsanov formula (see Appendix B), this can be expressed in terms of the probability law $\hat{P}^x$ of b_t as follows:

$$\hat{E}^x[u(b_\tau, \xi)\mathcal{E}(\tau, \xi)] = u(x, \xi) + \hat{E}^x\left[\int_0^\tau L^{(\xi)}u(b_t, \xi)\mathcal{E}(t, \xi)dt\right], \tag{4.6.14}$$

where

$$\mathcal{E}(t,z) = \exp\left[\frac{1}{2}\int_0^t \nabla\gamma(b_s,z)db_s - \frac{1}{8}\int_0^t (\nabla\gamma)^2(b_s,z)ds\right]. \qquad (4.6.15)$$

$\hat{E}^x$ denotes expectation with respect to $\hat{P}^x$, and

$$\tau = \tau_U(\hat{\omega}) = \inf\{t > 0; b_t(\hat{\omega}) \notin U\}.$$

Letting $U \uparrow D$, we get from (4.6.14) and (4.6.8) that

$$u(x,\xi) = \hat{E}^x\left[\int_0^\tau F(b_t,\xi)\mathcal{E}(t,\xi)dt\right]. \qquad (4.6.16)$$

By Itô's formula we have

$$\gamma(b_t,\xi) = \gamma(b_0,\xi) + \int_0^t \nabla\gamma(b_s,\xi)db_s + \frac{1}{2}\int_0^t \Delta\gamma(b_s,\xi)ds \qquad (4.6.17)$$

or

$$\frac{1}{2}\int_0^t \nabla\gamma(b_s,\xi)db_s = \frac{1}{2}\gamma(b_t,\xi) - \frac{1}{2}\gamma(b_0,\xi) - \frac{1}{4}\int_0^t \Delta\gamma(b_s,\xi)ds. \qquad (4.6.18)$$

Substituting (4.6.18) and (4.6.11) in (4.6.16) we get, with $\tau = \tau_D$,

$$u(x,\xi) = \frac{1}{2}\exp\left[-\frac{1}{2}\gamma(x,\xi)\right] \cdot \hat{E}^x\left[\int_0^\tau f(b_t) \cdot \exp\left[-\frac{1}{2}\gamma(b_t,\xi)\right.\right.$$

$$\left.\left. -\frac{1}{4}\int_0^t \frac{1}{2}(\nabla\gamma)^2(b_s,\xi) + \Delta\gamma(b_s,\xi)ds\right]dt\right] \qquad (4.6.19)$$

for all $\xi \in \mathbb{K}_q(R) \cap \mathbb{R}^\mathbb{N}$.

Since $\gamma(x,\xi) = \sum_k(\phi_x,\eta_k)\xi_k; \xi_k \in \mathbb{R}$ has an obvious analytic extension to $z_k \in \mathbb{C}$ given by $\gamma(x,z) = \sum_k(\phi_x,\eta_k)z_k$, similarly with

$$\nabla\gamma(x,z) = \sum_k \nabla_x(\phi_x,\eta_k)z_k,$$

$$\Delta\gamma(x,z) = \sum_k \Delta_x(\phi_x,\eta_k)z_k,$$

we see that $\xi \to u(x, \xi); \xi \in (\mathbb{R}^{\mathbb{N}})_c$ given by (4.6.19) has an analytic extension (also denoted by u) given by

$$u(x, z) = \frac{1}{2} \exp\left[-\frac{1}{2}\gamma(x, z)\right] \cdot \hat{E}^x\left[\int_0^\tau \tilde{f}(b_t) \cdot \exp\left[-\frac{1}{2}\gamma(b_t, z)\right.\right.$$

$$\left.\left. -\frac{1}{4}\int_0^t \frac{1}{2}(\nabla\gamma)^2(b_s, z) + \Delta\gamma(b_s, z)ds\right]dt,\right] \tag{4.6.20}$$

provided the expression converges. If $z \in \mathbb{K}_q(\delta)$, then

$$|\gamma(x, z)|^2 = \left|\sum_k (\phi_x, \eta_k)z_k\right|^2$$

$$\leq \left(\sum_k (\phi_x, \eta_k)^2\right) \sum_k |z_k|^2$$

$$\leq \|\phi\|^2 \sum_\alpha |z^\alpha|^2 (2\mathbb{N})^{q\alpha}$$

$$\leq \delta^2\|\phi\|^2 \quad \text{for all } q > 0,$$

similarly with $|\nabla\gamma(x, z)|$ and $|\Delta\gamma(x, z)|$.

This gives

$$|u(x, z)| \leq C_1 \exp\left[\frac{\delta}{2}\|\phi\|\right]\hat{E}^x\left[\int_0^\tau \exp\left[\frac{\delta}{2}\|\phi\|\right.\right.$$

$$\left.\left. +\frac{1}{4}\left(\frac{\delta^2}{2}\|\nabla\phi_x(\cdot)\|^2 + \delta\|\Delta\phi_x\|\right)t\right]dt\right]$$

where C_1 is a constant. Since D is bounded, there exists $\rho > 0$ such that

$$\hat{E}^x[\exp[\rho\tau]] < \infty.$$

Therefore, if we choose $\delta > 0$ such that

$$\frac{1}{4}\left(\frac{\delta^2}{2}\|\nabla\phi_x\|^2 + \delta\|\Delta\phi_x\|\right) < \rho,$$

we obtain that $u(x, z)$ is bounded for $z \in \mathbb{K}_q(\delta)$.

We must verify that $u(\cdot, z)$ satisfies equations (4.6.8). We know that this is the case when $z = \xi \in \mathbb{K}_q(\delta) \cap \mathbb{R}^{\mathbb{N}}$. Moreover, the solution $u(x, \xi)$ is real analytic in a neighborhood of $\xi = 0$, so we can write

$$u(x, \xi) = \sum_\alpha c_\alpha(x)\xi^\alpha.$$

Note that since $u(x, \xi) \in C^{2+\lambda}(D)$ with respect to x for all ξ, we must have $c_\alpha(x) \in C^{2+\lambda}(D)$ for all α. Moreover, by (4.6.8) we have $c_\alpha(x) = 0$ for $x \in \partial D$.

Similarly, we may write $F(x, z) = \sum a_\alpha(x)z^\alpha$, and we have

$$\nabla\gamma(x, z) = \sum_k \nabla(\phi_x, \eta_k)z_k.$$

Substituted in (4.6.8), this gives

$$\sum_\alpha \frac{1}{2}\Delta c_\alpha(x)\xi^\alpha + \sum_{\beta,k} \nabla(\phi_x, \eta_k) \cdot \nabla c_\beta(x)\xi^{\beta+\epsilon^{(k)}} = \sum_\alpha a_\alpha(x)\xi^\alpha, \quad (4.6.21)$$

i.e.,

$$\sum_\alpha \left(\frac{1}{2}\Delta c_\alpha(x) + \sum_{\substack{\beta,k \\ \beta+\epsilon^{(k)}=\alpha}} \nabla(\phi_x, \eta_k) \cdot \nabla c_\beta(x) \right) \xi^\alpha = \sum_\alpha a_\alpha(x)\xi^\alpha. \quad (4.6.22)$$

Since this holds for all ξ small enough, we conclude that

$$\frac{1}{2}\Delta c_\alpha(x) + \sum_{\substack{\beta,k \\ \beta+\epsilon^{(k)}=\alpha}} \nabla(\phi_x, \eta_k) \cdot \nabla c_\beta(x) = a_\alpha(x) \quad (4.6.23)$$

for all multi-indices α. But then (4.6.22), and hence (4.6.21), also holds when ξ is replaced by $z \in \mathbb{K}_q(\delta)$. In other words, the analytic extension $u(x, z)$ of $u(x, \xi)$ does indeed solve the first part of (4.6.8).

Next, since $c_\alpha(x) = 0$ for $x \in \partial D$ for all α, it follows that $u(x, z) = 0$ for $x \in \partial D$ for all $z \in \mathbb{K}_q(\delta)$. We conclude that $u(x, z)$ does indeed satisfy (4.6.8).

Moreover, we saw above that $u(x, z)$ is uniformly bounded for all $(x, z) \in \bar{D} \times \mathbb{K}_q(\delta)$. Furthermore, for all $\xi \in \mathbb{K}_q(\delta) \cap \mathbb{R}^{\mathbb{N}}$, we know that $u(x, \xi) \in C^{2+\lambda(\xi)}(D)$. This implies that $c_\alpha(x) \in C^{2+\lambda(\xi)}(D)$ for all α. So all partial derivatives of $c_\alpha(x)$ up to order two are continuous and uniformly bounded in D. By bounded convergence we conclude that,

$$\Delta u(x, z) = \sum_\alpha \Delta c_\alpha(x)z^\alpha,$$

is continuous and uniformly bounded in D for each $z \in \mathbb{K}_q(\delta)$. So by Theorem 4.1.1 we conclude that inverse Hermite transform of $u(x)$,

$$p(x) := \mathcal{H}^{-1}u(x)$$

satisfies equation (4.6.5). Moreover, from (4.6.20) we see that $p_\phi(x)$ is given by (4.6.6). $\qquad \square$

4.6.2 An Inductive Approximation Procedure

We emphasize that although our solution $p(x, \cdot)$ lies in the abstract space $(\mathcal{S})_{-1}$, it does have a physical interpretation. For example, by taking the *generalized expectation* E_μ of equation (4.6.5) (see Definition 2.6.13) and using (2.6.45) we get that the function

$$\bar{p}(x) = E_\mu[p(x, \cdot)] \qquad (4.6.24)$$

satisfies the classical deterministic Poisson problem

$$\begin{cases} \Delta\bar{p}(x) = -E_\mu[f(x)]; & x \in D \\ \bar{p}(x) = 0; & x \in \partial D, \end{cases} \qquad (4.6.25)$$

i.e., the equation we will obtain if we replace the stochastic permeability $K_\phi(x, \omega) = \exp^\diamond W_\phi(x, \omega)$ by its expectation

$$\bar{K}(x) = E_\mu[K_\phi(x, \omega)] = 1,$$

which corresponds to a completely homogeneous medium.

We may regard $\bar{p}(x)$ as the *best ω-constant approximation* to $p(x, \omega)$. This ω-constant coincides with the zeroth-order term $c_0(x)$ of the expansion for $p(x, \omega)$,

$$p(x, \omega) = \sum_\alpha c_\alpha(x) H_\alpha(\omega), \qquad (4.6.26)$$

where $c_\alpha(x)$ is given inductively by (4.6.23). Having found $\bar{p}(x) = c_0(x)$, we may proceed to find the *best Gaussian approximation $p_1(x, \omega)$* to $p(x, \omega)$. This coincides with the sum of all first order terms:

$$p_1(x, \omega) = \sum_{|\alpha| \leq 1} c_\alpha(x) H_\alpha(\omega)$$

$$= c_0(x) + \sum_{j=1}^\infty c_{\epsilon(j)}(x) \langle \omega, \eta_j \rangle. \qquad (4.6.27)$$

From (4.6.23) we can find $c_{\epsilon(j)}(x)$ when $c_0(x)$ is known from the equation

$$\begin{cases} \frac{1}{2}\Delta c_{\epsilon(j)}(x) + \nabla(\phi_x, \eta_j) \cdot \nabla c_0(x) = a_{\epsilon(j)}(x); & x \in D \\ c_{\epsilon(j)}(x) = 0; & x \in \partial D. \end{cases} \qquad (4.6.28)$$

Similarly, one can proceed by induction to find higher-order approximation to $p(x, \omega)$. This may turn out to be the most efficient way of computing $p(x, \cdot)$ numerically. See, however, Holden and Hu (1996), for a different approach based on finite differences.

4.6.3 The 1-Dimensional Case

When $d = 1$, it is possible to solve equation (4.6.5) directly, using Wick calculus.

Theorem 4.6.2 Holden et al. (1995). *Let $a, b \in \mathbb{R}, a < b$, and assume that $f \in L^1[a, b]$ is a deterministic function. Then for all $\phi \in \mathcal{S}(\mathbb{R})$ the unique solution $p(x, \cdot) \in (\mathcal{S})_{-1}$ of the 1-dimensional pressure equation*

$$\begin{cases} (\exp^\diamond[W_\phi(x)] \diamond p'(x, \cdot))' = -f(x); & x \in (a, b) \\ p(a, \cdot) = p(b, \cdot) = 0 \end{cases} \qquad (4.6.29)$$

is given by

$$p(x, \cdot) = A \diamond \int_a^x \exp^\diamond[-W_\phi(t)]dt - \int_a^x \int_a^t f(s)ds \exp^\diamond[-W_\phi(t)]dt, \quad (4.6.30)$$

where

$$A = A(\omega)$$

$$= \left(\int_a^b \exp^\diamond[-W_\phi(t)]dt \right)^{\diamond(-1)} \diamond \int_a^b \int_a^t f(s)ds \exp^\diamond[-W_\phi(t)]dt \in (\mathcal{S})_{-1}.$$

$$(4.6.31)$$

Proof   Integrating (4.6.29) we get

$$\exp^\diamond[W_\phi(x)] \diamond p'(x, \cdot) = A - \int_a^x f(t)dt; \; x \in (a, b),$$

where $A = A(\omega)$ does not depend on x. Since $\exp^\diamond[-X] \diamond \exp^\diamond[X] = 1$ for all $X \in (\mathcal{S})_{-1}$, we can write this as

$$p'(x, \cdot) = A \diamond \exp^\diamond[-W_\phi(x)] - \int_a^x f(s)ds \cdot \exp^\diamond[-W_\phi(x)]. \qquad (4.6.32)$$

Using the condition $p(a, \cdot) = 0$, we deduce from (4.6.32) that $p(x, \cdot)$ is given by the expression

$$p(x, \cdot) = A \diamond \int_a^x \exp^\diamond[-W_\phi(t)]dt - \int_a^x \int_a^t f(s)ds \exp^\diamond[-W_\phi(t)]dt. \qquad (4.6.33)$$

It remains to determine the random variable A. The condition $p(b, \cdot) = 0$ leads to the expression

$$A \diamond \int_a^b \exp^\diamond[-W_\phi(t)]dt = \int_a^b \int_a^t f(s)ds \exp[-W_\phi(t)]dt. \qquad (4.6.34)$$

Set

$$Y = \int_a^b \exp^\diamond[-W_\phi(t)]dt. \qquad (4.6.35)$$

We have $Y \in (\mathcal{S})_{-1}$ and $E[Y] = b - a \neq 0$. Therefore $Y^{\diamond(-1)} \in (\mathcal{S})_{-1}$ exists by Example 2.6.15(iii). So

$$A := Y^{\diamond(-1)} \diamond \int_a^b \int_a^t f(s)ds \exp^\diamond[-W_\phi(t)]dt \in (\mathcal{S})_{-1},$$

and with this choice of A in (4.6.33) we see that $p(x, \cdot)$ given by (4.6.33) solves (4.6.29). □

4.6.4 The Singular Positive Noise Case

Theorem 4.6.3. *Let D and f be as in Theorem 4.6.1. Then the (singular) stochastic pressure equation*

$$\begin{cases} div(\exp^\diamond[W(x)] \diamond \nabla p(x)) = -f(x); & x \in D \\ p(x) = 0; & x \in \partial D, \end{cases} \qquad (4.6.36)$$

has a unique $(\mathcal{S})_{-1}$-valued solution $p(x) \in C^2(D) \cap C(\bar{D})$ given by

$$p(x) = \frac{1}{2} \exp^\diamond \left[-\frac{1}{2}W(x)\right] \diamond \hat{E}^x \left[\int_0^{\tau_D} f(b_t) \diamond \exp^\diamond \left[-\frac{1}{2}W(b_t)\right. \right.$$

$$\left. \left. -\frac{1}{4} \int_0^t \left[\frac{1}{2}(\nabla W(y))^{\diamond 2} + \Delta W(y)\right]_{y=b_s} ds \right] dt \right], \qquad (4.6.37)$$

where b_t, $\hat{E}^x$ and τ_D are as in Theorem 4.6.1 and where, in addition, we have that $W(x) = W(x, \omega) = \sum_j \eta_j(x) H_{\epsilon^{(j)}}(\omega)$ is the singular white noise process.

Proof The proof follows word for word the proof of Theorem 4.6.1, except that the smoothed versions are replaced by the singular versions everywhere.

For example, $\gamma(x) = \gamma_\phi(x) = \sum_j (\phi_x, \eta_j) z_j$ in (4.6.10) is replaced by the function $\gamma(x) = \check{\gamma}(x) = \sum_j \eta_j(x) z_j$. Because of our choice of η_j (see (2.2.8)), this will not disturb any of the arguments nor any of the estimates we used in the smoothed case. □

Corollary 4.6.4. *Let $\rho \geq 0$ and $\phi_k; k = 1, 2, \ldots$ be as in $(3.6.2)$. Let $p_{(k)}(x)$ be the solution of the ϕ_k-smoothed stochastic pressure equation $(4.6.5)$ (i.e., with $\phi = \phi_k$), and let $p(x)$ be the solution of the singular stochastic pressure equation $(4.6.36)$. Then*

$$p_{(k)}(x) \to p(x) \quad in \quad (\mathcal{S})_{-1} \quad as \quad k \to \infty,$$

uniformly for $x \in \bar{D}$.

Proof This follows by inspecting the solution formulas (4.6.6) and (4.6.37) or, rather, their Hermite transforms. For example, we see that

$$\gamma_{(k)}(x) = \sum_j \int_{\mathbb{R}^d} \phi_k(t - x) \eta_j(t) dt z_j \to \sum_j \eta_j(x) z_j \quad as \quad k \to \infty,$$

since $\{\phi_k\}_{k=1}^\infty$ constitutes an approximate identity. Moreover, the convergence is uniform for $(x, z) \in D \times \mathbb{K}_q(\delta)$. Similar convergence and estimates are obtained for the other terms. We omit the details. □

Remark The approach used above has the advantage of giving (relatively) specific solution formulas – when it works. But its weakness is that it only works in some cases. There are other methods that can be applied to obtain existence and uniqueness more generally, without explicit solution formulas. One of these methods is outlined in the next section (see also Gjerde (1995a), (1995b)). Another interesting method is based on fundamental estimates for the Wick product in the Hilbert spaces $(\mathcal{S})_{-1,-k}$ (see, e.g., Proposition 3.3.2) combined with the Lax–Milgram theorem. See Våge (1995), (1996a), (1996b) for details. In Øksendal and Våge (1996), this approach is applied to the study of stochastic variational inequalities, with applications to the moving boundary problem in a stochastic medium, i.e., a stochastic version of equations (1.1.2)–(1.1.4) in Chapter 1.

4.7 The Heat Equation in a Stochastic, Anisotropic Medium

In the previous section the following method was used: When taking the Hermite transform we got an equation in $U(x, z)$ that could be solved for real values λ_k of the parameters z_k. Then, from the solution formula for $u(x, \lambda)$, it was apparent that it had an analytic extension to $u(x, z)$ for complex z. Finally, to prove that $u(x, z)$ was the Hermite transform of an element in $(\mathcal{S})_{-1}$, we proved boundedness for z in some $\mathbb{K}_q(\delta)$.

In other equations the extension from the real case $z_k = \lambda_k \in \mathbb{R}$ to the complex, analytic case $z_k \in \mathbb{C}$ may not be as obvious, and it is natural to ask if it is sufficient with good enough estimates in the real case alone to obtain the same conclusion. Of course, examples like

$$g(z) = \cos z = \frac{1}{2}(e^{iz} + e^{-iz}); \quad z \in \mathbb{C},$$

remind us that good estimates for $z = \lambda \in \mathbb{R}$ do not necessarily imply good estimates for $z \in \mathbb{C}$. Nevertheless, as discovered by Gjerde (1995a), good estimates for $z = \lambda \in \mathbb{K}_q(R) \cap \mathbb{R}^N$ do imply good estimates for complex z in a smaller neighborhood $\mathbb{K}_{\hat{q}}(\hat{R})$. Using this he could solve, for example, the heat equation in a stochastic medium.

We now explain this in more detail. Our presentation is based on Gjerde (1998).

Definition 4.7.1.   A function $f : \mathbb{K}_q(\delta) \cap \mathbb{R}^N \to \mathbb{R}$ is said to be *real analytic* if the restriction of f to $\mathbb{K}_q(\delta) \cap \mathbb{R}^k$ is real analytic for all natural numbers k, and there exist $M < \infty, \rho > 0$ *independent of* k such that

$$|\partial^\alpha f(0)| \leq M |\alpha|! \rho^{-|\alpha|} \qquad (4.7.1)$$

for all $\alpha \in \mathbb{N}_0^k$.

Lemma 4.7.2 Gjerde (1998). *Suppose f is real analytic on $\mathbb{K}_q(\delta) \cap \mathbb{R}^N$. Then there exist $\hat{q} < \infty, \hat{\delta} > 0$ and a bounded analytic function $F : \mathbb{K}_{\hat{q}}(\hat{\delta}) \to \mathbb{C}$ such that*

$$f(\lambda) = F(\lambda) \quad for \quad \lambda \in \mathbb{K}_{\hat{q}}(\hat{\delta}) \cap \mathbb{R}^N.$$

In short, f has an extension to a bounded analytic function on $\mathbb{K}_{\hat{q}}(\hat{\delta})$.

Proof   From the general theory of real analytic functions (see, e.g., John (1986), Chapter 3) we know that, with M, ρ as in (4.7.1), for all $k \in \mathbb{N}$, the power series expansion

$$f(\lambda) = \sum_{\alpha \in \mathbb{N}_0^k} \frac{\partial^\alpha f(0)}{\alpha!} \lambda^\alpha \qquad (4.7.2)$$

is valid for all $\lambda \in \mathbb{R}^k$ with $|\lambda_i| < \rho$. Moreover,

$$|\partial^\alpha f(0)| \leq M |\alpha|! \rho^{-|\alpha|} \quad \text{for all} \quad \alpha \in \mathbb{N}_0^k.$$

In particular, this expansion holds for

$$\lambda \in \mathbb{R}_k(\rho) = \left\{ x \in \mathbb{R}^k; \sum_{j=1}^{k} |x_j| < \rho \right\}.$$

It is clear that we can extend $f|_{\mathbb{R}_k(\rho)}$ analytically to the set

$$\mathbb{C}_k(\rho) = \left\{ z \in \mathbb{C}^k; \sum_{j=1}^{k} |z_j| < \rho \right\}$$

by defining

$$f(z) = \sum_{\alpha \in \mathbb{N}_0^k} \frac{\partial^\alpha f(0)}{\alpha!} z^\alpha; \quad z \in \mathbb{C}_k(\rho).$$

Now note that if $\sum_{j=1}^{k} |z_j| \le \rho_1 < \rho$, then

$$|f(z)| \le \sum_{\alpha \in \mathbb{N}_0^k} \frac{1}{\alpha!} |\partial^\alpha f(0)| |z^\alpha| \le \sum_{\alpha \in \mathbb{N}_0^k} \frac{1}{\alpha!} M|\alpha|! \rho^{-|\alpha|} |z^\alpha|$$

$$= M \sum_{j=0}^{\infty} \rho^{-j} \sum_{\substack{|\alpha|=j \\ \alpha \in \mathbb{N}_0^k}} \frac{j!}{\alpha!} |z_1|^{\alpha_1} \cdots |z_k|^{\alpha_k}$$

$$= M \sum_{j=0}^{\infty} \rho^{-j} (|z_1| + \cdots + |z_k|)^j \le M \sum_{j=0}^{\infty} \left(\frac{\rho_1}{\rho} \right)^j = \frac{M\rho}{\rho - \rho_1}.$$

From this we see that $f(z)$ is analytic and bounded, uniformly in k, on

$$\left\{ z \in \mathbb{C}^k; \sum_{j=1}^{k} |z_j| \le \rho_1 \right\} \quad \text{for all} \quad k.$$

Therefore it suffices to find $\hat{q} < \infty, \hat{\delta} > 0$ such that

$$\mathbb{K}_{\hat{q}}(\hat{\delta}) \subset \left\{ z \in \mathbb{C}^\mathbb{N}; \sum_{j=1}^{\infty} |z_j| < \rho \right\}. \tag{4.7.3}$$

To this end note that

$$\sum_{j=1}^{\infty} |z_j| \le \sum_{\alpha} |z^\alpha| \le \left(\sum_{\alpha} |z^\alpha|^2 (2\mathbb{N})^{q\alpha} \right)^{\frac{1}{2}} \cdot \left(\sum_{\alpha} (2\mathbb{N})^{-q\alpha} \right)^{\frac{1}{2}} < \delta A(q)^{\frac{1}{2}},$$

where by Proposition 2.3.3, $A(q) = \sum_\alpha (2\mathbb{N})^{-q\alpha} < \infty$ for $q > 1$. Therefore, if $\hat{\delta} \leq \delta$ and $\hat{q} \geq q$ are chosen such that

$$\hat{\delta} A(\hat{q})^{\frac{1}{2}} < \rho,$$

then (4.7.3) holds and the proof is complete.   □

Theorem 4.7.3.   *Consider an SPDE of the form (4.1.2), i.e.*

$$A^\circ(t, x, \partial_t, \nabla_x, U, \omega) = 0 \tag{4.7.4}$$

for $(t, x) \in G \subset \mathbb{R}^{d+1}$. Assume that for some $q < \infty, \delta > 0$ there exists a solution $u(t, x, \lambda)$ of the Hermite transformed equation

$$\widetilde{A}(t, x, \partial_t, \nabla_x, u, \lambda) = 0; \quad (t, x) \in G \tag{4.7.5}$$

for real $\lambda = (\lambda_1, \lambda_2, \ldots) \in \mathbb{K}_q(\delta) \cap \mathbb{R}^\mathbb{N}$. Moreover, assume the following:

For all $(t, x) \in G$ the function $\lambda \to u(t, x, \lambda)$ is real analytic on $\mathbb{K}_q(\delta) \cap \mathbb{R}^\mathbb{N}$
$$\tag{4.7.6}$$

and

$u(t, x, \lambda)$ and all its partial derivatives with respect to t and x which are involved in (4.7.5) are real analytic with respect to $\lambda \in \mathbb{K}_q(\delta) \cap \mathbb{R}^\mathbb{N}$. Moreover, $u(t, x, \lambda)$ is continuously differentiable in all the variables $(t, x, \lambda) \in G \times \mathbb{K}_q(\delta) \cap \mathbb{R}^\mathbb{N}$ for all orders with respect to λ and all the above orders with respect to (t, x).   (4.7.7)

Then there exists $U(t, x) \in (\mathcal{S})_{-1}$ such that $\mathcal{H}U(t, x) = u(t, x)$ and $U(t, x)$ solves (in the strong sense in $(\mathcal{S})_{-1}$) equation (4.7.4).

Proof   By assumption (4.7.6) and Lemma 4.7.2 we know that $\lambda \to u(t, x, \lambda)$ has an analytic extension to a bounded function $z \to u(t, x, z)$ for $z \in \mathbb{K}_{\hat{q}}(\hat{\delta})$ for some $\hat{q}, \hat{\delta}$. In order to apply Theorem 4.1.1 it suffices to prove that all partial derivatives of u involved in (4.7.5) are bounded and continuous on $G \times \mathbb{K}_{\hat{q}}(\hat{\delta})$. Fix $(t, x, z) \in G \times \mathbb{K}_{\hat{q}}(\hat{\delta})$, and let e_k be the kth unit vector in $\mathbb{R}^d, \varepsilon > 0$. Then by (4.7.2) we have

$$\frac{1}{\varepsilon}[u(t, x + \varepsilon e_k, z) - u(t, x, z)] = \sum_\alpha \frac{1}{\alpha!} \frac{1}{\varepsilon} (\partial_\lambda^\alpha u(t, x + \varepsilon e_k, 0) - \partial_\lambda^\alpha u(t, x, 0)) z^\alpha$$

$$= \sum_\alpha \frac{1}{\alpha!} \cdot \frac{\partial}{\partial x_k} (\partial_\lambda^\alpha u(t, x + \theta e_k, 0)) z^\alpha$$

$$\to \sum_\alpha \frac{1}{\alpha!} \partial_\lambda^\alpha \frac{\partial u}{\partial x_k}(t, x, 0) z^\alpha \quad \text{as} \quad \varepsilon \to 0,$$

by (4.7.7) and the mean value theorem ($\theta = \theta(\varepsilon) \to 0$ as $\varepsilon \to 0$). Here ∂_λ^α means that the derivatives of order α are taken with respect to $\lambda = (\lambda_1, \lambda_2, \ldots)$. This shows that $\partial/\partial x_k u(t, x, z)$ exists and is analytic and bounded with respect to z for z in some $\mathbb{K}_{q_1}(\delta_1)$. The same argument works for other derivatives with respect to t or x. By restricting ourselves to $(t, x) \in G_0 \subset\subset G$, we obtain boundedness. Hence Theorem 4.1.1 applies and the proof is complete. $\qquad\square$

As an illustration of Theorem 4.7.3 we give the following important application:

Theorem 4.7.4 Gjerde (1998). *Let* $\mathbf{K} = \exp^\diamond \mathcal{W}(x)$ *be the positive noise matrix defined in (2.6.61), (2.6.62). Let $T > 0$ and suppose there exists $\rho > 0$ such that*

$$(t, x) \to g(t, x) \in (\mathcal{S})_{-1} \quad \textit{belongs to} \quad C_b^{0+\rho}([0, T] \times \mathbb{R}^d) \qquad (4.7.8)$$

and

$$x \to f(x) \in (\mathcal{S})_{-1} \quad \textit{belongs to} \quad C_b^{2+\rho}(\mathbb{R}^d). \qquad (4.7.9)$$

Then the heat equation in a stochastic medium

$$\begin{cases} \frac{\partial U}{\partial t} = div(\mathbf{K}(x) \diamond \nabla U) + g(t, x); & (t, x) \in (0, T) \times \mathbb{R}^d \\ U(0, x) = f(x); & x \in \mathbb{R}^d \end{cases} \qquad (4.7.10)$$

has a unique $(\mathcal{S})_{-1}$-valued solution $U(t, x) \in C^{1,2}([0, T] \times \mathbb{R}^d)$.

Idea of proof Taking Hermite transforms we get the equation

$$\begin{cases} \frac{\partial u}{\partial t} = div(\widetilde{\mathbf{K}} \cdot \nabla u) + \widetilde{g}(t, x); & (t, x) \in (0, T) \times \mathbb{R}^n \\ u(0, x) = \widetilde{f}(x) \end{cases} \qquad (4.7.11)$$

in $u = u(t, x, z)$. We seek a solution for each z in some $\mathbb{K}_q(R)$. First choose $z = \lambda \in \mathbb{K}_q(R) \cap \mathbb{R}^\mathbb{N}$, with q, R to be specified later. Note that for $y \in \mathbb{R}^d$ with $|y| = 1$ we have

$$y^T \widetilde{\mathbf{K}}(x, \lambda) y = \sum_{n=0}^\infty \frac{1}{n!} y^T \widetilde{\mathcal{W}}(x, \lambda)^n y$$

$$= \sum_{n=0}^\infty \frac{1}{n!} (y^T \widetilde{\mathcal{W}}(x, \lambda) y)^n$$

$$= \exp[y^T \widetilde{\mathcal{W}}(x, \lambda) y].$$

Since the components of $\widetilde{\mathcal{W}}(x, \lambda)$ are uniformly bounded for $(x, \lambda) \in \mathbb{R}^d \times \mathbb{K}_2(1)$, we conclude that $\widetilde{\mathbf{K}}(x, \lambda)$ is *uniformly elliptic* for $x \in \mathbb{R}^d$ and has λ-uniform ellipticity constants, i.e., there exist C_1, C_2 such that

$$C_1 \leq y^T \widetilde{\mathbf{K}}(x, \lambda) y \leq C_2 \tag{4.7.12}$$

for all $(x, \lambda) \in \mathbb{R}^d \times (\mathbb{K}_2(1) \cap \mathbb{R}^N)$ and all $y \in \mathbb{R}^d$ with $|y| = 1$.

Hence from the general theory of deterministic uniformly elliptic Cauchy problems, we conclude that (4.7.11) has a unique $C^{2+\rho}$ solution $u(x, \lambda)$ for each λ.

The next step is to prove that the solution depends real–analytically on λ. This is achieved by applying known estimates from the theory of deterministic parabolic differential equations. These estimates also yield (4.7.7) and hence we can apply Theorem 4.7.3. We refer to Gjerde (1998), for the details and for applications to other SPDEs. □

4.8 A Class of Quasilinear Parabolic SPDEs

In Section 3.6 we saw how Gjessing's lemma (Theorem 2.10.7) could be used to solve quasilinear SDEs. This method was further developed in Benth and Gjessing (2000), to apply to a class of quasilinear SPDEs. To illustrate the main idea of the method, we consider the following general situation:

Let

$$L(t, x, \nabla_x)$$

be a partial differential operator operating on $x \in \mathbb{R}^d$. Let $W(t)$ be 1-parameter, 1-dimensional white noise, and consider the following SPDE:

$$\begin{cases} \frac{\partial}{\partial t} U(t, x) = L(t, x, \nabla U(t, x)) + \sigma(t) U(t, x) \diamond W(t); \quad t > 0, x \in \mathbb{R}^d, \\ U(0, x) = g(x); \quad x \in \mathbb{R}^d \end{cases}$$

$$\tag{4.8.1}$$

where $\sigma(t), g(x) = g(x, \omega)$ are given functions. Note that, in view of Theorem 2.5.9, this equation is a (generalized) Skorohod SPDE of the form

$$dU_t = L(t, x, \nabla U_t)dt + \sigma(t)U_t dB_t; \quad U_0(x) = g(x).$$

As in the proof of Theorem 3.6.1, we put

$$\sigma^{(t)}(s) = \sigma(s)\chi_{[0,t]}(s)$$

and

$$J_\sigma(t) = J_\sigma(t, \omega) = \exp^\diamond \left[-\int_0^t \sigma(s)dB(s) \right]$$

$$= \exp^\diamond \left[-\int_{\mathbb{R}} \sigma^{(t)}(s)dB(s) \right]. \tag{4.8.2}$$

The following result is a direct consequence of the method in Benth and Gjessing (1994):

Theorem 4.8.1. *Assume the following:*

$\sigma(t)$ *is a deterministic function bounded on bounded intervals in* $[0, \infty)$. $\hspace{2cm}$ (4.8.3)

For almost all ω *(fixed), the deterministic PDE* $\hspace{2cm}$ (4.8.4)

$$\begin{cases} \frac{\partial Y}{\partial t} = J_\sigma(t, \omega) \cdot L(t, x, J_\sigma^{-1}(t, \omega)\nabla Y); \quad t > 0, x \in \mathbb{R}^d \\ Y(0, x) = g(x, \omega) \end{cases} \hspace{1cm} (4.8.5)$$

has a unique solution $Y(t, x) = Y(t, x, \omega)$, *and there exists* $p > 1$ *such that*

$$Y(t, x, \cdot) \in L^p(\mu) \quad \text{for all} \quad t, x.$$

Then the quasilinear SPDE (4.8.1) has a unique solution $U(t, x, \omega)$ *with*

$$U(t, x, \cdot) \in L^q(\mu)$$

for all $q < p$. *Moreover, the solution is given by*

$$U(t, x, \cdot) = J_\sigma^{\diamond(-1)}(t) \diamond Y(t, x, \cdot) = \exp^\diamond \left[\int_0^t \sigma(s)dB_s(\cdot) \right] \diamond Y(t, x, \cdot). \hspace{0.5cm} (4.8.6)$$

Proof We proceed along the same lines as in the proof of Theorem 3.6.1. Regarding equation (4.8.1) as an equation in $(\mathcal{S})_{-1}$, we can Wick-multiply both sides of (4.8.1) by $J_\sigma(t)$. This yields, after rearranging,

$$J_\sigma(t) \diamond \frac{\partial U}{\partial t} - \sigma(t)J_\sigma(t) \diamond U \diamond W(t) = J_\sigma(t) \diamond L(t, x, \nabla U)$$

or

$$\frac{\partial}{\partial t}(J_\sigma(t) \diamond U) = J_\sigma(t) \diamond L(t, x, \nabla U). \hspace{1cm} (4.8.7)$$

Now define

$$Y(t, x) = J_\sigma(t) \diamond U(t, x). \hspace{1cm} (4.8.8)$$

By Theorem 2.10.7 we have, if $U \in L^p(\mu)$ for some $p > 1$,

$$Y(t, x) = J_\sigma(t) \cdot T_{-\sigma(t)}U(t, x) \hspace{1cm} (4.8.9)$$

and

$$J_\sigma(t) \diamond L(t, x, U) = J_\sigma(t) \cdot L(t, x, T_{-\sigma(t)}U). \hspace{1cm} (4.8.10)$$

Substituting this into (4.8.7), we get

$$\frac{\partial Y}{\partial t} = J_\sigma(t) \cdot L(t, x, J_\sigma^{-1}(t)\nabla Y) \tag{4.8.11}$$

with the initial value

$$Y(0, x) = U(0, x) = g(x). \tag{4.8.12}$$

This is an equation of the form (4.8.5) which we, by assumption, can solve for each ω, thereby obtaining a unique solution

$$Y(t, x) = Y(t, x, \omega)$$

for each $\omega \in \mathcal{S}'(\mathbb{R}^d)$. Since

$$Y(t, x, \cdot) \in L^p(\mu),$$

it follows from Theorem 2.10.7 that

$$U(t, x) = J_\sigma^{\diamond(-1)}(t) \diamond Y(t, x)$$

$$= \exp^\diamond \left[\int_0^t \sigma(s) dB_s(\cdot) \right] \diamond Y(t, x) \in L^q(\mu) \quad \text{for all} \quad q < p.$$

$$\square$$

As an example of an application of this general method, consider the SPDE

$$\begin{cases} \frac{\partial U}{\partial t}(t, x, \omega) + \nabla_x(f(t, x, U(t, x, \omega))) \\ \quad = \nu \Delta_x U(t, x, \omega) + \sigma(t)U(t, x, \omega) \diamond W(t, \omega) \\ U(0, x, \omega) = \phi_0(x, \omega), \end{cases} \tag{4.8.13}$$

where $f : \mathbb{R} \times \mathbb{R}^d \times \mathbb{R} \to \mathbb{R}, \sigma : \mathbb{R} \to \mathbb{R}$ and $\phi_0 : \mathbb{R}^d \times \mathcal{S}'(\mathbb{R}) \to \mathbb{R}$ are given (measurable) functions and ν is a positive constant.

Corollary 4.8.2 Benth and Gjessing (2000). *Assume the following:*

f is Lipschitz in u, in the sense there exists $C(t, x) > 0$ such that
$|f(t, x, u) - f(t, x, v)| \le C(t, x)|u - v|$ for all $u, v \in \mathbb{R}^d$ and that
$\sup_{t,x} C(t, x) < \infty.$ (4.8.14)

$f(t, x, 0) = 0$ *for all t, x.* (4.8.15)

$\sigma(t)$ *is bounded on bounded intervals.* (4.8.16)

There exists $p > 2$ and $A(\omega) \in L^p(\mu_1)$ such that (4.8.17)

$$\sup_x |\phi_0(x, \omega)| \le A(\omega).$$

Then for $1 \leq q < p/2$, there exists a unique solution $U(t, x, \cdot) \in L^q(\mu_1)$ of (4.8.13) (interpreted in the weak form with respect to t and x).

Idea of proof The assumptions are used to establish that there exists a solution of the corresponding deterministic equation (see (4.8.5)). Then one can apply Theorem 4.8.1. We refer to Benth and Gjessing (2000), for details and other results.

$\square$

4.9 SPDEs Driven by Poissonian Noise

So far we have discussed only SPDEs driven by Gaussian white noise. By this we mean that the underlying basic probability measure is the Gaussian measure $\mu = \mu_1$ defined by (2.1.3). From a modeling point of view one might feel that this is too special. One can easily envisage situations where the underlying noise has a different nature. For example, Kallianpur and Xiong (1994), and Gjerde (1996a), discuss stochastic models for pollution growth when the rate of increase of the concentration is a Poissonian noise.

It turns out, however, that there is a close mathematical connection between SPDEs driven by Gaussian and Poissonian noise, at least for Wick-type equations. More precisely, there is a unitary map between the two spaces, such that one can obtain the solution of the Poissonian SPDE simply by applying this map to the solution of the corresponding Gaussian SPDE. This fact seems to have evaded several researchers. On the other hand, versions of it have been known to some experts, see, e.g., Itô and Kubo (1988), Albeverio et al. (1996), Albeverio et al. (1993b), and Kondratiev et al. (1995b).

A nice, concise account of this connection was recently given by Benth and Gjerde (1995), and we will base our presentation on that paper.

Analogous to Theorem 2.1.1, we have

Theorem 4.9.1 (The Bochner–Minlos theorem II). *There exists a unique probability measure π on $\mathcal{B} = \mathcal{B}(\mathcal{S}'(\mathbb{R}^d))$ with the property*

$$\int_{\mathcal{S}'(\mathbb{R}^d)} e^{i\langle \omega, \phi \rangle} \, d\pi(\omega) = \exp\left[\int_{\mathbb{R}^d} (e^{i\phi(x)} - 1) dx \right] \qquad (4.9.1)$$

for all $\phi \in \mathcal{S}(\mathbb{R}^d)$.

The existence of the measure π follows from Theorem A.3 in Appendix A, because the function

$$g(\phi) = \exp\left[\int_{\mathbb{R}^d} (e^{i\phi(x)} - 1) dx \right]$$

satisfies the conditions (i), (ii), (iii) of this theorem. (For more information about positive definite functions, see, e.g., Berg and Forst (1975).)

Definition 4.9.2. The triple $(\mathcal{S}, \mathcal{B}, \pi)$ is called the *Poissonian white noise probability space* and π is called the *Poissonian white noise probability measure*.

Based on the measure π one can now develop a machinery similar to the one we constructed in Chapter 2 for μ. We only outline this construction here. For proofs see Hida et al. (1993), Itô (1988), and Us (1995). For simplicity we only consider the case with noise dimension $m = 1$ and state dimension $N = 1$.

Lemma 4.9.3. *Let $\phi \in \mathcal{S}(\mathbb{R}^d)$. Then*

$$E_\pi[\langle \cdot, \phi \rangle] := \int_{\mathcal{S}'(\mathbb{R}^d)} \langle \omega, \phi \rangle \, d\pi(\omega) = \int_{\mathbb{R}^d} \phi(x) \, dx \qquad (4.9.2)$$

and

$$E_\pi\left[\left(\langle \cdot, \phi \rangle - \int_{\mathbb{R}^d} \phi(x) \, dx\right)^2\right] = \|\phi\|_2^2, \qquad (4.9.3)$$

or

$$E_\pi[\langle \cdot, \phi \rangle^2] = \|\phi\|_2^2 + \left(\int_{\mathbb{R}^d} \phi(x) \, dx\right)^2.$$

Hence the map

$$J : \phi \to \langle \omega, \phi \rangle - \int_{\mathbb{R}^d} \phi(x) \, dx; \quad \phi \in \mathcal{S}(\mathbb{R}^d), \qquad (4.9.4)$$

can be extended to an isometry, denoted by J_π, from $L^2(\mathbb{R}^d)$ into $L^2(\pi)$, by the definition

$$J_\pi(\phi) = \lim_{n \to \infty} J(\phi_n) \qquad (4.9.5)$$

for $\phi \in L^2(\mathbb{R}^d)$, the limit being in $L^2(\pi)$, where $\{\phi_n\}$ is any sequence in $\mathcal{S}(\mathbb{R}^d)$ converging to ϕ in $L^2(\mathbb{R}^d)$. By (4.9.3) the limit in (4.9.5) exists and is independent of the sequence $\{\phi_n\}$.

In particular, for each $x = (x_1, \ldots, x_d) \in \mathbb{R}^d$ we can define

$$\widetilde{P}(x, \omega) := \langle \omega, \theta_x \rangle = J_\pi(\theta_x) + \prod_{j=1}^{d} x_j \in L^2(\pi), \qquad (4.9.6)$$

where $\theta_x(t_1, \ldots, t_d) = \theta_{x_1}(t_1) \cdots \theta_{x_d}(t_d)$, with

$$\theta_{x_j}(s) = \begin{cases} 1 & \text{if } 0 < s \leq x_j \\ -1 & \text{if } x_j < s \leq 0 \\ 0 & \text{otherwise.} \end{cases} \qquad (4.9.7)$$

Then $\widetilde{P}(x,\omega)$ has a right-continuous integer-valued version $P(x,\omega)$ called (*d-parameter*) *Poisson process*. The process

$$Q(x,\omega) = P(x,\omega) - \prod_{j=1}^{d} x_j = J_\pi(\chi_{[0,x_1]} \times \cdots \times \chi_{[0,x_d]}) \in L^2(\pi) \quad (4.9.8)$$

is called the *compensated (d-parameter) Poisson process*.

If $d = 1$ and $t \geq 0$, then $P(t,\cdot)$ has a Poisson distribution with mean t. Moreover, the process $\{P(t,\cdot)\}_{t\in\mathbb{R}}$ has independent increments. Define

$$Q(t) = Q(t,\omega) = P(t,\omega) - t; \ t \in \mathbb{R}.$$

Then Q is a martingale, so we can define the stochastic integral: in the usual way

$$\int_{\mathbb{R}} f(t,\omega)dQ(t)$$

of (t,ω)-measurable processes $f(t,\omega)$, adapted with respect to the filtration $\mathcal{G}_t$ generated by $Q(s,\cdot); s \leq t$, and satisfies

$$E\left[\int_{\mathbb{R}} f^2(t,\omega)dt\right] < \infty$$

(see Appendix B). Similar to the Brownian motion case (see Section 2.5), we can define the multiple stochastic integrals

$$\int_{\mathbb{R}^{nd}} g(x)dQ^{\otimes n}(x)$$

with respect to Q, for all $g \in \hat{L}^2(\mathbb{R}^{nd})$. Moreover, similar to (2.5.5) we have (see, e.g., Hida et al. (1993), Itô (1988), or Itô and Kubo (1988))

Theorem 4.9.4 (The Wiener–Itô chaos expansion). *Every $g \in L^2(\pi)$ has the (unique) representation*

$$g(\omega) = \sum_{n=0}^{\infty} \int_{\mathbb{R}^{nd}} g_n(x)dQ^{\otimes n}(x) \quad (4.9.9)$$

where $g_n \in \hat{L}^2(\mathbb{R}^{nd})$ for all n.

Moreover, we have the isometry

$$\|g\|_{L^2(\pi)}^2 = \sum_{n=0}^{\infty} n!\|g_n\|_{L^2(\mathbb{R}^{nd})}^2. \quad (4.9.10)$$

We have seen that for Gaussian white noise analysis the Hermite polynomial functionals $H_\alpha(\omega)$ defined in Section 2.2 play a fundamental role. In the Poissonian case this role is played by the *Charlier polynomial functionals* $C_\alpha(\omega)$, defined by

$$C_\alpha(\omega) = C_{|\alpha|}(\omega; \overbrace{\eta_1, \ldots, \eta_1}^{\alpha_1}, \ldots, \overbrace{\eta_k, \ldots, \eta_k}^{\alpha_k}) \qquad (4.9.11)$$

for $\alpha = (\alpha_1, \ldots, \alpha_k) \in \mathbb{N}_0^{\mathbb{N}}$ (with η_k as before, see (2.2.8)), where

$$C_n(\omega; \phi_1, \ldots, \phi_n) = \frac{\partial^n}{\partial u_1 \cdots \partial u_n} \exp\left[\left\langle \omega, \log\left(1 + \sum_{j=1}^n u_j \phi_j \right) \right\rangle \right.$$
$$\left. - \sum_{j=1}^n u_j \int_{\mathbb{R}^d} \phi_j(y) dy \right]\Bigg|_{u_1 = \cdots = u_n = 0} \qquad (4.9.12)$$

for $n \in \mathbb{N}, \phi_j \in \mathcal{S}(\mathbb{R}^d)$.

Thus, with $\bar{\phi} = \int \phi(x) dx$,

$$C_1(\omega; \phi) = \langle \omega, \phi \rangle - \bar{\phi}; \quad \phi \in \mathcal{S}(\mathbb{R}^d) \qquad (4.9.13)$$

and

$$C_2(\omega; \phi_1, \phi_2) = \langle \omega, \phi_1 \rangle \langle \omega, \phi_2 \rangle - \langle \omega, \phi_1 \phi_2 \rangle - \langle \omega, \phi_1 \rangle \bar{\phi}_2$$
$$- \langle \omega, \phi_2 \rangle \bar{\phi}_1 + \bar{\phi}_1 \bar{\phi}_2; \quad \phi_i \in \mathcal{S}(\mathbb{R}^d). \qquad (4.9.14)$$

Analogous to (2.2.29) we can express multiple integrals with respect to Q in terms of the Charlier polynomial functionals as follows:

$$\int_{\mathbb{R}^{nd}} \eta_1^{\hat{\otimes} \alpha_1} \hat{\otimes} \cdots \hat{\otimes} \eta_k^{\hat{\otimes} \alpha_k} dQ^{\otimes n} = C_\alpha(\omega), \qquad (4.9.15)$$

where $n = |\alpha|, \alpha = (\alpha_1, \ldots, \alpha_k)$.

Combined with Theorem 4.9.4 this gives the following (unique) representation of $g \in L^2(\pi)$:

$$g(\omega) = \sum_\alpha b_\alpha C_\alpha(\omega) \quad (b_\alpha \in \mathbb{R}) \qquad (4.9.16)$$

where

$$\|g\|_{L^2(\pi)}^2 = \sum_\alpha \alpha! b_\alpha^2. \qquad (4.9.17)$$

Corollary 4.9.5 Benth and Gjerde (1998a). *The map* $\mathcal{U} : L^2(\mu) \to L^2(\pi)$ *defined by*

$$\mathcal{U}\left(\sum_\alpha b_\alpha H_\alpha(\omega) \right) = \sum_\alpha b_\alpha C_\alpha(\omega) \qquad (4.9.18)$$

is isometric and surjective, i.e., unitary.

Analogous to Definition 2.3.2 we are now able to define the *Kondratiev spaces of Poissonian test functions* $(\mathcal{S})_{\rho;\pi}$ and *Poissonian distributions* $(\mathcal{S})_{-\rho;\nu}$ respectively, for $0 \le \rho \le 1$ as follows:

Definition 4.9.6 Benth and Gjerde (1998a). Let $0 \le \rho \le 1$.

a) Define $(\mathcal{S})_{\rho;\pi}$ to be the space of all $g(\omega) = \sum_\alpha b_\alpha C_\alpha(\omega) \in L^2(\pi)$ such that

$$\|g\|_{\rho,k;\pi}^2 := \sum_\alpha b_\alpha^2 (\alpha!)^{1+\rho} (2\mathbb{N})^{k\alpha} < \infty \qquad (4.9.19)$$

for *all* $k \in \mathbb{N}$.

b) Define $(\mathcal{S})_{-\rho;\pi}$ to be the space of all formal expansions

$$G(\omega) = \sum_\alpha a_\alpha C_\alpha(\omega)$$

such that

$$\|G\|_{-\rho,-k;\pi}^2 := \sum_\alpha a_\alpha^2 (\alpha!)^{1-\rho} (2\mathbb{N})^{-k\alpha} < \infty \qquad (4.9.20)$$

for some $k \in \mathbb{N}$.

As in the Gaussian case, i.e., for the spaces $(\mathcal{S})_\rho = (\mathcal{S})_{\rho;\mu}$ and $(\mathcal{S})_{-\rho} = (\mathcal{S})_{-\rho;\mu}$, the space $(\mathcal{S})_{-\rho;\pi}$ is the dual of $(\mathcal{S})_{\rho;\pi}$ when the spaces are equipped with the inductive (projective, respectively) topology given by the seminorms $\| \cdot \|_{\rho,k;\pi}$ ($\| \cdot \|_{-\rho,-k;\pi}$, respectively). If

$$G(\omega) = \sum_\alpha a_\alpha C_\alpha(\omega) \in (\mathcal{S})_{-\rho;\pi} \quad \text{and} \quad g(\omega) = \sum_\alpha b_\alpha C_\alpha(\omega) \in (\mathcal{S})_{\rho;\pi},$$

then the action of G on g is given by

$$\langle G, g \rangle = \sum_\alpha \alpha! a_\alpha b_\alpha. \qquad (4.9.21)$$

Corollary 4.9.7 Benth and Gjerde (1998a). *We can extend the map* $\mathcal{U}$ *defined in (4.9.18) to a map from* $(\mathcal{S})_{-1;\mu}$ *to* $(\mathcal{S})_{-1;\pi}$ *by putting*

$$\mathcal{U}\left(\sum_\alpha b_\alpha H_\alpha(\omega) \right) = \sum_\alpha b_\alpha C_\alpha(\omega) \qquad (4.9.22)$$

when

$$F := \sum_\alpha b_\alpha H_\alpha(\omega) \in (\mathcal{S})_{-1;\mu}.$$

Then $\mathcal{U}$ *is linear and an isometry, in the sense that*

$$\|\mathcal{U}(F)\|_{\rho,k;\pi} = \|F\|_{\rho,k;\mu} \qquad (4.9.23)$$

for all $F \in (S)_{\rho;\mu}$ *and all* $k \in \mathbb{Z}, \rho \in [-1,1]$. *Hence* $\mathcal{U}$ *maps* $(S)_{\rho;\mu}$ *onto* $(S)_{\rho;\pi}$ *for all* $\rho \in [-1,1]$.

Definition 4.9.8. If $F(\omega) = \sum_\alpha a_\alpha C_\alpha(\omega)$ and $G(\omega) = \sum_\beta b_\beta C_\beta(\omega)$ are two elements of $(S)_{-1;\pi}$, we define the *Poissonian Wick product* of F and G, $F \tilde{\diamond} G$, by

$$(F \tilde{\diamond} G)(\omega) = \sum_\gamma \left(\sum_{\alpha+\beta=\gamma} a_\alpha b_\beta \right) C_\gamma(\omega). \qquad (4.9.24)$$

As in the Gaussian case, one can now prove that the Poissonian Wick product is a commutative, associative and distributive binary operation on $(S)_{-1;\pi}$ and $(S)_{1;\pi}$.

From (4.9.24) and (4.9.22) we immediately get that the map $\mathcal{U}$ respects the Wick products.

Lemma 4.9.9. *Suppose* $F, G \in (S)_{-1;\mu}$. *Then*

$$\mathcal{U}(F \diamond G) = \mathcal{U}(F) \tilde{\diamond} \mathcal{U}(G). \qquad (4.9.25)$$

Definition 4.9.10. The *(d-parameter) Poissonian compensated white noise* $V(x) = V(x,\omega)$ is defined by

$$V(x,\omega) = \sum_{k=1}^\infty \eta_k(x) C_{\varepsilon_k}(\omega). \qquad (4.9.26)$$

Note that

$$V(x,\omega) = \mathcal{U}(W(x,\omega)) \qquad (4.9.27)$$

and

$$Q(x,\omega) = \mathcal{U}(B(x,\omega)). \qquad (4.9.28)$$

Using the isometry $\mathcal{U}$ we see that the results for the Gaussian case carry over to the Poissonian case. For example, we have

$$V(x,\omega) \in (S)_{-0;\pi} \quad \text{for all} \quad x \in \mathbb{R}^d \qquad (4.9.29)$$

and

$$V(x,\omega) = \frac{\partial^d}{\partial x_1 \cdots \partial x_d} Q(x,\omega). \qquad (4.9.30)$$

Definition 4.9.11. The *Poissonian Hermite transform* $\mathcal{H}_\pi(F)$ of an element $F(\omega) = \sum_\alpha a_\alpha C_\alpha(\omega) \in (S)_{-1;\pi}$ is defined by

$$\mathcal{H}_\pi(F) = \sum_\alpha a_\alpha z^\alpha; \ z = (z_1, z_2, \ldots) \in (\mathbb{C}^{\mathbb{N}})_c. \qquad (4.9.31)$$

(Compare with Definition 2.6.1.)

By the same proofs as in the Gaussian case (see Section 2.6), we get

Lemma 4.9.12. *If* $F, G \in (\mathcal{S})_{-1;\pi}$, *then*

$$\mathcal{H}_\pi(F \overset{\pi}{\diamond} G) = \mathcal{H}_\pi(F) \cdot \mathcal{H}_\pi(G).$$

Lemma 4.9.13. *Suppose* $g(z) = g(z_1, z_2, \ldots) = \sum_\alpha a_\alpha z^\alpha$ *is bounded and analytic on some* $\mathbb{K}_q(R)$. *Then there exists a unique* $G(\omega) \in (\mathcal{S})_{-1;\pi}$ *such that*

$$\mathcal{H}_\pi(G) = g,$$

namely

$$G(\omega) = \sum_\alpha a_\alpha C_\alpha(\omega).$$

(*Compare with Theorem 2.6.11b.*)

Lemma 4.9.14. *Suppose* $g(z) = \mathcal{H}_\pi(X)(z)$ *for some* $X \in (\mathcal{S})_{-1;\pi}$. *Let* $f : D \to \mathbb{C}$ *be an analytic function on a neighborhood* D *of* $g(0)$ *and assume that the Taylor expansion of* f *around* $g(0)$ *has real coefficients. Then there exists a unique* $Y \in (\mathcal{S})_{-1;\pi}$ *such that*

$$\mathcal{H}_\pi(Y) = f \circ g.$$

(*Compare with Theorem 2.6.12.*)

Thus we see that the machinery that has been constructed for Gaussian SPDE carries over word-for-word to a similar machinery for Poissonian SPDE.

Moreover, the operator $\mathcal{U}$ enables us to transform any Wick-type SPDE with Poissonian white noise into a Wick-type SPDE with Gaussian white noise and vice versa.

Theorem 4.9.15 Benth and Gjerde (1998a). *Let*

$$A^\diamond(t, x, \partial_t, \nabla_x, U, \omega) = 0$$

be a (Gaussian) Wick type SPDE with Gaussian white noise. Suppose $U(t, x, \omega) \in (\mathcal{S})_{-1;\mu}$ *is a solution of this equation. Then*

$$Z(t, x, \omega) := \mathcal{U}(U(t, x, \cdot)) \in (\mathcal{S})_{-1;\pi}$$

solves the Poissonian Wick type SPDE

$$A^{\overset{\pi}{\diamond}}(t, x, \partial_t, \nabla_x, Z, \omega) = 0$$

with Poissonian white noise.

Example 4.9.16 (The stochastic Poissonian Burgers equation).

Under the conditions of Theorem 4.5.4, where the spaces and the Hermite transforms are interpreted as Poissonian $((\mathcal{S})_{-1;\pi}$ and $\mathcal{H}_{\pi})$, we get that the unique gradient type solution $Z(t,x) \in (\mathcal{S})^d_{-1;\pi}$ of the Poissonian Burgers equation

$$\begin{cases} \frac{\partial Z_k}{\partial t} + \lambda \sum_{j=1}^{d} Z_j \diamond \frac{\partial Z_k}{\partial x_j} = \pi \Delta Z_k - \frac{\partial N}{\partial x_k}; & t > 0, x \in \mathbb{R}^d \\ Z_k(0,x) = -\frac{\partial R}{\partial x_k}(x); & x \in \mathbb{R}^d \end{cases} \tag{4.9.32}$$

is given by

$$Z_k = \mathcal{U}(U_k); \quad 1 \le k \le d,$$

where

$$U = (U_1, \ldots, U_d) \in (\mathcal{S})^d_{-1;\pi}$$

is the solution (4.5.24) of the Gaussian Burgers equation (4.5.23).

We refer to Benth and Gjerde (1998a), for more details and other applications.

Exercises

4.1 Prove the following special case of Theorem 4.1.1: Suppose there exist an open interval I, real numbers q, R and a function $u(x,z) : I \times \mathbb{K}_q(R) \to \mathbb{C}$ such that

$$\frac{\partial^2 u}{\partial x^2}(x,z) = \widetilde{F}(x,z) \text{ for } (x,z) \in I \times \mathbb{K}_q(R),$$

where $F(x) \in (\mathcal{S})_{-1}$ for all $x \in I$.

Suppose $\partial^2 u / \partial x^2(x,z)$ is bounded for $(x,z) \in I \times \mathbb{K}_q(R)$ and continuous with respect to $x \in I$ for each $z \in \mathbb{K}_q(R)$. Then there exists $U(x) \in (\mathcal{S})_{-1}$ such that

$$\frac{\partial^2 U}{\partial x^2} = F(x) \text{ in } (\mathcal{S})_{-1}, \text{ for all } x \in I.$$

4.2 Let σ_j be defined as in (4.4.18) and let $\widetilde{H}(s) = \int_0^s \widetilde{G}(b_r)dr$. Prove that

$$\lim_{j \to \infty} \frac{\hat{E}^x[\widetilde{H}(\sigma_j)]}{\hat{E}[\sigma_j]} = \widetilde{G}(x),$$

where $\widetilde{G}(x) = \exp \widetilde{W}_\phi(x)$.

4.3 Let $U(x) = \int_{\mathbb{R}^d} G(x,y)W(y)dy$ be the unique solution of the stochastic Poisson equation (4.2.1) on $D \subset \mathbb{R}^d$. Then $U(x) \in (\mathcal{S})^*$ for all d. For what values of d is $U(x) \in L^2(\mu)$?

4.4 Let $X(t)$ be an Itô diffusion in $\mathbb{R}^d$ with generator L. Assume that L is uniformly elliptic. For fixed deterministic functions $\phi \in L^2(\mathbb{R}^d)$, $f \in L^\infty(\mathbb{R}^d)$, consider the SPDE

$$\begin{cases} \frac{\partial U}{\partial t} = LU + W_\phi(x) \diamond U; & (t,x) \in \mathbb{R} \times \mathbb{R}^d \\ U(0,x) = f(x); & x \in D, \end{cases}$$

where $W_\phi(x)$ is smoothed white noise in $\mathbb{R}^d$ (noise in the space variables only).

a) Show that the solution is given by

$$U = U_\phi(t,x) = \hat{E}^x \left[f(X(t)) \exp^\diamond \left[\int_0^t W_\phi(X(s))ds \right] \right],$$

where $\hat{E}^x$ denotes the expectation with respect to the law Q^x of $X(t)$ when $X(0) = x$.

b) Find the limit of $U_\phi(t,x)$ in $(\mathcal{S})_{-1}$ as ϕ approaches the Dirac measure δ at 0 (in the weak star topology on the space of measures).

4.5 (Guitar string in a sand storm) In Walsh (1986), the following SDE is discussed as a model for the motion of a string of a guitar "carelessly left outdoors" and being exposed to a sand storm:

$$\begin{cases} \frac{\partial^2 U}{\partial t^2} - \frac{\partial^2 U}{\partial x^2} = W(t,x) & for(t,x) \in \mathbb{R}^+ \times \mathbb{R} \\ U(0,x) = \frac{\partial U}{\partial t}(0,x) = 0. \end{cases}$$

a) Show that the unique $(\mathcal{S})^*$ solution is given by

$$U(t,x) = \frac{1}{2} \int_0^t \int_{x+s-t}^{x+t-s} W(s,y)dyds.$$

b) In particular, if the noise only occurs in the space variable, we get

$$U(t,x) = \frac{1}{2} \int_0^t B(x+t-s) - B(x+s-t)ds.$$

4.6 Find the general solution of the SPDE

$$a\frac{\partial U}{\partial t} + b\frac{\partial U}{\partial x} = cW(t,x); \quad (t,x) \in \mathbb{R}^2,$$

where a, b, c are constants.

4.7 Study the 1-dimensional Schrödinger equation

$$\begin{cases} \frac{1}{2}U''(t) + V(t) \diamond U(t) = -f(t); & t \in [0,T] \\ U(t) = 0 & for \ t = 0, t = T, \end{cases}$$

where $V(t), f(t)$ are given stochastic distribution processes, by transforming the equation into a stochastic Volterra equation, as in Example 3.4.4.

4.8 Consider the 1-dimensional stochastic pressure equation

$$\begin{cases} (K(x) \diamond p'(x))' = -f(x); & x \in (a,b) \\ p(a) = p(b) = 0, \end{cases}$$

where $K(x) = \exp^\diamond[W_\phi(x)]$ and $f(x) \equiv 1$. Find $c_0(x)$ and $c_{\varepsilon_j}(x)$ in the chaos expansion

$$p(x,\omega) = \sum_\alpha c_\alpha(x) H_\alpha(\omega)$$

of the solution $p(x,\omega)$, by using (4.6.28).

4.9 Consider the heat equation in a 1-dimensional stochastic medium

$$\begin{cases} \frac{\partial U}{\partial t} = \frac{\partial}{\partial x}(K(x) \diamond \frac{\partial U}{\partial x}); & (t,x) \in \mathbb{R}^+ \times \mathbb{R} \\ U(0,x) = f(x); & x \in \mathbb{R}, \end{cases}$$

where f is a bounded deterministic function. Show that this equation has a unique $(\mathcal{S})_{-1}$ solution proceeding as in the sketch of the proof of Theorem 4.7.3.

4.10 a) Use the method of Theorem 4.8.1 to solve the SPDE

$$\begin{cases} \frac{\partial U}{\partial t} = \Delta U + W(t,\omega) \diamond U \\ U(0,x) = f(x), \end{cases}$$

where $f(x)$ is bounded, deterministic and $W(t) = W(t,\omega)$ is 1-parameter white noise.

b) Compare the result with the general solution (4.3.5) of the stochastic transport equation.

Chapter 5
Stochastic Partial Differential Equations Driven by Lévy Processes

5.1 Introduction

In the last decades there has been an increased interest in stochastic models based on other processes than the Brownian motion $B(t)$. In particular, the following two generalizations of Brownian motions as driving processes have been (and still are) studied:

(i) *Generalization 1: Fractional Brownian motion*

By definition a *fractional Brownian motion* with *Hurst parameter* $H \in (0,1)$, denoted by $B_H(t)$; $t \in \mathbb{R}$, is the continuous Gaussian process with mean

$$E[B_H(t)] = 0 = B_H(0); \quad t \geq 0,$$

and covariance

$$E[B_H(s)B_H(t)] = \frac{1}{2} \left(|t|^{2H} + |s|^{2H} - |t-s|^{2H} \right); \quad s,t \in \mathbb{R}.$$

If $H = 1/2$, we see that $B_H(t) = B(t)$, i.e., the classical Brownian motion. But for all other values of H the process $B_H(t)$ does not have independent increments. In fact, if $H > 1/2$ the increment $B_H(n+1) - B_H(n)$ for $n \geq 1$ is always *positively* correlated to $B_H(1) - B_H(0)$, (i.e., the process $B_H(t)$ is *persistent*). On the other hand, if $H < 1/2$, the increment $B_H(n+1) - B_H(n)$ for $n \geq 1$ is always *negatively* correlated to $B_H(1) - B_H(0)$, (i.e., the process $B_H(t)$ is *antipersistent*). This makes fractional Brownian motion a useful tool in many stochastic models, including turbulence ($H < 1/2$) and weather-related cases ($H > 1/2$).

We will not deal with stochastic partial differential equations driven by (multiparameter) fractional Brownian motion here, but we refer to the forthcoming book (Biagnini et al. (2006)) and the references therein.

H. Holden et al., *Stochastic Partial Differential Equations*, 2nd ed., Universitext, 213
DOI 10.1007/978-0-387-89488-1_5, © Springer Science+Business Media, LLC 2010

(ii) *Generalization 2: Lévy processes*

A Lévy process, denoted by $\eta(t)$; $t \geq 0$, has *stationary, independent* increments just like Brownian motion $B(t)$, but it differs from $B(t)$ in that it does not necessarily have continuous paths, (and in general it is not Gaussian). In fact, if we impose the condition that a Lévy process $\eta(t)$ is continuous, then it has the form

$$\eta(t) = a\,t + \sigma B(t); \quad t \geq 0$$

where a and σ are constants, so we are basically back in the Brownian motion case.

The possibility of jumps makes it possible to get more realistic models. For example, it has been pointed out that certain classes of processes based on (discontinuous) Lévy processes fit the stock prices data better than the classical Samuelson-Black-Scholes model based on Brownian motion. We refer to, e.g., Barndorff-Nielsen (1998), Eberlein (2001), Schoutens (2003), and Cont and Tankov (2003) for more information. Similarly, stochastic partial differential equations driven by (multiparameter) Lévy processes allow for more realistic models than in the classical Brownian motion case. It turns out that it is possible to develop a white noise theory for Lévy processes, even in the multiparameter case, and this theory shares many of the features of the classical white noise theory described in the previous chapters. It is the purpose of this chapter to explain this in detail and apply the theory to solve stochastic partial differential equations driven by (multiparameter) Lévy processes.

According to the Lévy–Itô representation (see (E.12)) any Lévy process $\eta(t)$ can be written on the form

$$\eta(t) = a\,t + \sigma B(t) + \int_0^t \int_{|z|<1} z\tilde{N}(ds,dz) + \int_0^t \int_{|z|\geq 1} zN(ds,dz), \qquad (5.1.1)$$

where $N(\cdot,\cdot)$ is the Poisson random jump measure of η and where $\tilde{N}(ds,dz) = N(ds,dz) - \nu(dz)ds$ is the compensated Poisson random measure of η, $\nu(dz)$ being the Lévy measure of η.

For simplicity we will from now on assume that

$$E[\eta^2(t)] < \infty \quad \text{for all } t \geq 0. \qquad (5.1.2)$$

This implies that $\eta(t)$ can be written on the form

$$\eta(t) = a_1\,t + \sigma B(t) + \int_0^t \int_{\mathbb{R}} z\tilde{N}(ds,dz),$$

where

$$a_1 = a + \int_{|z| \geq 1} z\nu(dz).$$

Since the Brownian motion case is covered by the previous chapters, we are now primarily interested in the so-called "pure jump" case, i.e., when $\eta(t)$ has the form

$$\eta(t) = \int_0^t \int_{\mathbb{R}} z\tilde{N}(ds, dz). \tag{5.1.3}$$

We consider also the *multiparameter* case

$$\eta(x) = \eta(x_1, \ldots, x_d); \quad x = (x_1, \ldots, x_d) \in \mathbb{R}^d$$

of a Lévy process and – more generally – of a compensated Poisson random measure $\tilde{N}(dx, dz)$.

5.2 The White Noise Probability Space of a Lévy Process ($d = 1$)

This presentation is based on [Di Nunno et al.] and [Øksendal et al.] Recall that by the Lévy–Khintchine formula a pure jump Lévy process $\eta(t) - \eta(t, \omega)$; $(t, \omega) \in [0, \infty) \times \Omega$, with $\mathrm{E}[\eta^2(t)] < \infty$ for all t, can be characterized as the (unique) càdlàg process $\eta(\cdot)$ such that

$$\mathrm{E}[\exp(iu\eta(t))] = \exp(t\Psi(u)), \tag{5.2.1}$$

where

$$\Psi(u) = \int_{\mathbb{R}} \left(e^{iuz} - 1 - iuz \right) \nu(dz); \quad u \in \mathbb{R} \tag{5.2.2}$$

where ν is the Lévy measure of $\eta(\cdot)$. This measure ν always satisfies

$$\int_{\mathbb{R}} z^2 \nu(dz) < \infty. \tag{5.2.3}$$

(See Theorem E.2.)

We will now prove a converse of the above. More precisely, given a measure ν on $\mathcal{B}(\mathbb{R}_0)$ such that (5.2.3) holds, we will construct a càdlàg stochastic process $\eta(t)$ such that (5.2.1)–(5.2.2) hold. This construction will be similar to the construction we did of Brownian motion in Section 2.1 (and of the Poisson process in Section 4.9).

Accordingly, let ν be a given measure on $\mathcal{B}(\mathbb{R}_0)$ such that

$$M := \int_{\mathbb{R}} z^2 \nu(dz) < \infty. \tag{5.2.4}$$

We now construct a pure jump Lévy process $\eta(t)$; $t \geq 0$ such that ν is the Lévy measure of $\eta(\cdot)$:

Theorem 5.2.1. *There exists a measure $\mu = \mu^{(\mathrm{L})}$ defined on the σ-algebra $\mathcal{B}(\Omega)$ of Borel subsets of $\Omega = \mathcal{S}'(\mathbb{R})$ such that*

$$\int_{\Omega} e^{i\langle \omega, f \rangle} d\mu(\omega) = \exp\left(\int_{\mathbb{R}} \Psi(f(y)) dy \right); \quad f \in \mathcal{S}(\mathbb{R}) \tag{5.2.5}$$

where

$$\Psi(w) = \int_{\mathbb{R}} \left(e^{iwz} - 1 - iwz \right) \nu(dz), \quad i = \sqrt{-1} \tag{5.2.6}$$

and $\langle \omega, f \rangle$ denotes the action of $\omega \in \mathcal{S}'(\mathbb{R})$ on $f \in \mathcal{S}(\mathbb{R})$.

Proof   The existence of $\mu^{(\mathrm{L})}$ follows from the Bochner–Minlos theorem (Appendix A). In order to apply this theorem we need to verify that the function

$$F : f \mapsto \exp\left(\int_{\mathbb{R}} \Psi(f(y)) dy \right); \quad f \in \mathcal{S}(\mathbb{R})$$

is positive definite, i.e., that

$$\sum_{j,k=1}^{n} z_j \bar{z}_k F(f_j - f_k) \geq 0 \tag{5.2.7}$$

for all complex numbers z_j and all $f_j \in \mathcal{S}(\mathbb{R})$, $n = 1, 2, \ldots$ (here $\bar{z}_k$ denotes the complex conjugate of $z_k \in \mathbb{C}$). We leave the proof of (5.2.7) to the reader (Exercise 5.1). □

Definition 5.2.2. The triple $(\Omega, \mathcal{B}(\Omega), \mu^{(\mathrm{L})})$ is called the (pure jump) *Lévy white noise probability space.*

Lemma 5.2.3. *Let $g \in \mathcal{S}(\mathbb{R})$. Then, with $\mu = \mu^{(\mathrm{L})}$, $\mathrm{E} = \mathrm{E}_\mu$ and $M = \int_{\mathbb{R}} z^2 \nu(dz)$, we have*

$$\mathrm{E}[\langle \cdot, g \rangle] = 0, \tag{5.2.8}$$

and

$$\text{Var}[\langle \cdot, g \rangle] := \text{E}[\langle \cdot, g \rangle^2] = M \int_{\mathbb{R}} g^2(y)dy. \tag{5.2.9}$$

Proof If we apply (5.2.5) to the function $f(y) = t\,g(y)$ for a fixed $t \in \mathbb{R}$, we get

$$\text{E}[\exp(i\,t\langle \omega, g \rangle)] = \exp\left(\int_{\mathbb{R}} \Psi(t\,g(y))dy \right)$$

$$= \exp\left(\int_{\mathbb{R}} \int_{\mathbb{R}} \left(e^{itzg(y)} - 1 - itzg(y) \right) \nu(dz)dy \right).$$

Assume for a moment that ν is supported on $[-R, R] \setminus \{0\}$ for some $R < \infty$ and that g has compact support. Then the expansion of the above in a Taylor series gives

$$\sum_{n=0}^{\infty} \frac{1}{n!} i^n t^n \, \text{E}[\langle \cdot, g \rangle^n] = \sum_{m=0}^{\infty} \frac{1}{m!} \left(\int_{\mathbb{R}} \int_{\mathbb{R}} \left(\sum_{k=2}^{\infty} \frac{1}{k!} i^k t^k z^k g^k(y) \right) \nu(dz)dy \right)^m$$

$$= \sum_{m=0}^{\infty} \frac{1}{m!} \left(\sum_{k=2}^{\infty} \frac{1}{k!} i^k t^k \int_{\mathbb{R}} z^k \nu(dz) \int_{\mathbb{R}} g^k(y)dy \right)^m.$$

Comparing the first order terms of t and then the second order terms (those containing t^2) we get, respectively,

$$i\,t\,\text{E}[\langle \cdot, g \rangle] = 0,$$

and

$$\frac{1}{2}(-1)t^2\text{E}[\langle \cdot, g \rangle^2] = \frac{1}{2}(-1)t^2 \int_{\mathbb{R}} z^2\nu(dz) \cdot \int_{\mathbb{R}} g^2(y)dy.$$

The case with general $g \in \mathcal{S}(\mathbb{R})$ and general ν now follows by an approximation argument. □

Using Lemma 5.2.3 we can extend the definition of $\langle \omega, f \rangle$ for $f \in \mathcal{S}(\mathbb{R})$ to any $f \in L^2(\mathbb{R})$ as follows:

If $f \in L^2(\mathbb{R})$, choose $f_n \in \mathcal{S}(\mathbb{R})$ such that $f_n \to f$ in $L^2(\mathbb{R})$. Then by (5.2.9) we see that $\{\langle \omega, f_n \rangle\}_{n=1}^{\infty}$ is a Cauchy sequence in $L^2(\mu)$ and hence convergent in $L^2(\mu)$. Moreover, the limit depends only on f and not the sequence $\{f_n\}_{n=1}^{\infty}$. We denote this limit by $\langle \omega, f \rangle$.

Now define

$$\tilde{\eta}(t) := \langle \omega, \chi_{[0,t]}(\cdot) \rangle; \quad t \in \mathbb{R} \tag{5.2.10}$$

where

$$\chi_{[0,t]}(s) = \begin{cases} 1 & \text{if } 0 \le s \le t, \\ -1 & \text{if } t \le s \le 0 \text{ except } t = s = 0, \\ 0 & \text{otherwise.} \end{cases} \tag{5.2.11}$$

Then we have

Theorem 5.2.4. *The stochastic process $\tilde{\eta}(t)$ has a càdlàg version, denoted by $\eta(t)$. This process $\eta(t)$; $t \ge 0$ is a pure jump Lévy process with Lévy measure ν.*

Proof We verify that $\tilde{\eta}$ satisfies (5.2.1), i.e., that

$$\mathrm{E}[\exp(i\,u\,\langle \omega, \chi_{[0,t]}(\cdot) \rangle)]$$

$$= \exp\left(t \int_{\mathbb{R}} \left(e^{iuz} - 1 - iuz \right) \nu(dz) \right); \quad u \in \mathbb{R}, t > 0 \tag{5.2.12}$$

By (5.2.5) we get, with $f(y) = u\,\chi_{[0,t]}(y)$, $t > 0$, $u \in \mathbb{R}$, that

$$\mathrm{E}\left[\exp\left(i\,u\,\langle \omega, \chi_{[0,t]}(\cdot) \rangle\right)\right]$$

$$= \exp\left(\int_{\mathbb{R}} \left(\int_{\mathbb{R}} \left(e^{iuz\chi_{[0,t]}(y)} - 1 - iuz\chi_{[0,t]}(y) \right) \nu(dz) \right) dy \right)$$

$$= \exp\left(t \int_{\mathbb{R}} \left(e^{iuz} - 1 - iuz \right) \nu(dz) \right)$$

which is (5.2.12).

It follows that $\tilde{\eta}$ has a càdlàg version η (see, e.g., Applebaum (2004), Theorem 2.1.7), which hence is a Lévy process with Lévy measure ν. By the Lévy–Khintchine formula (Theorem E.2) it follows that (by our choise of $\Psi(w)$ in (5.2.6)) that η is a *pure jump Lévy martingale*, i.e.,

$$\eta(t) = \int_0^t \int_{\mathbb{R}} z\tilde{N}(ds, dz). \tag{5.2.13}$$

$\square$

5.3 White Noise Theory for a Lévy Process $(d = 1)$

5.3.1 Chaos Expansion Theorems

Assume that $d = 1$ in this section, so that

$$\eta(t) = \int\limits_0^t \int\limits_{\mathbb{R}} z \tilde{N}(ds, dz); \quad t \geq 0, \qquad (5.3.1)$$

where $\tilde{N}(ds, dz) = N(ds, dz) - \nu(dz)ds$ is the compensated Poisson random measure of η. In the following we recall a chaos expansion for square integrable functionals of η, originally due to Itô (1956). The expansion is similar to the one for Brownian motion given in Theorem 2.2.7, but note that now the expansion is in terms of iterated integrals with respect to $\tilde{N}(ds, dz)$, not with respect to $d\eta(s)$. (The latter is in fact not possible in general. See Exercise 5.4).

Let λ denote the Lebesgue measure on $\mathbb{R}_+$ and let $L^2((\lambda \times \nu)^n)$ denote the space of all functions $f : (\mathbb{R}_+ \times \mathbb{R})^n \to \mathbb{R}$ such that

$$\|f\|_{(\lambda \times \nu)^n}^2 := \int\limits_{(\mathbb{R}_+ \times \mathbb{R})^n} f^2(t_1, z_1, \ldots, t_n, z_n) dt_1 \nu(dz_1) \cdots dt_n \nu(dz_n) < \infty.$$

$$(5.3.2)$$

If f is a (measurable) function from $(\mathbb{R}_+ \times \mathbb{R})^n \to \mathbb{R}$, we define its *symmetrization* $\hat{f}$ by

$$\hat{f}(t_1, z_1, \ldots, t_n, z_n) := \frac{1}{n} \sum_\sigma f(t_{\sigma(1)}, z_{\sigma(1)}, \ldots, t_{\sigma(n)}, z_{\sigma(n)}) \qquad (5.3.3)$$

where the sum is taken over all permutations σ of $\{1, 2, \ldots, n\}$. We call f *symmetric* if $f = \hat{f}$ and we define $\hat{L}^2((\lambda \times \nu)^n)$ to be the set of all symmetric functions $f \in L^2((\lambda \times \nu)^n)$. Put

$$S_n = \{(t, z) \in (\mathbb{R}_+ \times \mathbb{R})^n; \ 0 \leq t_1 \leq t_2 \leq \cdots \leq t_n < \infty\}.$$

Then if $f \in \hat{L}^2((\lambda \times \nu)^n)$, we have

$$\|f\|_{L^2((\lambda \times \nu)^n)}^2$$

$$= n! \int\limits_0^\infty \int\limits_{\mathbb{R}} \cdots \int\limits_0^{t_2} \int\limits_{\mathbb{R}} f^2(t_1, z_1, \ldots, t_n, z_n) dt_1 \nu(dz_1) \cdots dt_n \nu(dz_n)$$

$$= n! \|f\|_{L^2(S_n)}^2. \qquad (5.3.4)$$

If $g \in L^2(S_n)$, we define its *n-fold iterated integral with respect to* $\tilde{N}(\cdot, \cdot)$ *over S_n* by

$$J_n(g) := \int_0^\infty \int_{\mathbb{R}} \cdots \int_0^{t_2} \int_{\mathbb{R}} g(t_1, z_1, \ldots, t_n, z_n) \tilde{N}(dt_1, dz_1), \ldots, \tilde{N}(dt_n, dz_n).$$

(5.3.5)

If $f \in \hat{L}^2((\lambda \times \nu)^n)$, we define its *n-fold iterated integral with respect to* $\tilde{N}(\cdot, \cdot)$ *over $(\mathbb{R}_+ \times \mathbb{R})^n$* by

$$I_n(f) = n! J_n(f).$$

(5.3.6)

By applying the Itô isometry (E.11) inductively, we obtain the isometry

$$E[(I_n(f))^2] = E[(n!)^2 (J_n(f))^2] = (n!)^2 \|f_n\|^2_{L^2(S_n)}$$
$$= n! \|f_n\|^2_{L^2((\lambda \times \nu)^n)}; \quad f \in \hat{L}^2((\lambda \times \nu)^n).$$

(5.3.7)

Moreover we have the following orthogonality relation

$$E[I_n(f) I_m(g)] = \begin{cases} 0 & \text{if } n \neq m \\ n!(f, g)_{L^2((\lambda \times \nu)^n)} & \text{if } n = m \end{cases}$$

(5.3.8)

for $f, g \in \hat{L}^2((\lambda \times \nu)^n)$, where

$$(f, g)_{L^2((\lambda \times \nu)^n)} = \int_{(\mathbb{R}_+ \times \mathbb{R})^n} f(t, z) g(t, z) (\lambda \times \nu)^n (dt, dz)$$

(5.3.9)

is the inner product on $L^2((\lambda \times \nu)^n)$.

Theorem 5.3.1 (Itô (1956)) (Chaos expansion theorem I). *Let $F \in L^2(\mu)$ be measurable with respect to the σ-algebra $\mathcal{F}_\infty$ generated by $\{\eta(s); s \geq 0\}$. Then there exists a unique sequence of functions $f_n \in \hat{L}^2((\lambda \times \nu)^n)$ such that*

$$F = \sum_{n=0}^\infty I_n(f_n) \quad (\text{with } I_0(f_0) := E[F]).$$

(5.3.10)

Moreover, we have the isometry

$$E[F^2] = E[F]^2 + \sum_{n=1}^\infty n! \|f_n\|^2_{L^2((\lambda \times \nu)^n)}$$

(5.3.11)

Proof We have already outlined the proof that (5.3.10) implies (5.3.11). It remains to prove that the linear span of the family $\{I_n(f_n); f_n \in \hat{L}^2((\lambda \times \nu)^n), n = 0, 1, \dots\}$ is dense in $L^2(\mu)$. See Itô (1956) for details. For an alternative proof of this we refer to Løkka (2001). $\square$

Remark 5.3.2 If $F \in L^2(\mu)$ is measurable with respect to the σ-algebra $\mathcal{F}_T$ generated by $\{\eta(s); 0 \le s \le T\}$, then

$$\operatorname{supp} f_n(\cdot, z) \subseteq [0, T]^n \qquad \text{for all } z \in \mathbb{R}_0 \tag{5.3.12}$$

The proof of this is similar to the proof of Lemma 2.5.2. Thus in this case we get the expansion

$$F = \mathrm{E}[F]$$
$$+ \sum_{n=0}^{\infty} n! \int_0^T \int_{\mathbb{R}} \cdots \int_0^{t_2} \int_{\mathbb{R}} f_n(t_1, z_1, \dots, t_n, z_n) \tilde{N}(dt_1, dz_1) \cdots \tilde{N}(dt_n, dz_n).$$

$$\tag{5.3.13}$$

Example 5.3.3 Choose $F = \eta^2(T) = \left(\int_0^T \int_{\mathbb{R}} z \tilde{N}(dt, dz) \right)^2$. Then by the Itô formula (Theorem E.4)

$$d(\eta^2(t)) = \int_{\mathbb{R}} \left((\eta(t) + z)^2 - \eta^2(t) - 2\eta(t)z \right) \nu(dz) dt$$

$$+ \int_{\mathbb{R}} \left((\eta(t) + z)^2 - \eta^2(t) \right) \tilde{N}(dt, dz)$$

$$= \int_{\mathbb{R}} z^2 \nu(dz) dt + \int_{\mathbb{R}} (2\eta(t) + z) \, z \tilde{N}(dt, dz)$$

$$= \int_{\mathbb{R}} z^2 \nu(dz) dt + \int_{\mathbb{R}} \left(2 \int_0^t \int_{\mathbb{R}} \zeta \tilde{N}(ds, d\zeta) + z \right) z \tilde{N}(dt, dz)$$

$$= \int_{\mathbb{R}} z^2 \nu(dz) dt + \int_{\mathbb{R}} z^2 \tilde{N}(dt, dz)$$

$$+ \int_{\mathbb{R}} \left(\int_0^t \int_{\mathbb{R}} 2\zeta \, z \tilde{N}(ds, d\zeta) \right) \tilde{N}(dt, dz).$$

Hence

$$\eta^2(T) = T \int_{\mathbb{R}} z^2 \nu(dz)dt + \int_0^T \int_{\mathbb{R}} z^2 \tilde{N}(dz, dt)$$

$$+ \int_0^T \int_{\mathbb{R}} \left(\int_0^t \int_{\mathbb{R}} 2\zeta \, z \tilde{N}(ds, d\zeta) \right) \tilde{N}(dt, dz). \tag{5.3.14}$$

This is the chaos expansion of $\eta^2(T)$, with

$$I_0(f_0) = \mathrm{E}[\eta^2(T)] = T \int_{\mathbb{R}} z^2 \nu(dz),$$

$$f_1(t, z) = z^2, \quad f_2(s, \zeta, t, z) = 2\zeta \, z.$$

From now on we assume that the Lévy measure ν satisfies the following integrability condition:

For all $\epsilon > 0$ there exists $\lambda > 0$ such that

$$\int_{\mathbb{R}\backslash(-\epsilon,\epsilon)} \exp(\lambda|z|)\,\nu(dz) < \infty \tag{5.3.15}$$

This condition implies that η has finite moments of order n for all $n \geq 2$. It is trivially satisfied if ν is supported on $[-R, R]$ for some $R > 0$. The condition implies that the polynomials are dense in $L^2(\rho)$, where

$$\rho(dz) = d\rho(z) = z^2 \nu(dz) \tag{5.3.16}$$

(See [Nualart and Schoutens]). Now let $\{l_m\}_{m\geq 0} = \{1, l_1, l_2, \dots\}$ be the orthogonalization of $\{1, z, z^2, \dots\}$ with respect to the inner product of $L^2(\rho)$. Define

$$p_j(z) := \|l_{j-1}\|_{L^2(\rho)}^{-1} \, z l_{j-1}(z); \quad j = 1, 2, \dots \tag{5.3.17}$$

In particular, we have

$$p_1(z) = M^{-\frac{1}{2}} z \text{ or } z = M^{\frac{1}{2}} p_1(z)$$

Then $\{p_j(z)\}_{j=1}^\infty$ is an orthonormal basis for $L^2(\nu)$. Define the bijective map $\kappa : \mathbb{N} \times \mathbb{N} \to \mathbb{N}$ by

$$\kappa(i, j) = \frac{j + (i + j - 2)(i + j - 1)}{2} \tag{5.3.18}$$

Let $\{\xi_i(t)\}_{i=1}^\infty$ be the Hermite functions. Then if $k = \kappa(i, j)$, we define

$$\delta_k(t, z) = \delta_{\kappa(i,j)}(t, z) = \xi_i(t)p_j(z); \quad (i, j) \in \mathbb{N} \times \mathbb{N} \tag{5.3.19}$$

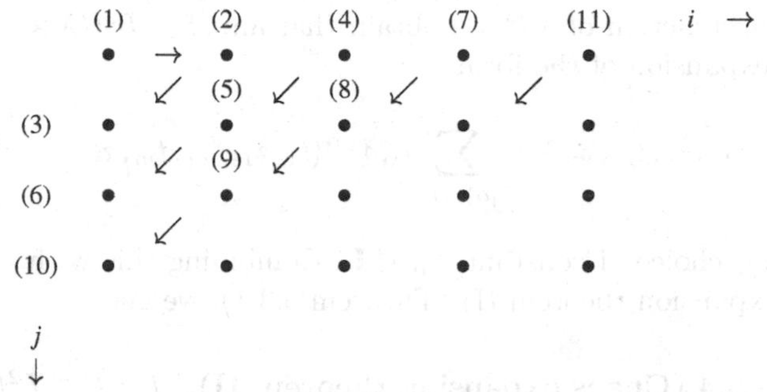

Fig. 5.1 The function k

If $\alpha \in \mathcal{J}$ with $\text{Index}(\alpha) = j$ and $|\alpha| = m$, we define the function $\delta^{\otimes \alpha}$ by

$$\delta^{\otimes \alpha}(t_1, z_1, \ldots, t_m, z_m) = \delta_1^{\otimes \alpha_1} \otimes \cdots \otimes \delta_j^{\otimes \alpha_j}(t_1, z_1, \ldots, t_m, z_m)$$

$$= \underbrace{\delta_1(t_1, z_1) \cdots \delta_1(t_{\alpha_1}, z_{\alpha_1})}_{\alpha_1 \text{ factors}} \cdots \underbrace{\delta_j(t_{m-\alpha_j+1}, z_{m-\alpha_j+1}) \cdots \delta_j(t_m, z_m)}_{\alpha_j \text{ factors}}.$$

$$\tag{5.3.20}$$

(The factors where $\alpha_i = 0$ are set equal to 1, i.e., $\delta_i^{\otimes 0} = 1$)

Finally we define the *symmetrized tensor product* of the δ_k's, denoted by $\delta^{\hat{\otimes} \alpha}$, by

$$\delta^{\hat{\otimes} \alpha}(t_1, z_1, \ldots, t_m, z_m) = \widehat{\delta^{\otimes \alpha}}(t_1, z_1, \ldots, t_m, z_m)$$

$$= \delta_1^{\otimes \alpha_1} \hat{\otimes} \cdots \hat{\otimes} \delta_j^{\otimes \alpha_j}(t_1, z_1, \ldots, t_m, z_m). \tag{5.3.21}$$

For $\alpha \in \mathcal{J}$ define

$$K_\alpha = K_\alpha(\omega) = I_{|\alpha|}\left(\delta^{\hat{\otimes} \alpha}\right)(\omega); \quad \omega \in \Omega, \tag{5.3.22}$$

where $I_{|\alpha|}$ is the iterated integral of order $m = |\alpha|$ with respect to $\tilde{N}(\cdot, \cdot)$, as defined in (5.3.5)–(5.3.6).

For example

$$K_{\epsilon(\kappa(i,j))} = I_1(\delta_{\kappa(i,j)})$$

$$= I_1(\xi_i(t)p_j(z))$$

$$= \int_0^\infty \int_{\mathbb{R}} \xi_i(t)p_j(z)\tilde{N}(dt, dz).$$

To simplify the notation we will from now on sometimes write

$$\epsilon^{(i,j)} = \epsilon^{(\kappa(i,j))} = (0, 0, \ldots, 1) \text{with 1 on place number } \kappa(i, j) \tag{5.3.23}$$

By our construction of $\delta^{\hat{\otimes}\alpha}$ we obtain that any $f \in \hat{L}^2((\lambda \times \nu)^m)$ has an orthogonal expansion of the form

$$f(t_1, z_1, \ldots, t_m, z_m) = \sum_{|\alpha|=m} c_\alpha \delta^{\hat{\otimes}\alpha}(t_1, z_1, \ldots, t_m, z_m) \qquad (5.3.24)$$

for a (unique) choice of constants $c_\alpha \in \mathbb{R}$. Combining this with (5.3.21) and the Chaos expansion theorem (I) (Theorem 5.3.1), we get

Theorem 5.3.4 (Chaos expansion theorem II). *Let $F \in L^2(\mu)$ be measurable with respect to the σ-algebra $\mathcal{F}_\infty$. Then there exists a unique sequence $\{c_\alpha\}_{\alpha \in \mathcal{J}}$ with $c_\alpha \in \mathbb{R}$ such that*

$$F(\omega) = \sum_{\alpha \in \mathcal{J}} c_\alpha K_\alpha(\omega).$$

Moreover, we have the isometry

$$\|F\|^2_{L^2(\mu)} = \sum_{\alpha \in \mathcal{J}} \alpha! c_\alpha^2. \qquad (5.3.25)$$

Example 5.3.5 Let $h \in L^2(\mathbb{R})$ be a deterministic function and define

$$F(\omega) = \int_\mathbb{R} h(s)d\eta(s) = I_1(h(s)z).$$

Since h has the expansion

$$h(s) = \sum_{i=1}^{\infty}(h, \xi_i)_{L^2(\mathbb{R})}\xi_i(s),$$

where

$$(h, \xi_i)_{L^2(\mathbb{R})} = \int_\mathbb{R} h(u)\xi_i(u)du,$$

we get the expansion

$$F(\omega) = \sum_{i=1}^{\infty}(h, \xi_i)_{L^2(\mathbb{R})}I_1(\xi_i(s)z) = M^{\frac{1}{2}}\sum_{i=1}^{\infty}(h, \xi_i)_{L^2(\mathbb{R})}K_{\epsilon(\kappa(i,1))}(\omega). \qquad (5.3.26)$$

In particular, choosing $h(s) = \mathcal{X}_{[0,t]}(s)$ for $t > 0$, we get the chaos expansion of $\eta(t)$:

$$\eta(t) = M^{\frac{1}{2}}\sum_{i=1}^{\infty}\int_0^t \xi_i(s)ds\, K_{\epsilon(\kappa(i,1))}(\omega). \qquad (5.3.27)$$

5.3.2 The Lévy–Hida–Kondratiev Spaces

We now proceed to define the Lévy analogues of the Kondratiev spaces and the Hida spaces introduced in Section 2.3:

Definition 5.3.6 (The Lévy–Hida–Kondratiev spaces of stochastic test functions and stochastic distributions) ($N = m = 1$)

a) *The Kondratiev stochastic test function spaces.* For $0 \leq \rho \leq 1$, let $(S)_\rho = (S)_\rho^{(L)}$ consist of those

$$\phi = \sum_{\alpha \in \mathcal{J}} c_\alpha K_\alpha(\omega) \in L^2(\mu^{(L)}) \quad (c_\alpha \in \mathbb{R} \text{ constants})$$

such that

$$||\phi||_{\rho,k}^2 := \sum_{\alpha \in \mathcal{J}} c_\alpha^2 (\alpha!)^{1+\rho} (2\mathbb{N})^{k\alpha} < \infty \quad \text{for } all \ k \in \mathbb{N}. \tag{5.3.28}$$

b) *The Kondratiev stochastic distribution spaces.* For $0 \leq \rho \leq 1$ let $(S)_{-\rho} = (S)_{-\rho}^{(L)}$ consist of all formal expansions

$$F = \sum_{\alpha \in \mathcal{J}} b_\alpha K_\alpha(\omega)$$

such that

$$||F||_{-\rho,-q}^2 := \sum_{\alpha \in \mathcal{J}} b_\alpha^2 (\alpha!)^{1-\rho} (2\mathbb{N})^{-q\alpha} < \infty \quad \text{for } some \ q \in \mathbb{N}. \tag{5.3.29}$$

c) The space $(S) = (S)_0^{(L)}$ is called the *Hida stochastic test function space* and the space $(S)^* = (S)_{-0}^{(L)}$ is called the *Hida distribution space*.

As in Section 2.3 we equip $(S)_\rho$ with the projective topology (intersection) defined by the norms $|| \cdot ||_{\rho,k}$; $k = 1, 2, \ldots$ and we equip $(S)_{-\rho}$ with the inductive topology (union) defined by the norms $|| \cdot ||_{-\rho,-q}$; $q = 1, 2, \ldots$. Then $(S)_{-\rho}$ becomes the dual of $(S)_\rho$ and the action of $F = \sum b_\alpha K_\alpha \in (S)_{-\rho}$ on $\phi = \sum a_\alpha K_\alpha \in (S)_\rho$ is given by

$$< F, \phi >= \sum_\alpha a_\alpha b_\alpha \alpha!. \tag{5.3.30}$$

We can now define the Lévy white noise process:

Definition 5.3.7 *The Lévy white noise process $\dot{\eta}(t)$ is defined by the expansion*

$$\dot{\eta}(t) = M^{\frac{1}{2}} \sum_{i=1}^{\infty} \xi_i(t) K_{\epsilon(\kappa(i,1))}(\omega); \ t \in \mathbb{R}. \tag{5.3.31}$$

Note that in this case $\dot\eta(t) = \sum_{\alpha \in \mathcal{J}} c_\alpha(t) K_\alpha(\omega)$, with

$$c_\alpha(t) = \begin{cases} M^{\frac{1}{2}} \xi_i(t) & if \ \alpha = (0, 0, \ldots, 1) = \epsilon^{(\kappa(i,1))} \\ 0 & \text{otherwise} \end{cases}$$

Hence

$$\sum_{\alpha \in \mathcal{J}} c_\alpha^2(t) \alpha! (2\mathbb{N})^{-q\alpha} = M \sum_{i=1}^{\infty} \xi_i^2(t) (\kappa(i,1))^{-q} < \infty$$

for all $q \geq 2$. Therefore $\dot\eta(t) \in (\mathcal{S})^*$ for all t. Note that by comparing (5.3.27) and (5.3.31) we have

$$\dot\eta(t) = \frac{d}{dt} \eta(t) \quad \text{in } (\mathcal{S})^* \tag{5.3.32}$$

which justifies the notation $\dot\eta(t)$ in (5.3.31). (Compare with (2.3.38).)

We can also define the white noise of the compensated Poisson random measure:

Definition 5.3.8 The white noise $\overset{\cdot}{\tilde{N}}(t, z)$ of the Poisson random measure $\tilde{N}(dt, dz)$ is defined by

$$\overset{\cdot}{\tilde{N}}(t, z) = \sum_{i,j \geq 1} \xi_i(t) p_j(z) K_{\epsilon^{(\kappa(i,j))}}. \tag{5.3.33}$$

Note that

$$\tilde{N}(t, U) = I_1 \left(\mathcal{X}_{[0,t]}(s) \mathcal{X}_U(z) \right)$$

$$= \sum_{i,j=1}^{\infty} (\mathcal{X}_{[0,t]}, \xi_i)_{L^2(\lambda)} (\mathcal{X}_U, p_j)_{L^2(\nu)} I_1(\xi_i(s) p_j(z))$$

$$= \sum_{i,j=1}^{\infty} \left(\int_0^t \xi_i(s) ds \right) \left(\int_U p_j(z) \nu(dz) \right) K_{\epsilon^{(\kappa(i,j))}}. \tag{5.3.34}$$

Therefore the random measure $\tilde{N}(dt, dz)$ can be given the representation

$$\tilde{N}(dt, dz) = \sum_{i,j=1}^{\infty} \xi_i(t) p_j(z) K_{\epsilon^{(\kappa(i,j))}} \nu(dz) dt. \tag{5.3.35}$$

We can therefore regard $\overset{\cdot}{\tilde{N}}(t, z)$ as the Radon–Nikodym derivative of $\tilde{N}(t, z)$ with respect to $dt \times \nu(dz)$, i.e.

$$\overset{\cdot}{\tilde{N}}(t, z) = \frac{\tilde{N}(dt, dz)}{dt \times \nu(dz)}. \tag{5.3.36}$$

Note that $\dot{\eta}(t)$ is related to $\dot{\tilde{N}}(t, z)$ by

$$\dot{\eta}(t) = \int_{\mathbb{R}} z \dot{\tilde{N}}(t, z) \nu(dz) \tag{5.3.37}$$

where the integral is an $(\mathcal{S})^*$-valued integral in the sense of Definition 2.5.5 (See Exercise 5.6).

Wick products and Skorohod integrals

Comparing the chaos expansion of Theorem 5.3.4 and Definition 5.3.6 with the expansions of Theorem 2.2.4 (with $m = 1$) and definition 2.3.2 (with $m = 1$), we see that - at least formally- the white noise theory for Brownian motion can be carried over to a white noise theory for the Lévy process $\eta(\cdot)$ through the correspondence

$$H_\alpha \leftrightarrow K_\alpha; \ \alpha \in \mathcal{J}.$$

We now explain this in more detail. We remark, however, that in spite of the close relation between the two cases, there are also significant differences.

Definition 5.3.9 (The Wick product) Let $F = \sum_{\alpha \in \mathcal{J}} a_\alpha K_\alpha \in (\mathcal{S})_{-1}^{(L)}$ and $G = \sum_{\beta \in \mathcal{J}} b_\alpha K_\beta \in (\mathcal{S})_{-1}^{(L)}$. Then the *Wick product* $F \diamond G$ of F and G is defined by the expansion

$$F \diamond G = \sum_{\alpha, \beta \in \mathcal{J}} a_\alpha b_\beta K_{\alpha+\beta} = \sum_{\gamma \in \mathcal{J}} \left(\sum_{\alpha + \beta = \gamma} a_\alpha b_\beta \right) K_\gamma. \tag{5.3.38}$$

Just as in Chapter 2 we can prove that the Wick product has the following properties:

$$F, G \in (\mathcal{S})_{-\rho} \Rightarrow F \diamond G \in (\mathcal{S})_{-\rho}; \quad 0 \le \rho \le 1 \tag{5.3.39}$$

$$F, G \in (\mathcal{S})_\rho \Rightarrow F \diamond G \in (\mathcal{S})_\rho; \quad 0 \le \rho \le 1 \tag{5.3.40}$$

$$\text{(commutative law)} \quad F \diamond G = G \diamond F \tag{5.3.41}$$

$$\text{(associative law)} \quad F \diamond (G \diamond H) = (F \diamond G) \diamond H \tag{5.3.42}$$

$$\text{(distributive law)} \quad F \diamond (G + H) = F \diamond G + F \diamond H, \ F, G, H \in (\mathcal{S})_{-1} \tag{5.3.43}$$

Similarly, in terms of the iterated integrals $I_n(\cdot)$ defined in (5.3.5)–(5.3.6), we have

$$I_n(f_n) \diamond I_m(g_m) = I_{n+m}(f_n \hat{\otimes} g_m), \tag{5.3.44}$$

where $f_n \hat{\otimes} g_m$ is the symmetrized (with respect to the (t, z)-variables) tensor product of $f_n(t_1, z_1, \ldots, t_n, z_n) \in \hat{L}^2((\lambda \times \nu)^n)$ and $g_m(t_1, z_1, \ldots, t_m, z_m) \in \hat{L}^2((\lambda \times \nu)^m)$ (see (5.3.3)).

This can be seen as follows: Suppose $f_n = \sum_{|\alpha|=n} c_\alpha \delta^{\hat{\otimes}\alpha} \in \hat{L}^2((\lambda \times \nu)^n)$ and $g_m = \sum_{|\beta|=m} b_\beta \delta^{\hat{\otimes}\beta} \in \hat{L}^2((\lambda \times \nu)^m)$. Then

$$f_n \hat{\otimes} g_m = \sum_{|\alpha|=n} \sum_{|\beta|=m} c_\alpha b_\beta \delta^{\hat{\otimes}(\alpha+\beta)} = \sum_{|\gamma|=n+m} \left(\sum_{\alpha+\beta=\gamma} c_\alpha b_\beta \right) \delta^{\hat{\otimes}\gamma}.$$

Therefore, by (5.3.22),

$$I_{n+m}(f_n \hat{\otimes} g_m) = \sum_{|\gamma|=n+m} \sum_{\alpha+\beta=\gamma} c_\alpha b_\beta K_\gamma$$

while, again by (5.3.22) and by (5.3.38)

$$I_n(f_n) \diamond I_m(g_m) = \left(\sum_{|\alpha|=n} c_\alpha K_\alpha \right) \diamond \left(\sum_{|\beta|=m} b_\beta K_\beta \right)$$

$$= \sum_{|\gamma|=n+m} \left(\sum_{\alpha+\beta=\gamma} c_\alpha b_\beta \right) K_\gamma.$$

This proves (5.3.44).

Example 5.3.10
(i) Choose $h \in L^2(\mathbb{R})$ and define

$$F(\omega) = \int_{\mathbb{R}} h(s) d\eta(s) = I_1(h(s)z).$$

Then by (5.3.44)

$$F \diamond F = I_2(h(s_1)z_1 h(s_2)z_2)$$

$$= 2 \int_0^\infty \int_{\mathbb{R}} \left(\int_0^{s_2} \int_{\mathbb{R}} h(s_1)z_1 h(s_2)z_2 \tilde{N}(ds_1, dz_1) \right) \tilde{N}(ds_2, dz_2)$$

$$= 2 \int_0^\infty \left(\int_0^{s_2} h(s_1) d\eta(s_1) \right) h(s_2) d\eta(s_2)$$

By the Itô formula (Theorem E.4) with $X(t) = \int_0^t h(s) d\eta(s)$

$$d(X^2(t)) = \int_{\mathbb{R}} \left((X(t) + h(t)z)^2 - X^2(t) - 2X(t)h(t)z \right) \nu(dz)dt$$

$$+ \int_{\mathbb{R}} \left((X(t) + h(t)z)^2 - X^2(t) \right) \tilde{N}(dt, dz)$$

$$= h^2(t) \int_{\mathbb{R}} z^2 \nu(dz) dt + \int_{\mathbb{R}} \left(2X(t)h(t)z + h^2(t)z^2 \right) \tilde{N}(dt, dz)$$

$$= 2X(t)dX(t) + h^2(t) \int_{\mathbb{R}} z^2 N(dt, dz). \qquad (5.3.45)$$

Therefore

$$F \diamond F = 2 \int_0^\infty X(s)dX(s) = F^2 - \int_0^\infty \int_{\mathbb{R}} h^2(s)z^2 N(ds, dz). \qquad (5.3.46)$$

In particular, choosing $h(s) = \mathcal{X}_{[0,t]}(s)$, we get

$$\eta(t) \diamond \eta(t) = \eta^2(t) - \int_0^t \int_{\mathbb{R}} z^2 N(ds, dz). \qquad (5.3.47)$$

Compare this to the Brownian motion case, where (see (2.4.14))

$$B(t) \diamond B(t) = B^2(t) - t.$$

(ii) It follows by the same method as above that

$$K_{\epsilon^{(i,1)}} \diamond K_{\epsilon^{(i,1)}} = K_{\epsilon^{(i,1)}} \cdot K_{\epsilon^{(i,1)}} - M^{-1} \int_0^\infty \!\!\! \int_{\mathbb{R}} \xi_i^2(s)z^2 N(ds, dz) \mathcal{X}_{i=j}, \qquad (5.3.48)$$

where, to simplify the notation, we have put

$$\epsilon^{(i,j)} := \epsilon^{(\kappa(i,j))} = (0, 0, \ldots, 1),$$

with 1 on place number $\kappa(i,j)$. (see Exercise 5.8).

The following definition of the *Skorohod integral* with respect to the Lévy process $\eta(t)$ is originally due to Kabanov (1975):

Definition 5.3.11 (The Skorohod integral)   Let $Y(t)$ be a measurable stochastic process such that

$$E[Y^2(t)] < \infty \quad \text{for all } t \geq 0$$

Then for each $t \geq 0$, $Y(t)$ has an expansion of the form

$$Y(t) = \sum_{n=0}^\infty I_n(f_n(\cdot, t))$$

where $f_n(\cdot, t) \in \hat{L}^2((\lambda \times \nu)^n)$ for $n = 1, 2, \ldots$ and $I_0(f_0(\cdot, t)) = \mathrm{E}[Y(t)]$. Let $\tilde{f}(t_1, z_1, \ldots, t_{n+1}, z_{n+1})$ be the symmetrization of

$$z_{n+1} f_n(t_1, z_1, \ldots, t_n, z_n, t_{n+1}).$$

Suppose

$$\sum_{n=0}^{\infty} (n+1)! \|\tilde{f}_n\|_{L^2((\lambda \times \nu)^{n+1})}^2 < \infty \tag{5.3.49}$$

Then we say that $Y(\cdot)$ is *Skorohod-integrable* with respect to $\eta(\cdot)$ and we define the *Skorohod integral* of $Y(\cdot)$ with respect to $\eta(\cdot)$ by

$$\int_0^\infty Y(t) \delta \eta(t) = \sum_{n=0}^{\infty} I_{n+1}(\tilde{f}_n). \tag{5.3.50}$$

Note that by (5.3.11) and (5.3.50) we have

$$\mathrm{E}\left[\left(\int_0^T Y(t) \delta \eta(t) \right)^2 \right] = \sum_{n=0}^{\infty} (n+1)! \|\tilde{f}_n\|_{L^2((\lambda \times \nu)^{n+1})}^2 < \infty \tag{5.3.51}$$

so $\int_0^T Y(t) \delta \eta(t) \in L^2(\mu)$. Moreover,

$$\mathrm{E}\left[\int_0^T Y(t) \delta \eta(t) \right] = 0. \tag{5.3.52}$$

Just as in the Brownian motion case one can now show that the Skorohod integral is an extension of the Itô integral, in the sense that if $Y(t)$ is $\mathcal{F}_t$-adapted and Skorohod-integrable, then the two integrals coincide:

Proposition 5.3.12. *Suppose $Y(t)$ is an $\mathcal{F}_t$-adapted process such that*

$$\mathrm{E}\left[\int_0^\infty Y^2(t) dt \right] < \infty. \tag{5.3.53}$$

Then $Y(\cdot)$ is both Skorohod-integrable and Itô integrable with respect to $\eta(\cdot)$, and the two integrals coincide.

Proof The proof is similar to the proof of Proposition 2.5.4 and is therefore omitted. □

Similarly, by replacing the basis elements H_α defined in Definition 2.2.1 by the basis elements K_α defined in (5.3.22) and using the chaos expansion in Definition 5.3.6, we can repeat the proof of Theorem 2.5.9 and obtain the following:

Theorem 5.3.13. *Assume that $Y(t) = \sum_{\alpha \in \mathcal{J}} c_\alpha(t) K_\alpha$ is a stochastic process which is Skorohod-integrable with respect to $\eta(\cdot)$. Then the process*

$$Y(t) \diamond \dot{\eta}(t)$$

is $(\mathcal{S})^*$-integrable and

$$\int_{\mathbb{R}} Y(t) d\eta(t) = \int_{\mathbb{R}} Y(t) \diamond \dot{\eta}(t) dt, \qquad (5.3.54)$$

where the integral on the right hand side is an $(\mathcal{S})^*$-valued integral in the sense of Definition 2.5.5.

Example 5.3.14 Let us compute the Skorohod integral

$$\int_0^T \eta(T) \delta\eta(t) = \int_0^\infty \eta(T) \mathcal{X}_{[0,T]}(t) \delta\eta(t)$$

in two ways:

(i) by using the chaos expansion (Definition 5.3.11)
(ii) by using Wick products, i.e., Theorem 5.3.13.

(i) The chaos expansion of $\eta(T)$ is

$$\eta(T) = \int_0^T \int_{\mathbb{R}} z\tilde{N}(dt, dz)$$

$$= \int_0^\infty \int_{\mathbb{R}} \mathcal{X}_{[0,T]}(t_1) z_1 \tilde{N}(dt_1, dz_1) = I_1(f_1),$$

where

$$F_1(t_1, z_1, t) = \mathcal{X}_{[0,T]}(t_1) z_1 \mathcal{X}_{[0,T]}(t)$$

Hence $\tilde{f}_1(t_1, z_1, t_2, z_2)$ is the symmetrization of $z_2 \mathcal{X}_{[0,T]}(t_1) z_1 \mathcal{X}_{[0,T]}(t_2)$ with respect to $(t_1, z_1), (t_2, z_2)$, i.e.

$$\tilde{f}_1(t_1, z_1, t_2, z_2) = z_1 z_2 \mathcal{X}_{[0,T]}(t_1) \mathcal{X}_{[0,T]}(t_2)$$

Hence

$$\int_0^T \eta(T)\delta\eta(t)$$

$$= I_2(\tilde{f}_1)$$

$$= 2 \int_0^\infty \int_{\mathbb{R}} \left(\int_0^{t_2} \int_{\mathbb{R}} z_1 z_2 \mathcal{X}_{[0,T]}(t_1) \mathcal{X}_{[0,T]}(t_2) \tilde{N}(dt_1, dz_1) \right) \tilde{N}(dt_2, dz_2)$$

$$= 2 \int_0^\infty \int_{\mathbb{R}} \left(\int_0^{t_2 \wedge T} z_1 \tilde{N}(dt_1, dz_1) \right) \mathcal{X}_{[0,T]}(t_2) z_2 \tilde{N}(dt_2, dz_2)$$

$$= 2 \int_0^\infty \int_{\mathbb{R}} \eta(t_2 \wedge T) \mathcal{X}_{[0,T]}(t_2) z_2 \tilde{N}(dt_2, dz_2)$$

$$= 2 \int_0^T \eta(t) d\eta(t) \qquad (5.3.55)$$

(ii) Using (5.3.55) and (5.3.47) we get

$$\int_0^T \eta(T)\delta\eta(t) = \int_0^T \eta(T) \diamond \dot{\eta}(t) dt$$

$$= \eta(T) \diamond \int_0^T \dot{\eta}(t) dt$$

$$= \eta(T) \diamond \eta(T) = \eta^2(T) - \int_0^T \int_{\mathbb{R}} z^2 N(ds, dz) \qquad (5.3.56)$$

This is the same as (5.3.56) by virtue of (5.3.45).

5.4 White Noise Theory for a Lévy Field ($d \geq 1$)

5.4.1 Construction of the Lévy Field

In this section we extend the definitions and results of Sections 5.2–5.3 to the multi-parameter case when time $t \in [0, \infty)$ is replaced by a point $x = (x_1, \ldots, x_d) \in \mathbb{R}^d$, for a fixed parameter dimension d. In this case the process

$\eta(x) = \eta(x, \omega)$; $(x, \omega) \in \mathbb{R}^d \times \Omega$ is called a *d-parameter (pure jump) Lévy process* or a *(pure jump) Lévy (random) field*.

We first extend the construction in Section 5.2 to arbitrary parameter dimension $d \geq 1$. As in the Brownian motion case, the construction for arbitrary $d \geq 1$ is basically the same as for $d = 1$. Nevertheless, we found it useful to go through the details for $d = 1$ first (Section 5.2), since this case is more familiar and has special interest. Now we only need to check that everything carries over to arbitrary d, mostly with only minor modifications. For completeness we give the details.

Let ν be a given measure on $\mathcal{B}_0(\mathbb{R}_0)$ such that

$$M := \int_{\mathbb{R}} z^2 \nu(dz) < \infty. \tag{5.4.1}$$

We will construct a *d-parameter Lévy process* $\eta(x)$; $x = (x_1, \ldots, x_d) \in \mathbb{R}^d$, such that ν is the *Lévy measure* of $\eta(\cdot)$, in the sense that

$$\nu(F) = \mathrm{E}[N(1, \ldots, 1; F)], \tag{5.4.2}$$

where $N(x; F) = N(x; F, \omega) : \mathbb{R}^d \times \mathcal{B}_0(\mathbb{R}_0) \times \Omega \to \mathbb{R}$ is the *jump measure* of $\eta(\cdot)$, defined by

$$
\begin{aligned}
N(x_1, \ldots, x_d; F) = \text{ } & \text{the number of jumps } \Delta\eta(u) = \eta(u) - \eta(u^-) \\
& \text{of size } \Delta\eta(u) \in F \text{ when } u_i \leq x_i; \ 1 \leq i \leq n, \\
& u = (u_1, \ldots, u_d) \in \mathbb{R}^d.
\end{aligned} \tag{5.4.3}
$$

As before let $\mathcal{S}(\mathbb{R}^d)$ denote the Schwartz space of rapidly decreasing smooth functions on $\mathbb{R}^d$ and let $\Omega = \mathcal{S}'(\mathbb{R}^d)$ be its dual, the space of tempered distributions. Then we define

Definition 5.4.1. The *d-parameter Lévy white noise probability measure* is the measure $\mu = \mu^{(L)}$ defined on the Borel σ-algebra $\mathcal{B}(\Omega)$ of subsets of Ω by

$$\int_{\Omega} e^{i \langle \omega, f \rangle} d\mu(\omega) = \exp\left[\int_{\mathbb{R}^d} \Psi(f(y)) dy \right]; \ f \in \mathcal{S}(\mathbb{R}^d) \tag{5.4.4}$$

where

$$\Psi(u) = \int_{\mathbb{R}} \left(e^{i \, u \cdot z} - 1 - i \, u \cdot z \right) \nu(dz); \ u \in \mathbb{R} \tag{5.4.5}$$

and $< \omega, f >$ denotes the action of $\omega \in \mathcal{S}'(\mathbb{R}^d)$ on $f \in \mathcal{S}(\mathbb{R}^d)$. The triple $(\Omega, \mathcal{B}(\Omega), \mu^{(L)})$ is called the *d-parameter Lévy white noise probability space*. For simplicity of notation we will write $\mu^{(L)} = \mu$ from now on.

Remark   The existence of μ follows from the Bochner–Minlos theorem (see Appendix A). In order to apply this theorem we need to verify that the map

$$F : f \mapsto \exp\left[\int_{\mathbb{R}^d} \Psi(f(y))dy\right]; \quad f \in \mathcal{S}(\mathbb{R}^d)$$

is positive definite on $\mathcal{S}(\mathbb{R}^d)$, i.e., that

$$\sum_{j,k=1}^{n} z_j \bar{z}_k F(f_j - f_k) \geq 0 \tag{5.4.6}$$

for all complex numbers z_j and all $f_j \in \mathcal{S}(\mathbb{R}^d)$, $n = 1, 2, \ldots$. We leave the proof of (5.4.6) to the reader (Exercise 5.1).

Lemma 5.4.2.   *Let $g \in \mathcal{S}(\mathbb{R}^d)$ and put $M = \int_{\mathbb{R}} z^2 \nu(dz) < \infty$. Then, with $E = E_\mu$,*

$$E\left[< \cdot, g >\right] = 0, \tag{5.4.7}$$

and

$$\mathrm{Var}_\mu\left[< \cdot, g >\right] := E\left[< \cdot, g >^2\right] = M \int_{\mathbb{R}^d} g^2(y)dy. \tag{5.4.8}$$

Proof   The proof is similar to the proof of Lemma 5.2.3: If we apply (5.4.4) to the function $f(y) = t\, g(y)$ for a fixed $t \in \mathbb{R}$ and for $y \in \mathbb{R}^d$ we get

$$E\left[\exp[i\, t < \omega, g >]\right] = \exp\left[\int_{\mathbb{R}^d} \Psi(t\, g(y))dy\right]$$

$$= \exp\left[\int_{\mathbb{R}^d} \int_{\mathbb{R}} \left(e^{itzg(y)} - 1 - itzg(y)\right)\nu(dz)dy\right].$$

Assume for a moment that ν is supported on $[-R, R]^d \setminus \{0\}$ for some $R < \infty$ and that g has compact support. Then by expansion of the above in a Taylor series we get

$$\sum_{n=0}^{\infty} \frac{1}{n!} i^n t^n E\left[< \cdot, g >^n\right]$$

$$= \sum_{m=0}^{\infty} \frac{1}{m!}\left\{\int_{\mathbb{R}^d}\left(\int_{\mathbb{R}}\left(\sum_{k=2}^{\infty}\frac{1}{k!}i^k t^k z^k g^k(y)\right)\nu(dz)\right)dy\right\}^m$$

$$= \sum_{m=0}^{\infty} \frac{1}{m!} \left\{ \sum_{k=2}^{\infty} \frac{1}{k!} i^k t^k \int_{\mathbb{R}} z^k \nu(dz) \int_{\mathbb{R}^d} g^k(y) dy \right\}^m.$$

Comparing the terms containing the first order term t and the second order term t^2, we get

$$i t \, \mathrm{E} \left[< \cdot, g > \right] = 0$$

and

$$\frac{1}{2}(-1)t^2 \mathrm{E} \left[< \cdot, g >^2 \right] = \frac{1}{2}(-1)t^2 \int_{\mathbb{R}} z^2 \nu(dz) \int_{\mathbb{R}^d} g^2(y) dy.$$

The case with general $g \in \mathcal{S}(\mathbb{R}^d)$ and general ν now follows by an approximation argument. $\qquad \square$

Using Lemma 5.4.2 we can extend the definition of $< \omega, f >$ from $f \in \mathcal{S}(\mathbb{R}^d)$ to any $f \in L^2(\mathbb{R}^d)$ as follows:

If $f \in L^2(\mathbb{R}^d)$ choose $f_n \in L^2(\mathbb{R}^d)$ such that $f_n \to f$ in $L^2(\mathbb{R}^d)$. Then by (5.4.8) we see that $\{ < \omega, f_n > \}_{n=1}^{\infty}$ is a Cauchy sequence in $L^2(\mu)$ and hence convergent in $L^2(\mu)$. Moreover, the limit depends only on f and not on the sequence $\{f_n\}_{n \geq 1}$. We denote this limit by $< \omega, f >$.

In general, if $\zeta(x) = \zeta(x, \omega)$; $x \in \mathbb{R}^d$ is a multiparameter stochastic process we define its *increments* $\Delta_h \zeta(x)$ for $h = (h_1, \ldots, h_d) \in \mathbb{R}_+^d$ as follows:

$\Delta_h \zeta(x)$

$$= \zeta(x + h) - \sum_{i=1}^{d} \zeta(x_1 + h_1, \ldots, x_i + \hat{h}_i, \ldots, x_d + h_d)$$

$$+ \sum_{\substack{i,j=1 \\ i \neq j}}^{d} \zeta(x_1 + h_1, \ldots, x_i + \hat{h}_i, \ldots, x_j + \hat{h}_j, \ldots, x_d + h_d) - \cdots + (-1)^d \zeta(x)$$

where the notation $\hat{}$ means that this term is deleted. In other words, if we regard $\zeta(x)$ as a random set function $\check{\zeta}$ by defining

$$\check{\zeta}\big((-\infty, x_1] \times \cdots \times (-\infty, x_d] \big) = \zeta(x_1, \ldots, x_d) \qquad (5.4.10)$$

and extending to all rectangles $(a_1, b_1] \times \cdots \times (a_d, b_d]$ by additivity, then

$$\Delta_h \zeta(x) = \check{\zeta}\big((x_1, x_1 + h_1] \times \cdots \times (x_d, x_d + h_d] \big). \qquad (5.4.11)$$

For example, if $d = 2$ we get

$$\Delta_h \zeta(x) = \zeta(x_1 + h_1, x_2 + h_2) - \zeta(x_1, x_2 + h_2) - \zeta(x_1 + h_1, x_2) + \zeta(x_1, x_2). \qquad (5.4.12)$$

Theorem 5.4.3 (Construction of a multiparameter pure jump Lévy process).
For $x = (x_1, \ldots, x_d) \in \mathbb{R}^d$ define

$$\tilde{\eta}(x) = \tilde{\eta}(x_1, \ldots, x_d) = \; <\omega, \mathcal{X}_{[0,x]}(\cdot)> \qquad (5.4.13)$$

where

$$\mathcal{X}_{[0,x]}(y) = \mathcal{X}_{[0,x_1]}(y_1) \cdots \mathcal{X}_{[0,x_d]}(y_d); \; y = (y_1, \ldots, y_d) \in \mathbb{R}^d \qquad (5.4.14)$$

with

$$\mathcal{X}_{[0,x_i]}(y_i) = \begin{cases} 1 & if \; 0 \le y_i \le x_i \; or \; x_i \le y_i \le 0, except \; x_i = y_i = 0 \\ 0 & otherwise \end{cases}$$

$$(5.4.15)$$

Then $\tilde{\eta}(x)$ has the following properties

$$\tilde{\eta}(x) = 0 \; \textit{if one of the components of } x \textit{ is } 0, \qquad (5.4.16)$$
$$\tilde{\eta} \textit{ has independent increments (see below)}, \qquad (5.4.17)$$
$$\tilde{\eta} \textit{ has stationary increments}, \qquad (5.4.18)$$
$$\tilde{\eta} \textit{ has a càdlàg version, denoted by } \eta. \qquad (5.4.19)$$

Proof The property (5.4.16) follows directly from (5.4.13)–(5.4.15). To prove (5.4.17) it suffices to prove that if $f, g \in \mathcal{S}(\mathbb{R}^d)$ have disjoint supports, then

$$<\omega, f> \; \text{and} \; <\omega, g> \; \text{are independent.}$$

To this end, it suffices to prove that, for all $\alpha, \beta \in \mathbb{R}$,

$$\mathrm{E}\left[e^{i\alpha<\omega,f>}e^{i\beta<\omega,g>}\right] = \mathrm{E}\left[e^{i\alpha<\omega,f>}\right]\mathrm{E}\left[e^{i\beta<\omega,g>}\right]. \qquad (5.4.20)$$

By (5.4.4) we have

$$\mathrm{E}\left[e^{i\alpha<\omega,f>}e^{i\beta<\omega,g>}\right] = \mathrm{E}\left[e^{i<\omega,\alpha f+\beta g>}\right]$$

$$= \exp\left[\int_{\mathbb{R}^d}\left(\int_{\mathbb{R}}\left\{e^{i(\alpha f+\beta g)z}-1-i(\alpha f+\beta g)z\right\}\nu(dz)\right)dy\right]$$

$$= \exp\left[\int_{\mathbb{R}}\left(\int_{\text{supp } f}\left\{e^{i\alpha f(y)z}-1-i\alpha f(y)z\right\}dy\right.\right.$$

$$+ \int_{\text{supp } g}\left.\left.\left\{e^{i\beta g(y)z}-1-i\beta g(y)z\right\}dy\right)\nu(dz)\right]$$

$$= \exp\left[\int_{\mathbb{R}}\left(\int_{\text{supp } f}\left\{e^{i\alpha f(y)z} - 1 - i\alpha f(y)z\right\}dy\right)\nu(dz)\right]$$

$$\cdot \exp\left[\int_{\mathbb{R}}\left(\int_{\text{supp } g}\left\{e^{i\beta g(y)z} - 1 - i\beta g(y)z\right\}dy\right)\nu(dz)\right]$$

$$= \mathrm{E}\left[e^{i\alpha<\omega,f>}\right] \cdot \mathrm{E}\left[e^{i\beta<\omega,g>}\right],$$

which proves (5.4.20).

To prove (5.4.18) it suffices to prove that if $f \in \mathcal{S}(\mathbb{R}^d)$, $h \in \mathbb{R}^d$, and we define $f_h(x) = f(x+h)$; $x \in \mathbb{R}^d$, then $< \omega, f >$ and $< \omega, f_h >$ have the same distribution. To this end it suffices to prove that, for all $\alpha \in \mathbb{R}$,

$$\mathrm{E}\left[e^{i\alpha<\omega,f>}\right] = \mathrm{E}\left[e^{i\alpha<\omega,f_h>}\right].$$

By (5.4.4) this is equivalent to the equation

$$\int_{\mathbb{R}^d}\left(\int_{\mathbb{R}}\left\{e^{i\alpha f(y)z} - 1 - i\alpha f(y)z\right\}\nu(dz)\right)dy$$

$$= \int_{\mathbb{R}^d}\left(\int_{\mathbb{R}}\left\{e^{i\alpha f(y+h)z} - 1 - i\alpha f(y+h)z\right\}\nu(dz)\right)dy$$

which follows from the translation invariance of Lebesgue measure dy.

The existence of a càdlàg version $\eta(x)$ of $\tilde{\eta}(x)$; $x \in \mathbb{R}^d$ follows from Theorem 2.1.7 in Applebaum (2004). The proof that (5.4.4) implies that ν is the Lévy measure of η is left for the reader (Exercise 5.10). □

In view of Theorem 5.4.4 it is natural to call the process $\eta(x)$ a *multiparameter* (pure jump) *Lévy process/martingale* or a (pure jump) *Lévy random field*. If $d = 1$, then $\eta(x) = \eta(t)$ is the classical pure jump Lévy martingale (see (5.2.13)).

The process $\eta(x)$; $x \in \mathbb{R}^d$ is the pure jump Lévy field that we will work with from now on.

By our choice (5.4.5) of the function $\Psi(u)$, it follows by the (multi-parameter) Lévy-Khintchine formula (see Theorem E.2 for the $d = 1$ case) that $\eta(x)$ is a pure jump Lévy martingale of the form

$$\eta(x) = \int_0^x\int_{\mathbb{R}} z\tilde{N}(dy, dz); \quad x \in \mathbb{R}^d \tag{5.4.21}$$

Note that if $\{\Delta_k\}_{k=1}^m$ is a disjoint family of disjoint rectangeles (boxes) in $\mathbb{R}^d$ and $c_k \in \mathbb{R}$ are constants, then

$$< \omega, \sum_k c_k \mathcal{X}_{\Delta_k}(\cdot) > = \sum_k c_k < \omega, \mathcal{X}_{\Delta_k}(\cdot) >$$

$$= \sum_k c_k \eta(\Delta_k) = \int_{\mathbb{R}} \left(\sum_k c_k \mathcal{X}_{\Delta_k}(x) \right) d\eta(x).$$

Thus by an approximation argument we obtain that

$$< \omega, h > = \int_{\mathbb{R}^d} h(x) d\eta(x)$$

$$= \int_{\mathbb{R}^d} \int_{\mathbb{R}} h(x) \, z \, \tilde{N}(dx, dz) \qquad (5.4.22)$$

for all (deterministic) $h \in L^2(\lambda)$, where λ denotes Lebesgue measure on $\mathbb{R}^d$.

5.4.2 Chaos Expansions and Skorohod Integrals $(d \geq 1)$

If $f(x^{(1)}, z_1, \ldots, x^{(n)}, z_n)$ is a function from $(\mathbb{R}^d \times \mathbb{R})^n$ into $\mathbb{R}$, we define its *symmetrization* $\hat{f}$ by

$$\hat{f}(x^{(1)}, z_1, \ldots, x^{(n)}, z_n) = \frac{1}{n!} \sum_\sigma f(x^{(\sigma_1)}, z_{\sigma_1}, \ldots, x^{(\sigma_n)}, z_{\sigma_n}) \qquad (5.4.23)$$

the sum being taken over all permutations σ of $\{1, 2, \ldots, n\}$. In other words, $\hat{f}$ is the symmetrization of f with respect to the n variables

$$y_1 = (x^{(1)}, z_1), \ldots, y_n = (x^{(n)}, z_n).$$

We let $\hat{L}^2((\lambda \times \nu)^n)$ denote the set of all symmetric functions $f \in L^2((\lambda \times \nu)^n)$.
Put

$$G = \left\{ (x^{(1)}, z_1, \ldots, x^{(n)}, z_n); \ x_j^{(1)} \leq x_j^{(2)} \leq \cdots \leq x_j^{(n)} \text{ for all } j = 1, 2, \ldots, d \right\}.$$
$$(5.4.24)$$

For $f \in \hat{L}^2((\lambda \times \nu)^n)$ define the n times iterated integral of f by

$$I_n(f) = n! \int_{G_n} f(x^{(1)}, z_1, \ldots, x^{(n)}, z_n) \tilde{N}(dx^{(1)}, dz_1) \cdots \tilde{N}(dx^{(n)}, dz_n).$$
$$(5.4.25)$$

By proceeding along the same lines as in Section 5.3, we obtain the following result (compare with Theorem 5.3.1):

Theorem 5.4.4 (Chaos expansion I for Lévy fields).
(i) *Every $F \in L^2(P)$ has a unique expansion*

$$F = \sum_{n=0}^{\infty} I_n(f_n); \quad f_n \in \hat{L}^2 \left((\lambda \times \nu)^n \right) \qquad (5.4.26)$$

where, by convention $I_0(f_0) = f_0$ when f_0 is a constant.

(ii) *Moreover, we have the isometry*

$$\|F\|^2_{L^2(P)} = \sum_{n=0}^{\infty} n! \|f_n\|^2_{L^2((\lambda \times \nu)^n)}. \qquad (5.4.27)$$

Example 5.4.5 Fix $x \in \mathbb{R}^d$. Then $F := \eta(x) \in L^2(P)$ has the expansion

$$\eta(x) = \int_0^x \int_{\mathbb{R}} z \tilde{N}(dy, dz) = I_1(f_1),$$

with
$$f_1(y, z) = \mathcal{X}_{[0,x]}(y) z = \mathcal{X}_{[0,x_1]}(y_1) \cdots \mathcal{X}_{[0,x_d]}(y_d),$$

where $x = (x_1, \ldots, x_d)$, $y = (y_1, \ldots, y_d)$.

We can now proceed to define Skorohod integrals in the same way as in Section 5.3 (see Definition 5.3.11):

Definition 5.4.6 (Skorohod integrals ($d \geq 1$)) Let $Y(x)$; $x \in \mathbb{R}^d$ be a stochastic process such that

$$\mathrm{E}\left[Y^2(x)\right] < \infty \text{ for all } x \in \mathbb{R}^d. \qquad (5.4.28)$$

Then for each $x \in \mathbb{R}^d$ $Y(x)$ has an expansion of the form

$$Y(x) = \sum_{n=0}^{\infty} I_n(f_n(\cdot, x)); \qquad (5.4.29)$$

where $f_n(\cdot, x) \in \hat{L}^2 \left((\lambda \times \nu)^n \right)$, with x as a parameter. Suppose that

$$\sum_{n=0}^{\infty} (n+1)! \|\tilde{f}_n\|^2_{L^2((\lambda \times \nu)^{n+1})} < \infty \qquad (5.4.30)$$

where $\tilde{f}_n(x^{(1)}, z_1, \ldots, x^{(n)}, z_n, x, z)$ is the symmetrization of

$$z f_n(x^{(1)}, z_1, \ldots, x^{(n)}, z_n, x)$$

with respect to the $n + 1$ variables

$$y_1 = (x^{(1)}, z_1), \ldots, y_n = (x^{(n)}, z_n), y_{n+1} = (x, z) := (x^{(n+1)}, z_{n+1}).$$

Then the *Skorohod integral* of Y with respect to η is defined by

$$\int_{\mathbb{R}^d} Y(x) \delta \eta(x) = \sum_{n=0}^{\infty} I_{n+1}(\tilde{f}_n). \tag{5.4.31}$$

Just as in the case $d = 1$ we can now prove that if $Y(x)$ is Skorohod-integrable, then

$$E\left[\left(\int_{\mathbb{R}^d} Y(x) \delta \eta(x) \right)^2 \right] = \sum_{n=0}^{\infty} (n+1)! \|\tilde{f}_n\|^2_{L^2((\lambda \times \nu)^{n+1})} < \infty \tag{5.4.32}$$

and

$$E\left[\int_{\mathbb{R}^d} Y(x) \delta \eta(x) \right] = 0. \tag{5.4.33}$$

See (5.3.52) and (5.3.53). Moreover, if $Y(\cdot)$ is *adapted*, in the sense that for all x the random variable $Y(x)$ is measurable with respect to the σ-algebra $\mathcal{F}_x$ generated by

$$\{\eta(y); \; y_1 \leq x_1, \ldots, y_d \leq x_d\}$$

and $E\left[\int_{\mathbb{R}^d} Y^2(x) dx \right] < \infty$, then $Y(x)$ is Skorohod-integrable and

$$\int_{\mathbb{R}^d} Y(x) \delta \eta(x) = \int_{\mathbb{R}^d} Y(x) d\eta(x),$$

where the integral on the right hand side is the Itô integral.

Proceeding as in Section 5.3 we assume from now on that ν satisfies condition (5.3.15). And we let $\{l_m\}_{m \geq 0} = \{1, l_1, l_2, \ldots\}$ be the orthogonalization of the polynomials $\{1, z, z^2, \ldots\}$ with respect to the inner product of $L^2(\rho)$, where

$$d\rho(z) = z^2 \nu(dz); \; z \in \mathbb{R}.$$

Define, as in (5.3.17),

$$p_j(z) := \|l_{j-1}\|^{-1}_{L^2(\rho)} z \, l_{j-1}(z); \; j = 1, 2, \ldots \tag{5.4.34}$$

where

$$M = \int_{\mathbb{R}} z^2 \nu(dz) < \infty.$$

Then $\{p_j(z)\}_{j=1}^{\infty}$ is an orthonormal basis of $L^2(\nu)$. Note that with this definition we have

$$p_1(z) = M^{-\frac{1}{2}}z \text{ or } z = M^{\frac{1}{2}}p_1(z); \ z \in \mathbb{R}. \tag{5.4.35}$$

As before let $\{\xi_i(t)\}_{i=1}^{\infty}$ be the Hermite functions on $\mathbb{R}$ and for

$$\gamma = (\gamma_1, \ldots, \gamma_d) \in \mathbb{N}^d$$

let

$$\xi_\gamma = \xi_{\gamma_1} \otimes \xi_{\gamma_2} \otimes \cdots \otimes \xi_{\gamma_d} \tag{5.4.36}$$

i.e.

$$\xi_\gamma(x) = \xi_{\gamma_1}(x_1)\xi_{\gamma_2}(x_2)\cdots\xi_{\gamma_d}(x_d); \ x = (x_1, \ldots, x_d) \in \mathbb{R}^d$$

Then $\{\xi_\gamma\}_{\gamma \in \mathbb{N}^d}$ is an orthonormal basis for $L^2(\mathbb{R}^d)$. As in (2.2.7) we may assume that $\mathbb{N}^d$ is *ordered*, $\mathbb{N}^d = \{\gamma^{(1)}, \gamma^{(2)}, \ldots \}$, in such a way that

$$i < j \ \Rightarrow \ \gamma_1^{(i)} + \cdots + \gamma_d^{(i)} \leq \gamma_1^{(j)} + \cdots + \gamma_d^{(j)} \tag{5.4.37}$$

and from now on we write (with abuse of notation)

$$\xi_i(x) := \xi_{\gamma(i)}(x); \ i = 1, 2, \ldots; \ x \in \mathbb{R}^d \tag{5.4.38}$$

With $\kappa : \mathbb{N} \times \mathbb{N} \to N$ as in (5.3.18), we define

$$\delta_{\kappa(i,j)}(x, z) = \xi_i(x)p_j(z); \ (i,j) \in \mathbb{N} \times \mathbb{N} \tag{5.4.39}$$

for $(x, z) \in \mathbb{R}^d \times \mathbb{R}$.

As before let $\mathcal{I}$ be the set of all multi-indices $\alpha = (\alpha_1, \ldots, \alpha_m) \in \mathcal{I}$ with

$$\alpha_i \in \mathbb{N} \cup \{0\}, \text{ for } i = 1, \ldots, m, \ m = 1, 2, \ldots$$

For $\alpha = (\alpha_1, \ldots, \alpha_m) \in \mathcal{I}$ with $\text{Index}(\alpha) := \max\{i; \ \alpha_i \neq 0\} = j$ and

$$|\alpha| := \alpha_1 + \cdots + \alpha_j = m$$

we define the function $\delta^{\otimes\alpha}$ by

$$\delta^{\otimes\alpha}(x^{(1)}, z_1, \ldots, x^{(m)}, z_m) = \delta_1^{\otimes\alpha_1} \otimes \cdots \otimes \delta_j^{\otimes\alpha_j}(x^{(1)}, z_1, \ldots, x^{(m)}, z_m)$$
$$= \underbrace{\delta_1(x^{(1)}, z_1)\cdots\delta_1(x^{(\alpha_1)}, z_{\alpha_1})}_{\alpha_1 \text{ factors}}\cdots\underbrace{\delta_j(x^{(m-\alpha_j+1)}, z_{m-\alpha_j+1})\cdots\delta_j(x^{(m)}, z_m)}_{\alpha_j \text{factors}}$$

$$\tag{5.4.40}$$

As usual we set $\delta_i^{\otimes 0} = 1$.

We thus define the symmetrized tensor product of the $\delta_k's$, denoted by $\delta^{\hat{\otimes}\alpha}$, as follows:

$$\delta^{\hat{\otimes}\alpha}(x^{(1)}, z_1, \ldots, x^{(m)}, z_m) = \widehat{(\delta^{\otimes\alpha})}(x^{(1)}, z_1, \ldots, x^{(m)}, z_m)$$
$$= \delta_1^{\hat{\otimes}\alpha_1} \hat{\otimes} \cdots \hat{\otimes} \delta_j^{\hat{\otimes}\alpha_j}(x^{(1)}, z_1, \ldots, x^{(m)}, z_m),$$
$$(5.4.41)$$

where the symbol $\hat{\ }$ denotes symmetrization.

Definition 5.4.7 For $\alpha \in \mathcal{I}$ define

$$K_\alpha = K_\alpha(\omega) := I_{|\alpha|}\left(\delta^{\hat{\otimes}\alpha}\right). \qquad (5.4.42)$$

As in Section 5.3 we simplify the notation and put

$$\epsilon^{(i,j)} = \epsilon^{(\kappa(i,j))} = (0, 0, \ldots, 1), \qquad (5.4.43)$$

with 1 on place number $\kappa(i,j)$, where $\kappa(i,j)$ is defined as in (5.3.18).

Example 5.4.8 By (5.4.40) and (5.4.35) we have

$$K_{\epsilon^{(i,j)}} = K_{\epsilon^{\kappa(i,j)}} = I_1\left(\delta^{\hat{\otimes}\epsilon^{\kappa(i,j)}}\right)$$
$$= I_1\left(\delta_{\kappa(i,j)}\right) = I_1\left(\xi_i(x)p_j(z)\right)$$
$$= \int_{\mathbb{R}^d} \int_{\mathbb{R}} \xi_i(x)p_j(z)\tilde{N}(dx, dz)$$
$$= M^{-\frac{1}{2}} \int_{\mathbb{R}^d} \int_{\mathbb{R}} \xi_i(x)z\tilde{N}(dx, dz). \qquad (5.4.44)$$

Let $f_n = \sum_{|\alpha|=n} a_\alpha \delta^{\hat{\otimes}\alpha} \in \hat{L}^2\left((\lambda \times \nu)^n\right)$ and $g_m = \sum_{|\beta|=m} b_\beta \delta^{\hat{\otimes}\beta} \in \hat{L}^2\left((\lambda \times \nu)^n\right)$. Then

$$f_n \hat{\otimes} g_m = \sum_{|\alpha|=n} \sum_{|\beta|=m} a_\alpha b_\beta \delta^{\hat{\otimes}(\alpha+\beta)}$$
$$= \sum_{|\gamma|=n+m} \left(\sum_{\alpha+\beta=\gamma} a_\alpha b_\beta\right) \delta^{\hat{\otimes}\gamma}.$$

Hence, by Definition 5.4.7,

$$I_n(f_n) = \sum_{|\alpha|=n} a_\alpha I_n\left(\delta^{\hat{\otimes}\alpha}\right) = \sum_{|\alpha|=n} a_\alpha K_\alpha \qquad (5.4.45)$$

and

$$I_{n+m}\left(f_n \hat{\otimes} g_m\right) = \sum_{|\gamma|=n+m} \left(\sum_{\alpha+\beta=\gamma} a_\alpha b_\beta\right) K_\gamma \qquad (5.4.46)$$

The identity (5.4.45) gives the link between Chaos expansion I (Theorem 5.4.4) and the following expansion:

Theorem 5.4.9 (Chaos expansion II for Lévy fields).

(i) *Every* $F \in L^2(P)$ *has a unique representation*

$$F = \sum_{\alpha \in \mathcal{I}} c_\alpha K_\alpha; \ c_\alpha \in \mathbb{R}. \qquad (5.4.47)$$

(ii) *Moreover, we have the isometry*

$$\|F\|_{L^2(P)}^2 = \sum_{\alpha \in \mathcal{I}} \alpha! c_\alpha^2. \qquad (5.4.48)$$

Example 5.4.10 Choose $F = \eta(x) = \int_0^x \int_{\mathbb{R}} z \tilde{N}(dy, dz)$, where $x \in \mathbb{R}^d$ is fixed. Then

$$F = \int_{\mathbb{R}^d} \int_{\mathbb{R}} \sum_{i=1}^\infty \left(\mathcal{X}_{[0,x]}(\cdot), \xi_i\right)_{L^2(\lambda)} \xi_i(y) z \tilde{N}(dy, dz)$$

$$= \sum_{i=1}^\infty \int_0^{x_d} \cdots \int_0^{x_1} \xi_i(u) du_1 \cdots du_d \left(\int_{\mathbb{R}^d} \int_{\mathbb{R}} \xi_i(y) z \tilde{N}(dy, dz)\right)$$

$$= M^{\frac{1}{2}} \sum_{i=1}^\infty \int_0^{x_d} \cdots \int_0^{x_1} \xi_i(u) du_1 \cdots du_d K_{\epsilon(\kappa(i,1))}$$

$$= M^{\frac{1}{2}} \sum_{i=1}^\infty \int_0^x \xi_i(u) du K_{\epsilon(i,1)}. \qquad (5.4.49)$$

We can now proceed word by word as in the case $d = 1$ (Section 5.3) to define the general d-dimensional Lévy field version of the *Hida–Kondratiev stochastic test function spaces*

$$(\mathcal{S}) := (\mathcal{S})_0 \text{ and } (\mathcal{S})_\rho; \ 0 \leq \rho \leq 1$$

and the *Hida–Kondratiev stochastic distributions spaces*

$$(\mathcal{S})^* := (\mathcal{S})_{-0} \text{ and } (\mathcal{S})_{-\rho}; \ 0 \leq \rho \leq 1.$$

(See Definition 5.3.6). As before we have

$$(\mathcal{S})_1 \subset (\mathcal{S})_\rho \subset (\mathcal{S}) \subset L^2(P) \subset (\mathcal{S})^* \subset (\mathcal{S})_{-\rho} \subset (\mathcal{S})_{-1}.$$

Example 5.4.11 The (*d-parameter*) *Lévy white noise* $\dot{\eta}(x)$ of the Lévy field $\eta(x)$ is defined by the expansion

$$\dot{\eta}(x) = M^{\frac{1}{2}} \sum_{i=1}^{\infty} \xi_i(x) K_{\epsilon(\kappa(i,1))} \qquad (5.4.50)$$

Just as in the case $d = 1$ (see Definition 5.3.7) we can verify that

$$\dot{\eta}(x) \in (\mathcal{S})^* \text{ for each } x \in \mathbb{R}^d$$

and (see (5.4.49))

$$\dot{\eta}(x) = \frac{\partial^d}{\partial x_1 \cdots \partial x_d} \eta(x); \quad x = (x_1, \ldots, x_d). \qquad (5.4.51)$$

5.4.3 The Wick Product

Based on the chaos expansion II (Theorem 5.4.9), we can now proceed to define the Wick product along the same lines as in Chapter 2:

Definition 5.4.12 (The Wick product) Let $F = \sum_{\alpha \in \mathcal{I}} a_\alpha K_\alpha$ and $G = \sum_{\beta \in \mathcal{I}} b_\beta K_\beta$ be two elements of $(\mathcal{S})_{-1}$. Then we define their *Wick product* $F \diamond G$ by the expression

$$F \diamond G = \sum_{\alpha, \beta \in \mathcal{I}} a_\alpha b_\beta K_{\alpha+\beta} = \sum_{\gamma \in \mathcal{I}} \left(\sum_{\alpha+\beta=\gamma} a_\alpha b_\beta \right) K_\gamma. \qquad (5.4.52)$$

As in Chapter 3 we can prove that all the spaces $(\mathcal{S})_\rho$ and $(\mathcal{S})_{-\rho}$; $0 \le \rho \le 1$ are closed under Wick multiplication.

Example 5.4.13 Note that by (5.4.45)–(5.4.46) we have, for $f_n \in \hat{L}^2((\lambda \times \nu)^n)$ and $g_m \in \hat{L}^2((\lambda \times \nu)^m)$

$$I_n(f_n) \diamond I_m(g_m) = \sum_{|\alpha|=n} a_\alpha K_\alpha \diamond \sum_{|\beta|=m} b_\beta K_\beta = \sum_{\alpha, \beta \in \mathcal{I}} a_\alpha b_\beta K_{\alpha+\beta}$$

$$= \sum_{|\gamma|=n+m} \left(\sum_{\alpha+\beta=\gamma} a_\alpha b_\beta \right) K_\gamma = I_{n+m}(f_n \hat{\otimes} g_m). \qquad (5.4.53)$$

In particular, by (5.4.44),

$$
\begin{aligned}
K_{\epsilon(\kappa(i,1))} \diamond K_{\epsilon(\kappa(i,1))} &= M^{-1} I_1 \left(\xi_i(x) z \right) \diamond I_1 \left(\xi_i(x) z \right) \\
&= M^{-1} I_2 \left(\xi_i(x^{(1)}) \xi_i(x^{(2)}) z_1 z_2 \right).
\end{aligned} \tag{5.4.54}
$$

Example 5.4.14 Let us compute

$$
\begin{aligned}
\eta^{\diamond 2}(x) &:= \eta(x) \diamond \eta(x) \\
&= I_1 \left(\mathcal{X}_{[0,x]}(y^{(1)}) z_1 \right) \diamond I_1 \left(\mathcal{X}_{[0,x]}(y^{(2)}) z_2 \right) \\
&= I_2 \left(\mathcal{X}_{[0,x]}(y^{(1)}) \mathcal{X}_{[0,x]}(y^{(2)}) z_1 z_2 \right) \\
&= 2 \int_0^x \int_{\mathbb{R}} \left(\int_0^{y^{(2)}} \int_{\mathbb{R}} z_1 z_2 \tilde{N}(dy^{(1)}, dz_1) \right) \tilde{N}(dy^{(2)}, dz_2) \\
&= 2 \int_0^x \int_{\mathbb{R}} z_2 \eta(y^{(2)}) \tilde{N}(dy^{(2)}, dz_2) = 2 \int_0^x \eta(y) d\eta(y).
\end{aligned} \tag{5.4.55}
$$

This is the Lévy analog (and d-dimensional extension) of the identity

$$
\int_0^t B(s) \delta B(s) = \frac{1}{2} B^{\diamond 2}(t)
$$

obtained in Example 2.5.11.

Remark Surprisingly, the relation in $d = 1$ between the Skorohod integrals and Wick product in Theorem 5.3.13, namely

$$
\int_{\mathbb{R}} Y(t) \delta \eta(t) = \int_{\mathbb{R}} Y(t) \diamond \dot{\eta}(t) dt,
$$

does *not* extend to $d > 1$. To see this, choose $d = 2$ and $Y(x) = \eta(x)$. By Wick calculus we have

$$
\frac{\partial}{\partial x_1} \eta^{\diamond 2}(x) = 2\eta(x) \diamond \frac{\partial}{\partial x_1} \eta(x)
$$

and

$$
\begin{aligned}
\frac{\partial^2}{\partial x_1 \partial x_2} \eta^{\diamond 2}(x) &= 2\eta(x) \diamond \frac{\partial^2}{\partial x_1 \partial x_2} \eta(x) + 2 \frac{\partial \eta}{\partial x_1} \diamond \frac{\partial \eta}{\partial x_2} \\
&= 2\eta(x) \diamond \dot{\eta}(x) + 2 \frac{\partial \eta}{\partial x_1} \diamond \frac{\partial \eta}{\partial x_2}.
\end{aligned}
$$

Integrating over the rectangle $[0, x_1] \times [0, x_2]$ we get

$$\eta^{\diamond 2}(x) = \int\limits_0^{x_2} \int\limits_0^{x_1} \frac{\partial^2}{\partial y_1 \partial y_2} \eta^{\diamond 2}(y) dy_1 dy_2$$

$$= 2 \int\limits_0^x \eta(y) \diamond \dot{\eta}(y) dy + 2 \int\limits_0^x \left(\frac{\partial \eta}{\partial y_1} \diamond \frac{\partial \eta}{\partial y_2} \right) (y) dy. \qquad (5.4.56)$$

Comparing this with (5.4.55) we obtain

$$\frac{1}{2} \eta^{\diamond 2}(x) = \int\limits_0^x \eta(y) d\eta(y)$$

$$= \int\limits_0^x \eta(y) \diamond \dot{\eta}(y) dy + \int\limits_0^x \left(\frac{\partial \eta}{\partial y_1} \diamond \frac{\partial \eta}{\partial y_2} \right) (y) dy. \qquad (5.4.57)$$

For more information about this, see Løkka and Proske (2006). On the other hand, if $Y : \mathbb{R}^d \to \mathbb{R}$ is *deterministic* and $\int_{\mathbb{R}^d} Y^2(y) dy < \infty$, we do have that

$$\int\limits_{\mathbb{R}^d} Y(y) \diamond \dot{\eta}(y) dy = \int\limits_{\mathbb{R}^d} Y(y) \dot{\eta}(y) dy = \int\limits_{\mathbb{R}^d} Y(y) d\eta(y). \qquad (5.4.58)$$

5.4.4 The Hermite Transform

Just as we have seen for the Brownian motion case in Chapter 3, we can define a correspondence between the elements of the stochastic distribution space $(\mathcal{S})_{-1}$ and a space of analytic functions of several complex variables. The correspondence is given in terms of the *Hermite transform*, defined as follows:

Definition 5.4.15 Let $F = \sum_{\alpha \in \mathcal{I}} a_\alpha K_\alpha \in (\mathcal{S})_{-1}$. Then the (Lévy) *Hermite transform* of F, denoted by $\mathcal{H}F$, is the function from the set $(\mathbb{C}^{\mathbb{N}})_c$ of all finite sequences of complex numbers into $\mathbb{C}$, defined by

$$\mathcal{H}F(\zeta_1, \zeta_2, \dots) = \sum_{\alpha \in \mathcal{I}} a_\alpha \zeta^\alpha \in \mathbb{C}, \qquad (5.4.59)$$

where $\zeta = (\zeta_1, \zeta_2, \dots) \in \mathbb{C}^{\mathbb{N}}$ and

$$\zeta^\alpha := \zeta_1^{\alpha_1} \cdot \zeta_2^{\alpha_2} \cdots \zeta_m^{\alpha_m} \text{ if } \alpha = (\alpha, \alpha_2, \dots, \alpha_m) \in \mathcal{I}.$$

Example 5.4.16 Let $F = \dot{\eta}(x) = M^{\frac{1}{2}} \sum_{j=1}^{\infty} \xi_j(x) K_{\epsilon^{(\kappa(j,1))}} \in (\mathcal{S})^*$. Then

$$(\mathcal{H}F)(\zeta) = M^{\frac{1}{2}} \sum_{j=1}^{\infty} \xi_j(x) \zeta^{\epsilon^{(\kappa(j,1))}}$$

$$= M^{\frac{1}{2}} \sum_{j=1}^{\infty} \xi_j(x) \zeta_{\kappa(j,1)}. \tag{5.4.60}$$

The following useful result is a direct consequence of the definition:

Proposition 5.4.17. *If $F, G \in (\mathcal{S})_{-1}$, then*

$$\mathcal{H}(F \diamond G)(\zeta) = (\mathcal{H}F)(\zeta) \cdot (\mathcal{H}G)(\zeta); \quad \zeta \in (\mathbb{C}^N)_c. \tag{5.4.61}$$

We end this section with a characterization theorem for Hermite transforms. Again the proof is similar to the Brownian motion case and is therefore omitted:

Theorem 5.4.19 (Characterization theorem for Hermite transforms).
Define, for $0 < q, R < \infty$, the infinite-dimensional neighborhood $N_q(R)$ of 0 in $\mathbb{C}^N$ by

$$N_q(R) = \left\{ (\zeta_1, \zeta_2, \dots) \in \mathbb{C}^N; \; \sum_{\alpha \neq 0} |\zeta^{\alpha}|^2 (2\mathbb{N})^{q\alpha} < R^2 \right\} \tag{5.4.62}$$

where $\zeta^{\alpha} = \zeta_1^{\alpha_1} \cdots \zeta_m^{\alpha_m}$ if $\alpha = (\alpha_1, \dots, \alpha_m) \in \mathcal{I}$.

(i) *If $F = \sum_{\alpha \in \mathcal{I}} a_\alpha K_\alpha \in (\mathcal{S})_{-1}$, then there exist $q, M_q < \infty$ such that*

$$|\mathcal{H}F(\zeta)| \leq \sum_{\alpha \in \mathcal{I}} |a_\alpha| |\zeta^{\alpha}| \leq M_q \left(\sum_{\alpha \in \mathcal{I}} (2\mathbb{N})^{q\alpha} |\zeta^{\alpha}|^2 \right)^{\frac{1}{2}} \tag{5.4.63}$$

for all $\zeta \in (\mathbb{C}^N)_c$. In particular, $\mathcal{H}F$ is a bounded analytic function on $N_q(R)$ for all $R < \infty$.

(ii) *Conversely, assume that $g(\zeta) := \sum_{\alpha \in \mathcal{I}} b_\alpha \zeta^{\alpha}$ is a power series in*

$$\zeta = (\zeta_1, \zeta_2, \dots) \in (\mathbb{C}^N)_c$$

such that there exist $q < \infty, \delta > 0$ with $g(\zeta)$ absolutely convergent and bounded on $N_q(\delta)$. Then there exists a unique $G \in (\mathcal{S})_{-1}$ such that

$$\mathcal{H}G = g,$$

namely

$$G = \sum_{\alpha \in \mathcal{I}} b_\alpha K_\alpha. \tag{5.4.64}$$

We now have the machinery needed to solve stochastic partial differential equations driven by Lévy white noise. Examples of such will be given in the next section.

5.5 The Stochastic Poisson Equation

We now use the white noise theory developed in the previous sections of this chapter to study the stochastic Poisson equation

$$\Delta U(x) = -\dot{\eta}(x); \quad x \in D \quad U(x) = 0; \quad x \in \partial D \qquad (5.5.1)$$

driven by Lévy white noise $\dot{\eta}(x)$. (See Section 4.2 for the Brownian white noise case.) As before $\Delta = \sum_{k=1}^{d} \partial^2/\partial x_k^2$ is the Laplace operator on $\mathbb{R}^d$ and $D \subset \mathbb{R}^d$ is a given bounded domain with regular boundary. This equation is a natural model for the temperature $U(x)$ at the point $x \in D$ if the boundary temperature is kept equal to 0 and there is a Lévy white noise heat source in D. As in Chapter 4 we regard (5.5.1) as an equation in $(\mathcal{S})_{-1}$ and the derivatives are defined in the topology of $(\mathcal{S})_{-1}$.

To solve (5.5.1) we proceed as in Chapter 4:

Step 1: We transform the equation into a deterministic partial differential equation with complex parameters ζ by applying the Hermite transform defined in the previous section.

Step 2: We solve this deterministic partial differential equation by classical methods for each parameter value $\zeta \in (\mathbb{C}^{\mathbb{N}})_c$.

Step 3: Finally, we use the characterization theorem for Hermite transforms (Theorem 5.4.19) to show that the solution found in Step 2 is the transform of an $(\mathcal{S})_{-1}$-valued function, which solves the original equation.

Applied to equation (5.5.1) this procedure gives the following:
Re. Step 1: Define

$$u(x; \zeta) = (\mathcal{H}U(x))(\zeta); \quad \zeta = (\zeta_1, \zeta_2, \dots) \in (\mathbb{C}^{\mathbb{N}})_c$$

Since

$$\mathcal{H}(\dot{\eta}(x))(\zeta) = M^{\frac{1}{2}} \sum_{j=1}^{\infty} \xi_j(x) \zeta_{\kappa(j,1)}$$

where $\zeta_{\kappa(j,1)} = \zeta^{\epsilon^{(\kappa(j,1))}} = \zeta^{\epsilon^{(j,1)}} \in \mathbb{C}$ for all j (see (5.4.59), the $\mathcal{H}$-transform of (5.5.1) is

$$\Delta u(x, \zeta) = -M^{\frac{1}{2}} \sum_{j=1}^{\infty} \xi_j(x) \zeta_{(j,1)}; \quad x \in D$$

$$u(x; \zeta) = 0; \quad x \in \partial D \qquad (5.5.2)$$

Re. Step 2: Let $G(x, y)$ be the classical Green function for the Laplace operator on D with 0 boundary conditions. Then for a given $\zeta \in (\mathbb{C}^{\mathbb{N}})_c$ the solution of (5.5.2) is

$$u(x; \zeta) = \int_D \mathcal{H}(\dot{\eta}(y))(\zeta)G(x, y)dy = M^{\frac{1}{2}} \sum_{j=1}^{\infty} \left(\int_D \xi_j(y)G(x, y)dy \right) \zeta_{(j,1)}.$$
(5.5.3)

Re. Step 3: Note that for all $\zeta \in (\mathbb{C}^{\mathbb{N}})_c$, $x \in D$ we have, using that $\zeta_{(j,1)} = \zeta_{\epsilon(\kappa(i,j))}$

$$|u(x; \zeta)| \leq M^{\frac{1}{2}} \sum_{j=1}^{\infty} \left(\int_D |\xi_j(y)| |G(x, y)dy \right) |\zeta_{(j,1)}| \leq \text{const} \sum_{j=1}^{\infty} |\zeta_{(j,1)}|$$

$$\leq \text{const} \left(\sum_{j=1}^{\infty} |\zeta_{(j,1)}|^2 (2\mathbb{N})^{2\epsilon^{(j,1)}} \right)^{\frac{1}{2}} \left(\sum_{j=1}^{\infty} (2\mathbb{N})^{-2\epsilon^{(j,1)}} \right)^{\frac{1}{2}}$$

$$\leq \text{const} \cdot R \left(\sum_{j=1}^{\infty} (2\mathbb{N})^{-2\epsilon^{(j,1)}} \right)^{\frac{1}{2}}$$

if $\zeta \in N_2(\mathbb{R})$ (see (5.4.62)). From (5.3.18) we see that

$$\sum_{j=1}^{\infty} (2\mathbb{N})^{-2\epsilon^{(j,1)}} = \sum_{j=1}^{\infty} (2\kappa(j, 1))^{-2} \leq \sum_{j=1}^{\infty} (2j)^{-2} < \infty. \qquad (5.5.4)$$

It follows that the series (5.5.3) converges absolutely for $\zeta \in N_2(\mathbb{R})$, for all $R < \infty$. Hence $u(x; \zeta)$ is an analytic function of $\zeta \in N_2(R)$ for all $R < \infty$ and we can apply part (ii) of Theorem 5.4.19 to conclude that for all $x \in D$ there exist a unique $U(x) \in (\mathcal{S})_{-1}$ such that

$$\mathcal{H}U(x) = u(x), \qquad (5.5.5)$$

namely

$$U(x) = M^{\frac{1}{2}} \sum_{j=1}^{\infty} \left(\int_D \xi_j(y)G(x, y)dy \right) K_{\epsilon(j,1)}. \qquad (5.5.6)$$

It remains to verify that this U satisfies

$$\Delta U(x) = -\dot{\eta}(x); \; x \in D. \qquad (5.5.7)$$

To this end we proceed as outlined in Theorem 4.1.1: From the general theory of deterministic elliptic partial differential equations (see, e.g., Bers et al. (1964), Theorem 3, p. 232) it is known that for all open and relatively compact sets $V \subset D$ and for all $\alpha \in (0,1)$ there exists a constant C such that

$$\|u(\cdot;\zeta)\|_{C^{2+\alpha}} \leq C \left(\|\Delta u(\cdot;\zeta)\|_{C^\alpha(V)} + \|u(\cdot;\zeta)\|_{C(V)} \right)$$

for all $\zeta \in (\mathbb{C}^N)_c$. Since both $\Delta u = -\mathcal{H}\dot{\eta}$ and u are bounded on $D \times K_2(R)$, we conclude that $\Delta u(x;\zeta) = \Delta(\mathcal{H}U(x))(\zeta)$ is uniformly bounded for $(x,\zeta) \in D \times K_2(R)$, it is x-continuous in D for each $\zeta \in K_2(R)$ and analytic with respect to $\zeta \in K_2(R)$ for all $x \in D$. Hence (5.5.7) follows, by (the Lévy analog of) Theorem 4.1.1. We have proved

Theorem 5.5.1 Løkka et al. (2004). *There exists a unique stochastic distribution process* $U\colon \overline{D} \to (\mathcal{S})^*$ *solving (5.5.1). The solution is twice continuously differentiable in* $(\mathcal{S})^*$ *and is given by*

$$U(x) = \int_D \dot{\eta}(y)G(x,y)dy$$

$$= M^{\frac{1}{2}} \sum_{k=1}^\infty \int_D \xi_k(y)G(x,y)dy K_{\epsilon(k,1)} \tag{5.5.8}$$

Remark Although we have worked in the stochastic distribution space $(\mathcal{S})_{-1}$, it is easy to see that the solution $U(x)$ given by (5.5.8) actually belongs to $(\mathcal{S})^*$ for each x. (Exercise 5.14.) As in the Brownian white noise case, the interpretation of such an $(\mathcal{S})^*$-valued (resp. $(\mathcal{S})_{-1}$-valued) solution $U(x)$ is the following:

For each $x \in \overline{D}$, $U(x)$ is a stochastic distribution whose action on a stochastic test function $f \in (\mathcal{S})$ (resp. $f \in (\mathcal{S})_1$) is

$$\langle U(x), f \rangle = \int_D G(x,y)\langle \dot{\eta}(y), f \rangle dy, \tag{5.5.9}$$

where

$$\langle \dot{\eta}(y), f \rangle = \left\langle M^{\frac{1}{2}} \sum_{j=1}^\infty \xi_j(y) K_{\epsilon(\kappa(j,1))}, f \right\rangle$$

$$= M^{\frac{1}{2}} \sum_{j=1}^\infty \xi_j(y) \mathrm{E}[K_{\epsilon(\kappa(j,1))} f]$$

$$= \sum_{j=1}^\infty \xi_j(y) \mathrm{E}\left[f \int_{\mathbb{R}^d} \int_{\mathbb{R}} \xi_j(x)\, z\, \tilde{N}(dx, dz) \right].$$

In this sense we may interpret $U(x) = U(x,\omega)$ as a solution which takes ω-averages for each x.

If the dimension d is low, we can prove that $U(x)$ is actually a classical $L^2(P)$ process:

Corollary 5.5.2. Løkka et al. (2004) *Suppose $d \leq 3$. Then $U(x) \in L^2(P)$ for all $x \in \overline{D}$, and $x \to U(x)$ is continuous in $L^1(P)$.*

Proof Since the singularity of $G(x,y)$ at $y = x$ has the order of magnitude $|x - y|^{2-d}$ for $d \geq 3$ and $\log[1/|x - y|]$ for $d = 2$ (with no singularity for $d = 1$), we see by using polar coordinates that

$$\int_D G^2(x,y)dy \leq \text{const} \cdot \int_0^1 r^{2(2-d)} r^{d-1} dr = \text{const} \cdot \int_0^1 r^{3-d} dr < \infty$$

for $d \leq 3$. Hence, using (5.4.58),

$$E[U^2(x)] = E\left[\left(\int_D G(x,y)\dot{\eta}(y)dy\right)^2\right]$$

$$= E\left[\left(\int_D G(x,y)d\eta(y)\right)^2\right]$$

$$= E\left[\left(\int_D \left(\int_{\mathbb{R}} G(x,y)\, z\, \tilde{N}(dy, dz)\right)\right)^2\right]$$

$$= \int_D \int_{\mathbb{R}} G^2(x,y)\, z^2\, \nu(dz)dy = M \int_D G^2(x,y)dy < \infty$$

This proves that $U(x) \in L^2(P)$ for each $x \in D$. Moreover, since

$$\sup_{x \in \overline{D}} \int_D G^2(x,y)dy < \infty,$$

the family $\left\{\int_D G(x,y)d\eta(y)\right\}_{x \in \overline{D}}$ is uniformly P-integrable and hence

$$E\left[\left|\int_D G(x^{(n)},y)d\eta(y) - \int_D G(x,y)d\eta(y)\right|\right] \to 0$$

for all sequences $\{x^{(n)}\}_{n=1}^{\infty}$ converging to $x \in \overline{D}$. Hence $U(x)$ is continuous in $L^1(P)$. □

5.6 Waves in a Region with a Lévy White Noise Force

Let $D \subset \mathbb{R}^m$ be a bounded domain with a C^1 boundary. Consider the stochastic wave equation

$$\frac{\partial^2 U}{\partial t^2}(t,x) - \Delta U(t,x) = F(t,x) \in C^{\frac{m+1}{2}}(\mathbb{R}_+ \times \mathbb{R}^m; (\mathcal{S})_{-1})$$

$$U(0,x) = G(x) \in C^{\frac{m+3}{2}}(\mathbb{R}^m; (\mathcal{S})_{-1}) \tag{5.6.1}$$

$$\frac{\partial U}{\partial t}(0,x) = H(x) \in C^{\frac{m+1}{2}}(\mathbb{R}^m; (\mathcal{S})_{-1})$$

Here

$$F(\cdot,\cdot) : \mathbb{R}_+ \times \mathbb{R}^m \to (\mathcal{S})_{-1} \quad \text{(corresponding to } d = m+1)$$
$$G(\cdot) : \mathbb{R}^m \to (\mathcal{S})_{-1} \quad \text{(corresponding to } d = m)$$

and

$$H(\cdot) : \mathbb{R}^m \to (\mathcal{S})_{-1}$$

are given stochastic distribution processes.

By applying the Hermite transform, then solving the corresponding deterministic PDE for each value of the parameter $\zeta \in (\mathbb{C}^{\mathbb{N}})_c$ and finally taking inverse Hermite transform as in the previous example, we get an $(\mathcal{S})_{-1}$-valued solution (in any dimension m). To illustrate this we just give the solution in the case $m = 1$ and we refer to Øksendal et al. (2006) for a solution in the general dimension.

Theorem 5.6.1 [Øksendal et al. (2006), $m = 1$ case]. *If $m = 1$ then the unique solution $U(t,x)$ of equation (5.6.1) is*

$$U(t,x) = \frac{1}{2}(G(x+t) - G(x-t)) + \frac{1}{2} \int\limits_{x-t}^{x+t} H(s)ds$$

$$+ \frac{1}{2} \int\limits_{0}^{t} \int\limits_{x-(t-s)}^{x+(t-s)} F(s,y)dy\,ds.$$

Here the integrals are $(\mathcal{S})^*$-valued integrals.

In particular, if $F(s,y) = \dot{\eta}(s,y)$, then the last term can be written

$$\frac{1}{2}\eta(D_{t,x})$$

where $D_{t,x} = \{(s,y); x - t + s \le y \le x + t - s, 0 \le s \le t\}$ is the domain of dependence of the point (t,x).

5.7 Heat Propagation in a Domain with a Lévy White Noise Potential

Consider the stochastic heat equation

$$\frac{\partial U}{\partial t}(t,x) = \frac{1}{2}\Delta U(t,x) + U(t,x) \diamond \dot{\eta}(t,x); \quad (t,x) \in [0,T] \times \mathbb{R}^d$$

$$U(0,x) = f(x); \quad x \in \mathbb{R}^d \ (f \text{ deterministic})$$

(5.7.1)

We take the Hermite transform and get the following deterministic heat equation in $u(t,x;\zeta)$ with $\zeta \in (\mathbb{C}^N)_c$ as a parameter:

$$\frac{\partial}{\partial t}u(t,x;\zeta) = \frac{1}{2}\Delta u(t,x;\zeta) + u(t,x;\zeta)\mathcal{H}\dot{\eta}(t,x;\zeta)$$

$$u(0,x;\zeta) = f(x).$$

(5.7.2)

This equation can be solved by using the Feynman–Kac formula, as follows: Let $\hat{B}(t)$ be an auxiliary Brownian motion on a filtered probability space $(\hat{\Omega}, \hat{\mathcal{F}}, \{\mathcal{F}_t\}_{t\geq 0}, \hat{P})$, independent of $B(\cdot)$.

Then the solution of (5.7.1) can be written

$$u(t,x;\zeta) = \hat{E}^x\left[f(\hat{B}(t))\exp\left[\int_0^t \mathcal{H}\dot{\eta}(s,\hat{B}(s);\zeta)ds\right]\right]$$

(5.7.3)

where $\hat{E}^x$ denotes expectation with respect to $\hat{P}$ when $\hat{B}(0) = x$. Taking inverse Hermite transforms we get:

Theorem 5.7.1. *The unique $(\mathcal{S})_{-1}$-solution of (5.7.1) is*

$$U(t,x) = \hat{E}^x\left[f(\hat{B}(t))\exp^\diamond\left[\int_0^t \dot{\eta}(s,\hat{B}(s))ds\right]\right],$$

(5.7.4)

where $\exp^\diamond[\cdot]$ denotes the Wick exponential, defined in general by

$$\exp^\diamond[F] = \sum_{n=0}^\infty \frac{1}{n!}F^{\diamond n}; \quad F \in (\mathcal{S})_{-1},$$

where

$$F^{\diamond n} = F \diamond F \diamond \cdots \diamond F \quad (n \text{ times}).$$

Exercises

5.1 Prove (5.2.9) and (5.4.6).

5.2 ($d = 1$) Find the chaos expansion $F = \sum_{m=0}^{\infty} I_m(f_m)$ (of the form (5.3.10)) for the random variables

(i) $F(\omega) = \eta^3(T)$ (Hint: See Example 5.3.3)
(ii) $F(\omega) = \exp[\eta(T)]$.

5.3 ($d = 1$) The Itô representation theorem (Itô (1956)) for pure jump Lévy processes states that if $F \in L^2(\mu)$ is $\mathcal{F}_T$-measurable, then there exists a unique predictable process $\phi(t, z)$ such that

$$
E\left[\int_0^T \int_{\mathbb{R}} \phi^2(t, z)\nu(dz)dt \right] < \infty
$$

and

$$
F = E[F] + \int_0^T \int_{\mathbb{R}} \phi(t, z)\tilde{N}(dt, dz).
$$

a) Find $\phi(t, z)$ when $F = \eta^2(T)$. (Hint: Use (5.3.14))
b) Prove that there does *not* exist a predictable process $\Psi(t)$ such that

$$
E\left[\int_0^T \Psi^2(t)dt \right] < \infty
$$

and

$$
\eta^2(T) = E[\eta^2(T)] + \int_0^T \Psi(t)d\eta(t)
$$

5.4 ($d = 1$) Prove that a chaos expansion with respect to $d\eta(t)$, i.e.

$$
F = E[F] + \sum_{n=1}^{\infty} n! \int_0^{\infty} \cdots \int_0^{t_2} f_n(t_1, \ldots, t_n)d\eta(t_1)\cdots d\eta(t_n)
$$

(similar to Theorem 2.2.7) with $f_n \in \hat{L}^2(\lambda^n)$ is not possible for $F = \eta^2(T)$. (Hint: Use (5.5.14))

5.5 ($d = 1$) Find the chaos expansion $F(\omega) = \sum_{\alpha \in \mathcal{I}} c_\alpha K_\alpha(\omega)$ (of the form (5.3.24)) for the following random variables:

(i) $F(\omega) = \eta^2(T)$
(ii) $F(\omega) = \exp[\eta(T)]$.

5.6 ($d = 1$) Prove (5.3.37) (Hint: Use (5.3.33) and (5.3.35)) to get

$$\int_0^t \int_{\mathbb{R}} z \,\mathring{\tilde{N}}(s,z)\nu(dz)ds = M^{\frac{1}{2}} \int_0^t \int_{\mathbb{R}} \sum_{i,j=1}^{\infty} z\,\xi_i(s)p_j(z)K_{\epsilon(\kappa(i,j))}\nu(dz)ds$$

$$= \int_0^t \int_{\mathbb{R}} z\,\tilde{N}(ds,dz) = \eta(t)$$

5.7 $(d=1)$ Show that

$$\eta(t_1) \diamond \eta(t_2) = \eta(t_1)\cdot\eta(t_2) - \int_0^{t_1 \wedge t_2} \int_{\mathbb{R}} z^2\,N(ds,dz) \quad \text{for } t_1, t_2 \geq 0.$$

(Hint: Use bilinearity of the Wick product to extend (5.3.45) to $F_1 \diamond F_2$), where

$$F_i(\omega) = \int_{\mathbb{R}} h_i(s)d\eta(s); \quad h_i \in L^2(\mathbb{R}); \quad i = 1,2$$

5.8 $(d=1)$ Prove (5.3.48).
(Hint: Use the extension of (5.3.46) in Exercise 5.7 and that $K_{\epsilon(\kappa(i,1))} = I_1(\xi_i(t)z)$.)

5.9 $(d=1)$ Find the Skorohod integral

$$\int_0^T \eta^2(T)\delta\eta(t)$$

in two ways:

  (i) By using chaos expansion (Definition 5.3.11)
  (ii) By using Wick products (Theorem 5.3.13) (Hint: Proceed as in Example 5.3.14.)

5.10 Let $\eta(x)$ be the pure jump Lévy field constructed from the measure ν, as in Theorem 5.4.3. Show that ν is the Lévy measure of $\eta(\cdot)$, in the sense of (5.4.2)–(5.4.3).

5.11 Show that for $x \in \mathbb{R}^d$ we have

$$\int_0^x \eta(x)\delta\eta(y) = 2\int_0^x \eta(y)d\eta(y)$$

(Hint: Proceed as in Example 5.3.14.)

5.12 Show that for $x \in \mathbb{R}^d$ and for $F = F(\omega)$ a random variable not depending on y we have

$$\int\limits_0^x F\delta\eta(y) = F \diamond \int\limits_0^x d\eta(y) = F \diamond \eta(x)$$

5.13 Prove that if $Y : \mathbb{R}^d \to \mathbb{R}$ is *deterministic* and $Y \in L^2(\mathbb{R}^d)$, then

$$\int\limits_{\mathbb{R}^d} Y(x)d\eta(x) = \int\limits_{\mathbb{R}^d} Y(x) \diamond \dot{\eta}(x)dx = \int\limits_{\mathbb{R}^d} Y(x)\dot{\eta}(x)dx.$$

5.14 Let $U(x)$ be as in (5.5.8). Prove that $U(x) \in (\mathcal{S})^*$ for all dimensions d and all $x \in \overline{D}$.

Appendix A
The Bochner–Minlos Theorem

As our approach to stochastic partial differential equations is completely based on the existence of the white noise measure μ_1 on $\mathcal{S}'(\mathbb{R}^d)$, we include a proof of its existence.

There are by now several extensions of the classical Bochner's theorem on $\mathbb{R}^d$. Instead of choosing the elegant and abstract approach as in Hanche–Olsen (1992), we will present a more analytic proof taken from Simon (1979), and Reed and Simon (1980). This means that we will first prove the Bochner–Minlos theorem on a space of sequences. The idea of the proof is simple. The existence of a measure on finite-dimensional subspaces is just the classical Bochner's theorem. Kolmogorov's theorem is used to obtain the existence of a measure on the full infinite-dimensional space, and we are left to prove that our space of sequences has full measure. The result is carried over to the set of tempered distributions using the Hermite functions as a basis. We will in fact prove a more general version of the Bochner–Minlos theorem than needed in Theorem 2.1.1 in that the right hand side of (2.1.3), $e^{-1/2\|\phi\|^2}$, is replaced by a positive definite functional.

For simplicity of notation we consider only $\mathcal{S}(\mathbb{R}^d)$ with $d = 1$. Let s be the space of sequences

$$s = \{a = \{a_n\}_{n \in \mathbb{N}_0}; \lim_{n \to \infty} n^p a_n = 0 \quad \text{for all} \quad p \in \mathbb{N}\}, \qquad (\text{A.1})$$

and let

$$s_m = \left\{a = \{a_n\}_{n \in \mathbb{N}_0}; \|a\|_m^2 := \sum_{n=0}^{\infty} (1 + n^2)^m |a_n|^2 < \infty\right\}, \quad m \in \mathbb{Z}. \quad (\text{A.2})$$

Clearly

$$s = \bigcap_{m \in \mathbb{Z}} s_m, \qquad (\text{A.3})$$

H. Holden et al., *Stochastic Partial Differential Equations*, 2nd ed., Universitext, DOI 10.1007/978-0-387-89488-1, © Springer Science+Business Media, LLC 2010

and s is a Fréchet space. The topological dual space to s, denoted by s', is given by

$$s' = \bigcup_{m \in \mathbb{Z}} s_m, \tag{A.4}$$

and the natural pairing of elements from s and s', denoted by $\langle \cdot, \cdot \rangle$, is given by

$$\langle a', a \rangle = \sum_{n=0}^{\infty} a'_n a_n \quad \text{for} \quad a' \in s', a \in s.$$

We equip s' with the cylinder topology. We will, whenever convenient, consider s and s' as subsets of $\mathbb{R}^{\mathbb{N}_0}$, the set of all sequences.

Assume now that we have a probability measure μ on s', and consider the functional g on s defined by

$$g(a) = \int_{s'} e^{i\langle a', a \rangle} d\mu(a'). \tag{A.5}$$

Then g has the following three properties:

(i) $g(0) = 1$;
(ii) g is positive definite, i.e.,

$$\sum_{j,l=1}^{n} z_j \bar{z}_l g(a_j - a_l) \geq 0 \quad \text{for any} \quad z_j \in \mathbb{C}, a_j \in s, j = 1, \ldots, n;$$

(iii) g is continuous in the Fréchet topology.

The Bochner–Minlos theorem states that conditions (i)–(iii) are not only necessary but also sufficient for the existence of μ.

Theorem A.1. *A necessary and sufficient condition for the existence of a probability measure μ on s' and a functional g on s satisfying (A.5) is that g satisfies conditions (i)–(iii).*

Proof (Simon (1979), p. 11f). We have already observed that conditions (i)–(iii) are necessary. Taking the classical Bochner's theorem for granted (see, e.g., Reed and Simon (1975), p. 13), we know that for any finite index set $N \subset \mathbb{N}$ we have a unique measure μ_N on $\mathbb{R}^{\#(N)}$ such that

$$g(a) = \int_{\mathbb{R}^{\#(N)}} e^{i\langle a', a \rangle} d\mu_N(a'), \tag{A.6}$$

where $\#(N)$ denotes the cardinality of N. We now use Kolmogorov's theorem (Simon (1979), p. 9) to conclude that there exists a measure μ on cylinder sets on $\mathbb{R}^{\mathbb{N}}$ such that (A.6) holds for all $a \in s$ with finite support. Considering

s as a subset of $\mathbb{R}^N$, it remains to show that $\mu(s') = 1$. To that end, let $\epsilon > 0$. Using the continuity of g, we can find $m \in \mathbb{N}$ and $\delta > 0$ such that $\|a\|_m < \delta$ implies $|g(a) - 1| < \epsilon$. We see that

$$Re\ g(a) = \int_{s'} \cos\langle a', a\rangle d\mu(a') \geq -1 \quad \text{for all} \quad a.$$

By considering the cases $\|a\|_m < \delta$ and $\|a\|_m \geq \delta$ separately, we obtain that

$$Re\ g(a) \geq 1 - \epsilon - 2\delta^{-2}\|a\|_m^2. \tag{A.7}$$

Let $\alpha = \{\alpha_n\}, \alpha_n > 0$, be some fixed sequence and define measures on $\mathbb{R}^{N+1}$ by

$$d\mu_{N,\sigma}(a) = \prod_{n=0}^{N} (2\pi\sigma\alpha_n)^{-\frac{1}{2}} \exp\left[-\frac{a_n^2}{2\sigma\alpha_n}\right] da_n, \tag{A.8}$$

where $\sigma > 0$ is a constant. Then $\mu_{N,\sigma}(\mathbb{R}^{N+1}) = 1$ and

$$\int_{\mathbb{R}^{N+1}} a_j a_l d\mu_{N,\sigma}(a) = \sigma\alpha_j \delta_{j,l},$$

$$\int_{\mathbb{R}^{N+1}} e^{i\langle a', a\rangle} d\mu_{N,\sigma}(a) = \exp\left[-\frac{\sigma}{2}\sum_{n=0}^{N} \alpha_n a_n'^2\right]. \tag{A.9}$$

By integrating (A.7) we find that

$$\int_{\mathbb{R}^{N+1}} \exp\left[-\frac{\sigma}{2}\sum_{n=0}^{N} \alpha_n a_n'^2\right] d\mu(a')$$

$$\geq 1 - \epsilon - 2\delta^{-2}\sigma \sum_{n=0}^{N} (1+n^2)^m \alpha_n. \tag{A.10}$$

Let $\alpha_n = (1+n^2)^{-m-1}$. Then $\sum_{n=1}^{\infty} \alpha_n (1+n^2)^m = c < \infty$. Monotone convergence implies that

$$\int_{\mathbb{R}^N} \exp\left[-\frac{\sigma}{2}\sum_{n=0}^{\infty} \alpha_n a_n'^2\right] d\mu(a') \geq 1 - \epsilon - 2\delta^{-2}\sigma C. \tag{A.11}$$

Let $\sigma \to 0$. Then $\mu(s_{-m-1}) \geq 1 - \epsilon$. Hence

$$\mu(s') \geq 1 - \epsilon,$$

which finally proves that $u(s') = 1$. $\qquad\square$

Our next result relates the sequence spaces s and s' with the Schwartz space, $\mathcal{S}(\mathbb{R})$, and the set of tempered distributions, $\mathcal{S}'(\mathbb{R})$, respectively. For that purpose we use the Hermite functions. Recall from (2.2.2) that the Hermite functions given by

$$\xi_n(x) = \frac{e^{-\frac{x^2}{2}}}{\sqrt{(n-1)!}\pi^{\frac{1}{4}}} h_{n-1}(\sqrt{2}x), \quad n = 1, 2, \ldots, \tag{A.12}$$

where h_n are the Hermite polynomials, constitute an orthonormal basis in $L^2(\mathbb{R})$. Furthermore, defining the operator

$$H = \frac{1}{2}\left(x - \frac{d}{dx}\right)\left(x + \frac{d}{dx}\right) : \mathcal{S}(\mathbb{R}) \to \mathcal{S}(\mathbb{R}), \tag{A.13}$$

we find that

$$H\xi_{n+1} = n\xi_{n+1} \quad n = 0, 1, \ldots. \tag{A.14}$$

For $f \in \mathcal{S}(\mathbb{R})$ we define the norm

$$\|f\|_m := \|(H^2 + 1)^m f\|_2, \tag{A.15}$$

where we use the $L^2(\mathbb{R})$-norm on the right hand side.

The relation between s and $\mathcal{S}(\mathbb{R})$ is the one induced by the Hermite functions; given a sequence in s we form a function in $\mathcal{S}(\mathbb{R})$ by using the elements as coefficients in an expansion along the basis $\{\xi_n\}$.

Theorem A.2. **a)** *The map $\mathcal{K} : \mathcal{S}(\mathbb{R}) \to s$ given by*

$$f \to \{(\xi_{n+1}, f)\}_{n=0}^{\infty} \tag{A.16}$$

is a one-to-one map, and $\|f\|_m = \|\{(\xi_{n+1}, f)\}\|_m$. Here $(\cdot, \cdot)$ denotes inner product in $L^2(\mathbb{R})$.

b) *The map $\mathcal{K}' : \mathcal{S}'(\mathbb{R}) \to s'$ given by*

$$\widetilde{f} \to \{\widetilde{f}(\xi_{n+1})\}_{n=0}^{\infty} \tag{A.17}$$

is a one-to-one map.

Proof (Reed and Simon (1975), p. 143f)

a) Let $f \in \mathcal{S}(\mathbb{R})$. Then $H^m f \in \mathcal{S}(\mathbb{R})$, which implies that $a_n = (\xi_{n+1}, f)$ satisfies $\sum_{n=0}^{\infty} a_n n^m \xi_{n+1} \in L^2(\mathbb{R})$; or $\sum_{n=0}^{\infty} |a_n|^2 n^{2m} < \infty$, which implies that we have $\lim_{n\to\infty} |a_n| n^m = 0$. Hence $a = \{a_n\} \in s$. By direct computation we see that $\|f\|_m = \|\mathcal{K}f\|$, thereby proving injectivity. If $a = \{a_n\} \in s$, we define $f_N = \sum_{n=0}^{N} a_n \xi_{n+1}$. We easily see that $\{f_N\}$ is a Cauchy sequence in each of the norms $\|\cdot\|_m$, and hence $f_N \to f$ as $N \to \infty$.

b) Consider now $\widetilde{f} \in \mathcal{S}'(\mathbb{R})$. Let $a_n' = \widetilde{f}(\xi_{n+1})$. Then $|a_n'| = |\widetilde{f}(\xi_{n+1})| \leq C\|\xi_{n+1}\|_m = C(n^2+1)^m$, and hence $a' = \{a_n'\} \in s'$. If $a' = \{a_n'\} \in s'$, then $|a_n'| \leq C(1+n^2)^m$ for the same m. Define

$$\widetilde{f}\left(\sum_{n=0}^{\infty} a_n \xi_{n+1}\right) = \sum_{n=0}^{\infty} a_n a_n'$$

for $a = \{a_n\} \in s$. Then

$$\left|\widetilde{f}\left(\sum_{n=0}^{\infty} a_n \xi_{n+1}\right)\right| \leq \sum |a_n||a_n'|$$

$$\leq C\left(\sum_{n=0}^{\infty}(1+n^2)^{2m+2}|a_n|^2\right)^{\frac{1}{2}} \left(\sum_{n=0}^{\infty}(1+n^2)^{-2}\right)^{\frac{1}{2}}$$

$$\leq \widetilde{C}\|a\|_{m+1},$$

which proves that $\widetilde{f} \in \mathcal{S}'(\mathbb{R})$. □

We now obtain the Bochner–Minlos theorem for $\mathcal{S}'(\mathbb{R})$.

Theorem A.3. *A necessary and sufficient condition for the existence of a probability measure μ on $\mathcal{S}'(\mathbb{R})$ and a functional g on $\mathcal{S}(\mathbb{R})$ such that*

$$g(\phi) = \int_{\mathcal{S}'(\mathbb{R})} e^{i\langle \omega, \phi \rangle} d\mu(\omega), \quad \phi \in \mathcal{S}(\mathbb{R}), \tag{A.18}$$

is that g satisfies

(i) $g(0) = 1$,
(ii) *g is positive definite,*
(iii) *g is continuous in the Fréchet topology.*

In order to conclude that Theorem 2.1.1 follows from Theorem A.3, it remains to show that

$$g(\phi) = e^{-\frac{1}{2}\|\phi\|^2}$$

is positive definite. To that end, it suffices to show that the matrix

$$[e^{-\frac{1}{2}|u_j - u_k|^2}]_{j,k=1}^n, \tag{A.19}$$

with elements $u_j, u_k \in \mathbb{R}^d$, is positive definite. Let $\xi_j \in \mathbb{C}$. Then

$$\sum_{j,k=1}^n e^{-\frac{1}{2}|u_j - u_k|^2} \bar{\xi}_j \xi_k = (2\pi)^{-\frac{d}{2}} \int e^{-\frac{1}{2}|\eta|^2} \left|\sum_j \xi_j e^{i\eta u_j}\right|^2 d\eta, \tag{A.20}$$

proving that indeed $[e^{-\frac{1}{2}|u_j - u_k|^2}]_{j,k=1}^n$ is positive definite. □

Appendix B
Stochastic Calculus Based on Brownian Motion

Here we recall some of the basic concepts and results from Brownian motion based Itô calculus that are used and implicitly assumed known in the text. General references for this material are, e.g., Ikeda and Watanabe (1989), Karatzas and Shreve (1991), and Øksendal (2003).

First we recall the definition of d-dimensional (1-parameter) *Brownian motion* (or Wiener process) $B(t, \omega)$. We saw in Section 2.1 that such a stochastic process can be constructed explicitly using the probability space $(\mathcal{S}'(\mathbb{R}), \mathcal{B}, \mu_1)$, and this is indeed a useful way of looking at Brownian motion, for several reasons. However, the Brownian motion process need not be linked directly to $(\mathcal{S}'(\mathbb{R}), \mathcal{B}, \mu_1)$ but can be linked to a general probability space $(\Omega, \mathcal{F}, P^x)$, where Ω is a set, $\mathcal{F}$ is a σ-algebra of subsets of Ω, and P^x is a probability measure on $\mathcal{F}$ for each $x \in \mathbb{R}^d$:

Definition B.1. A d-dimensional *Brownian motion* (*Wiener process*) is a family $\{B(t, \cdot)\}_{t \geq 0}$ of random variables $B(t, \cdot)$ on Ω with values in $\mathbb{R}^d$ such that the following, (B.1)–(B.4), hold:

$$P^x[B(0, \cdot) = x] = 1, \text{ i.e., } B(t) \text{ starts at } x \text{ a.s. } P^x; \qquad (B.1)$$

$\{B(t, \cdot)\}$ has *independent increments*, i.e., if $0 = t_0 < t_1 < \cdots < t_n$, then the increments

$$\{B(t_1) - B(t_0), B(t_2) - B(t_1), \ldots, B(t_n) - B(t_{n-1})\} \qquad (B.2)$$

are independent with respect to P^x;

For all $0 \leq t < s$ the random variable $B(s) - B(t) \in \mathbb{R}^d$ has a multinormal distribution with mean

$$E^x[B(s) - B(t)] = 0 \qquad (B.3)$$

263

and covariance matrix

$$E^x[(B_i(s) - B_i(t))(B_j(s) - B_j(t))] = \begin{cases} 0 & \text{if } i \neq j \\ n(s-t) & \text{if } i = j, \end{cases}$$

where E^x denotes expectation with respect to P^x;
 $B(t)$ has continuous paths a.s., i.e., the map

$$t \rightarrow B(t, \omega); \quad t \in [0, \infty) \tag{B.4}$$

is continuous, for almost all $\omega \in \Omega$ with respect to P^x.

There are also other (equivalent) descriptions of Brownian motion. (The definition given here can easily be seen as equivalent to the one given in Chapter2.) The first construction of Brownian motion was done by Wiener in 1923. He also constructed the *Wiener integral*, which was later generalized by Itô in 1942 and 1944 to what is now known as the *Itô integral*. The basic idea of the construction of this integral is the following:
 Assume that $B(t, \omega) = B(t)$ is a 1-dimensional starting at 0, and set $P^\circ = P, E^\circ = E$. For $t \geq 0$, let $\mathcal{F}_t$ be the σ-algebra generated by the random variables $\{B(s)\}_{s \leq t}$. Then for $0 \leq t < s < \infty$ we have

$$\{\emptyset, \Omega\} = \mathcal{F}_0 \subseteq \mathcal{F}_t \subseteq \mathcal{F}_s \subseteq \mathcal{F}.$$

We now define the Itô integral

$$I(f) := \int_S^T f(t, \omega) dB(t), \quad \text{where} \quad 0 \leq S < T,$$

for the following class $\mathcal{U}(S, T)$ of integrands $f(t, \omega)$ satisfying (B.5)–(B.7):

$(t, \omega) \rightarrow f(t, \omega)$ is $\mathcal{B} \times \mathcal{F}$-measurable, where $\mathcal{B}$ is the Borel

$$\sigma\text{-algebra on}[0, \infty); \tag{B.5}$$

f is $\mathcal{F}_t$-*adapted*, i.e., $f(t, \cdot)$ is $\mathcal{F}_t$-measurable, for all $t \in [S, T]$; (B.6)

$$E\left[\int_S^T f^2(t, \omega) dt\right] < \infty. \tag{B.7}$$

To construct $I(f)$ for $f \in \mathcal{U}(S, T)$, we first consider the case when $f \in \mathcal{U}(S, T)$ is *elementary*, i.e.,

$$f = \phi(t, \omega) = \sum_{j=1}^m e_j(\omega) \chi_{[t_j, t_{j+1})}(t),$$

where $t_j = j \cdot 2^{-n} \in [S, T]$; $j, n \in \mathbb{N}$ and e_j is $\mathcal{F}_{t_j}$-measurable, $E[e_j^2] < \infty$.

For such ϕ we define the *Itô integral* $I(\phi)$ by

$$I(\phi) = \int_S^T \phi(t,\omega)dB_t(\omega) = \sum_{j=1}^m e_j(\omega)(B(t_{j+1}) - B(t_j)). \qquad (B.8)$$

For general $f \in \mathcal{U}(S,T)$, we define the *Itô integral* $I(f)$ by

$$I(f) = \lim_{k \to \infty} I(\phi_k) \quad (\text{limit in } L^2(P)), \qquad (B.9)$$

where $\{\phi_k\}$ is a sequence of elementary processes such that

$$E\left[\int_S^T (f - \phi_k)^2 dt\right] \to 0 \quad \text{as} \quad k \to \infty. \qquad (B.10)$$

One can show that such a sequence $\{\phi_k\}$ satisfying (B.10) exists. Moreover, (B.10) implies that $\lim_{k \to \infty} I(\phi_k)$ exists in $L^2(P)$. Furthermore, this limit does not depend on the actual sequence $\{\phi_k\}$ chosen. The proofs of these statements follow from the fundamental *Itô isometry*

$$E\left[\left(\int_S^T \phi dB\right)^2\right] = E\left[\int_S^T \phi^2 dt\right], \qquad (B.11)$$

which is first (easily) verified for all elementary ϕ and hence holds for all $\phi \in \mathcal{U}(S,T)$.

This construction can be generalized in several ways. Here we just mention that the requirement (B.7) can be relaxed to

$$P\left[\int_S^T f^2(t,\omega)dt < \infty\right] = 1, \qquad (B.12)$$

in which case the convergence in (B.9) will be in measure rather than in $L^2(P)$. Next, the requirement (B.6) can be relaxed to

There exists an increasing family $\mathcal{H}_t; t \geq 0$ of σ-algebras such that

a) $B(t)$ is a martingale with respect to $\mathcal{H}_t$ and
b) $f(t,\omega)$ is $\mathcal{H}_t$-adapted.

This last extension allows us to define the Itô integral

$$\int_S^T f(t,\omega)dB_k(t,\omega) \qquad (B.13)$$

with respect to component number $k, B_k(t, \omega)$, of d-dimensional Brownian motion when $f(t, \omega)$ satisfies (B.5), (B.12) and is measurable with respect to

$$\mathcal{H}_t = \mathcal{F}^d := \text{the } \sigma\text{-algebra generated by } \{B_k(s, \cdot); k = 1, \ldots, d, s \leq t\}. \tag{B.14}$$

One can show that, as a function of the upper limit t, the Itô integral process

$$M(t, \omega) := \int_0^t f(s, \omega)dB_k(s, \omega)$$

has a *continuous* version. As is customary, we will assume from now on that this continuous version is chosen.

The Itô Formula

Similar to the situation for deterministic integration, definition (B.9) is not very useful for the actual computation of Itô integrals. There is no fundamental theorem of stochastic calculus, but the *Itô formula* is a good substitute. To describe it, consider an *Itô process* $X(t, \omega) \in \mathbb{R}^d$, i.e., a sum $X(t, \omega)$ of an Itô integral and a Lebesgue integral:

$$X(t, \omega) = x + \int_0^t u(s, \omega)ds + \int_0^t v(s, \omega)dB(s, \omega); \quad 0 \leq t \leq T. \tag{B.15}$$

Here u, v are processes in $\mathbb{R}^d, \mathbb{R}^{d \times p}$ respectively, $x \in \mathbb{R}^d$ is a constant and $B(s, \omega)$ is p-dimensional Brownian motion, and both u and v are $\mathcal{F}_t$-adapted (with $\mathcal{F}_t = \mathcal{F}_t^d$ as in (B.14)) and satisfy suitable growth conditions, like (B.12).

A shorthand (differential) notation for (B.15) is

$$dX(t, \omega) = u(t, \omega)dt + v(t, \omega)dB(t, \omega); \quad t \in [0, T], \quad X(0, \omega) = x. \tag{B.16}$$

Here and in (B.15) we use the matrix notation

$$[v(t, \omega)dB(t, \omega)]_i = \sum_{j=1}^{p} v_{ij}(t, \omega)dB_j(t, \omega); \quad 1 \leq i \leq d, \tag{B.17}$$

where, as usual, a_i denotes the ith component of the vector $a \in \mathbb{R}^d$.

The Itô formula states that if $X(t, \omega)$ is an Itô process as in (B.15) and

$$g(t, x) : \mathbb{R}^+ \times \mathbb{R}^d \to \mathbb{R}$$

is a function in $C^{1,2}(\mathbb{R}^+ \times \mathbb{R}^d)$, then the process

$$Y(t, \omega) := g(t, X(t, \omega))$$

is again an Itô process, described explicitly by the formula

$$dY(t, \omega) = \frac{\partial g}{\partial t}(t, X)dt + \sum_{i=1}^{d} \frac{\partial g}{\partial x_i}(t, X)dX_i + \frac{1}{2} \sum_{i,j=1}^{d} \frac{\partial^2 g}{\partial x_i \partial x_j}(t, X)dX_i dX_j,$$

(B.18)

where dX_i is component number i of dX given by (B.16) and (B.17) and

$$dX_i dX_j = [vv^T]_{ij}dt, \quad 1 \le i, j \le d,$$

(B.19)

where v^T is the transposed of the matrix v. In other words, $dX_i dX_j$ is computed from (B.16), (B.17) by using the distributive law plus the "multiplication" rules

$$\begin{cases} dtdt = dtdB_i = dB_i dB_j = 0 & \text{for } i \ne j; 1 \le i, j \le d \\ dB_i^2 = dt; \quad 1 \le i \le d. \end{cases}$$

(B.20)

In particular, if $X(t)$ and $Y(t)$ are two Itô processes, then Itô's formula gives that

$$d(X(t)Y(t)) = X(t)dY(t) + Y(t)dX(t) + dX(t)dY(t),$$

(B.21)

where $dX(t)dY(t)$ is computed from (B.20).

If we apply this to the special case when $Y(t)$ is absolutely continuous, i.e., $Y(t)$ has the form

$$dY(t, \omega) = w(t, \omega)dt,$$

we get the following *integration by parts formula*:

$$\int_0^T Y(t)dX(t) = X(T)Y(T) - X(0)Y(0) - \int_0^T X(t)w(t)dt,$$

(B.22)

because $dX(t)dY(t) = 0$.

Stochastic Differential Equations

If $b : \mathbb{R} \times \mathbb{R} \to \mathbb{R}^d$ and $\sigma : \mathbb{R} \times \mathbb{R}^d \to \mathbb{R}^{d \times p}$ are given deterministic functions and $Z(\omega) \in \mathbb{R}^d$ is a given random variable, then the equation

$$X(t) = Z + \int_0^t b(s, X(s))ds + \int_0^t \sigma(s, X_s)dB(s)$$

(B.23)

(with $B(s) \in \mathbb{R}^p$) is called a *stochastic differential equation* (SDE) (strictly speaking, a stochastic *integral* equation would be a better name). A fundamental result from the theory of SDEs is the following:

Theorem B.2.   *Assume that Z is independent of $B(t, \cdot); t \geq 0$ and $E[Z^2] < \infty$. Moreover, assume that b and σ satisfy the conditions*

$$|b(t, x)| + |\sigma(t, x)| \leq C(1 + |x|) \tag{B.24}$$

and

$$|b(t, x) - b(t, y)| + |\sigma(t, x) - \sigma(t, y)| \leq D|x - y| \tag{B.25}$$

for all $t \in [0, T]; x, y \in \mathbb{R}^d$, where C and D do not depend on t, x, y. Then there exists a unique stochastic process $X(t, \omega)$ which is $\mathcal{F}_t$-adapted and satisfies (B.23). Moreover, we have $E[X^2(t, \cdot)] < \infty$ for all $t \in [0, T]$.

We call $X(t, \omega)$ the (strong) solution of the SDE (B.23).

The Girsanov Theorem

This important result relates the probability law of one Itô process $Y(t) \in \mathbb{R}^d$ of the form

$$dY(t) = \beta(t, \omega)dt + \theta(t, \omega)dB(t); \quad 0 \leq t \leq T \text{ (constant)} \tag{B.26}$$

to the law of the related process $X(t) \in \mathbb{R}^d$ of the form

$$dX(t) = \alpha(t, \omega)dt + \theta(t, \omega)dB(t), \quad 0 \leq t \leq T, \tag{B.27}$$

where $\alpha \in \mathbb{R}^d, \beta \in \mathbb{R}^d$ and $\theta \in \mathbb{R}^{d \times p}$ are $\mathcal{F}_t$-adapted processes, each component of which satisfies (B.7): If there exists an $\mathcal{F}_t$-adapted process $u(t, \omega) \in \mathbb{R}^p$ satisfying

$$\theta(t, \omega)u(t, \omega) = \beta(t, \omega) - \alpha(t, \omega) \tag{B.28}$$

and such that

$$E\left[\exp\left[\frac{1}{2}\int_0^T u^2(s, \omega)ds\right]\right] < \infty, \tag{B.29}$$

then the *Girsanov theorem* states that the Q-law of $\{Y(t)\}_{t \leq T}$ coincides with the P-law of $\{X(t)\}_{t \leq T}$, where the measure Q is defined by

$$dQ(\omega) = M_T(\omega)dP(\omega) \quad \text{on} \quad \mathcal{F}_T, \tag{B.30}$$

with

$$M_T(\omega) = \exp\left[-\int_0^T u(t)dB(t) - \frac{1}{2}\int_0^T u^2(t)dt\right].$$

(B.31)

In particular, $Q \ll P$ and $P \ll Q$.

For an important special case, choose $\theta(t,\omega) \equiv I_d, \alpha(t,\omega) \equiv 0$. Then we get $u(t,\omega) = \beta(t,\omega)$, so if (B.29) holds, we conclude that the Q-law of

$$Y(t) := \int_0^t \beta(s,\omega)ds + B(t); \quad 0 \le t \le T$$

(B.32)

coincides with the P-law of $X(t) := B(t); \ 0 \le t \le T$. In other words, $\hat{B}(t) := Y(t)$ is a *Brownian motion* with respect to Q, where

$$dQ(\omega) = \exp\left[-\int_0^T \beta dB(t) - \frac{1}{2}\int_0^T \beta^2 dt\right] dP(\omega) \quad \text{on} \quad \mathcal{F}_T.$$

(B.33)

In particular, for any bounded function $g : \mathbb{R}^d \to \mathbb{R}$ and all $t \le T$ we have

$$E_Q[g(Y(t))] = E_P[g(B(t))],$$

or

$$\int_\Omega g\left(\int_0^t \beta ds + B(t)\right) \exp\left[-\int_0^T \beta dB(t) - \frac{1}{2}\int_0^T \beta^2 dt\right] dP(\omega)$$

$$= \int_\Omega g(B(t))dP(\omega).$$

(B.34)

If $\beta(t,\omega) = \beta(t)$ is deterministic, and we define $\beta(t) = 0$ for $t > T$, then we can write

$$\exp\left[-\int_0^T \beta dB(t) - \frac{1}{2}\int_0^T \beta^2 dt\right] = \exp^\diamond[-w(\beta)]$$

(see Lemma 2.6.16), and (B.34) obtains the form

$$\int_\Omega g\left(\int_0^t \beta ds + B(t)\right) \exp^\diamond[-w(\beta)]dP(\omega) = \int_\Omega g(B(t))dP(\omega).$$

(B.35)

This can also be written

$$\int_{\Omega} g(B(t)) \exp^{\diamond}[w(\beta)] dQ(\omega) = \int_{\Omega} g\left(B(t) + \int_0^t \beta ds\right) dQ(\omega), \qquad \text{(B.36)}$$

and in this form we can see the resemblance to formula (2.10.11) in Corollary 2.10.5, but note that the settings are different in these two formulas.

Appendix C

Properties of Hermite Polynomials

The Hermite polynomials $h_n(x)$ are defined by

$$h_n(x) = (-1)^n e^{\frac{1}{2}x^2} \frac{d^n}{dx^n} (e^{-\frac{1}{2}x^2}); \quad n = 0, 1, 2, \ldots \tag{C.1}$$

or, alternatively,

$$h_n(x) = \sum_{k=0}^{[\frac{n}{2}]} \left(-\frac{1}{2} \right)^k \frac{n!}{k!(n-2k)!} x^{n-2k}. \tag{C.2}$$

Thus the first Hermite polynomials can be calculated very easily, for example,

$$h_0(x) = 1, \quad h_1(x) = x, \quad h_2(x) = x^2 - 1, \quad h_3(x) = x^3 - 3x$$
$$h_4(x) = x^4 - 6x^2 + 3, \quad h_5(x) = x^5 - 10x^3 + 15x, \ldots \tag{C.3}$$

In fact, any polynomial can also be expressed in terms of Hermite polynomials. The formula is

$$x^n = \sum_{k=0}^{[\frac{n}{2}]} \binom{n}{2k} (2k-1)!! h_{n-2k}(x), \tag{C.4}$$

where $(2k-1)!! = (2k-1)(2k-3)(2k-5)\cdots 1$.

The generating function of Hermite polynomials is (see Exercise 2.15)

$$\sum_{n=0}^{+\infty} \frac{t^n}{n!} h_n(x) = \exp \left[-\frac{t^2}{2} + tx \right]. \tag{C.5}$$

The following relations may be deduced from (C.5):

$$\left(\frac{d^2}{dx^2} - x\frac{d}{dx} + n\right)h_n(x) = 0, \tag{C.6}$$

$$h_{n+1}(x) - xh_n(x) + nh_{n-1}(x) = 0, \tag{C.7}$$

$$\frac{dh_n(x)}{dx} = nh_{n-1}(x). \tag{C.8}$$

Besides, one has the following integral representation and orthogonal property:

$$h_n(x) = \int_{-\infty}^{+\infty} (x+iy)^n \frac{1}{\sqrt{2\pi}} e^{-\frac{y^2}{2}} dy \tag{C.9}$$

$$\int_{-\infty}^{\infty} h_n(x)h_m(x) \exp\left[-\frac{x^2}{2}\right] dx = \begin{cases} \sqrt{2\pi}\, n! & \text{if } n = m \\ 0 & \text{if } n \neq m \end{cases}. \tag{C.10}$$

Thus $\{h_n(x), n \geq 1\}$ forms an orthogonal basis for $L^2(\mathbb{R}, \mu(dx))$ if $\mu(dx) = 1/\sqrt{2\pi}e^{-x^2/2}dx$.

The Hermite functions $\xi_n(x)$ are defined by

$$\xi_n(x) = \pi^{-\frac{1}{4}}((n-1)!)^{-\frac{1}{2}} e^{-\frac{1}{2}x^2} h_{n+1}(\sqrt{2}x); \quad n \geq 1. \tag{C.11}$$

The most important properties of ξ_n used in this book are

$$-\frac{d^2\xi_n}{dx^2} + x^2\xi_n(x) = 2n\xi_n(x) \tag{C.12}$$

and

$$\xi_n \in \mathcal{S}(\mathbb{R}) \quad \text{for all} \quad n \geq 1. \tag{C.13}$$

The collection $\{\xi_n\}_{n=1}^{\infty}$ constitutes an orthonormal basis for $L^2(\mathbb{R})$.

$$\tag{C.14}$$

$$\sup_x |\xi_n(x)| = O(n^{-\frac{1}{12}}) \tag{C.15}$$

$$\xi_n(u) = O(n^{-\frac{1}{4}}) \quad \text{for all} \quad u \in \mathbb{R} \tag{C.16}$$

$$\|\xi_n\|_{L^1(\mathbb{R})} = O(n^{\frac{1}{4}}). \tag{C.17}$$

We refer to Hille and Phillips (1957), Hida (1980), and Hida et al. (1993), and the references therein for more information.

Appendix D

Independence of Bases in the Wick Products

It may appear that the definition of the Wick product depends on the choice of basis elements $\{e^{(k)}\}_{k=1}^{\infty}$ for $\oplus_{k=1}^{m} L^2(\mathbb{R}^d)$. In this appendix we will prove directly that this is not the case.

First we establish some properties of Hermite polynomials. Recall that Hermite polynomials are defined by the relation

$$h_n(x) = (-1)^n e^{\frac{x^2}{2}} \frac{d^n}{dx^n} \left(e^{-\frac{x^2}{2}} \right).$$

We adopt the convention that $h_{-1}(x) = 0$. Then for $n = 0, 1, 2 \ldots$ we have

$$h_{n+1}(x) = h_1(x) h_n(x) - n h_{n-1}(x). \tag{D.1}$$

If $x = (x_1, x_2, \ldots, x_N)$ is a vector, and $\alpha = (\alpha_1, \alpha_2, \ldots, \alpha_M)$ is a multi-index, we define

$$h_\alpha(x) = h_{\alpha_1}(x_1) h_{\alpha_2}(x_2) \cdots h_{\alpha_M}(x_M).$$

Formulated in this language, (D.1) takes the form

$$h_{\alpha+\beta}(x) = h_\alpha(x) h_\beta(x) - \frac{\alpha!}{(\alpha - \beta)!} h_{\alpha-\beta}(x) \quad \text{if } |\beta| = 1. \tag{D.2}$$

Lemma D.1.   *If $a^2 + b^2 = 1$, then for all $n = 0, 1, 2 \ldots$ we have*

$$h_n(ax + by) = \sum_{k=0}^{n} \binom{n}{k} a^k b^{n-k} h_k(x) h_{n-k}(y).$$

Proof   By induction. The cases $n = 0, 1$ are trivial. We use (D.1) to get

$$h_{n+1}(ax + by) = (ax + by) h_n(ax + by) - n h_{n-1}(ax + by)$$

$$= (ax + by) \sum_{k=0}^{n} \binom{n}{k} a^k b^{n-k} h_k(x) h_{n-k}(y)$$

$$- n \sum_{k=0}^{n-1} \binom{n-1}{k} a^k b^{n-1-k} h_k(x) h_{n-1-k}(y)$$

$$= \sum_{k=0}^{n} \binom{n}{k} a^{k+1} b^{n-k} x h_k(x) h_{n-k}(y)$$

$$+ \sum_{k=0}^{n} \binom{n}{k} a^k b^{n+1-k} h_k(x) y h_{n-k}(y)$$

$$- \sum_{k=0}^{n-1} \binom{n-1}{k} n a^k b^{n-1-k} h_k(x) h_{n-1-k}(y).$$

Using the equation (D.1) backwards, we have the following:

$$h_{n+1}(ax + by) = \sum_{k=0}^{n} \binom{n}{k} a^{k+1} b^{n-k} h_{k+1}(y)$$

$$+ \sum_{k=0}^{n-1} \binom{n}{k} a^k b^{n+1-k} h_k(x) h_{n+1-k}(y)$$

$$+ \sum_{k=1}^{n} \binom{n}{k} a^{k+1} b^{n-k} h_{k-1}(x) h_{n-k}(y)$$

$$+ \sum_{k=0}^{n-1} \binom{n}{k} a^k b^{n+1-k} h_k(x) y h_{n-1-k}(y)$$

$$- \sum_{k=0}^{n-1} \binom{n-1}{k} n a^k b^{n-1-k} h_k(x) h_{n-1-k}(y).$$

The sum of the two first terms gives the required expression. As for the three last terms, when we let S denote the sum of these, we have

$$S = \sum_{k=0}^{n-1} \left[\binom{n}{k+1} (k+1) a^2 + \binom{n}{k} (n-k) b^2 \right.$$

$$\left. - \binom{n-1}{k} n \right] a^k b^{n-1-k} h_k(x) h_{n-1-k}(y).$$

Here

$$\binom{n}{k+1}(k+1)a^2 + \binom{n}{k}(n-k)b^2 - \binom{n-1}{k} n = (a^2 + b^2 - 1) \binom{n-1}{k} n = 0,$$

and this proves the lemma. $\square$

Proposition D.2. *If $a = (a_1, a_2, \ldots, a_M)$ with $\sum_{i=1}^{M} a_i^2 = 1$, we have*

$$h_n\left(\sum_{i=1}^{M} a_i x_i\right) = \sum_{\substack{\alpha=(\alpha_1,\alpha_2,\ldots,\alpha_M) \\ |\alpha|=n}} \frac{n!}{\alpha!} a^\alpha h_\alpha(x).$$

Proof Trivial if $M = 1$. If $M \geq 1$, we set

$$\sum_{i=1}^{M+1} a_i x_i = a_1 x_1 + \sqrt{\sum_{i=2}^{M+1} a_i^2} \sum_{i=2}^{M+1} \frac{a_i}{\sqrt{\sum_{i=2}^{M+1} a_i}} x_i =: a_1 x_1 + by,$$

i.e.,

$$b = \sqrt{\sum_{i=2}^{M+1} a_i^2} \qquad y = \sum_{i=2}^{M+1} \left(\frac{a_i}{b}\right) x_i.$$

Then $a_1^2 + b^2 = 1$, so by Lemma D.1

$$h_n\left(\sum_{i=1}^{M+1} a_i x_i\right) = \sum_{k=0}^{n} \binom{n}{k} a_1^k b^{n-k} h_k(x_1) h_{n-k}(y).$$

The induction hypothesis applies to $h_{n-k}(y)$. Hence

$$h_n\left(\sum_{i=1}^{M+1} a_i x_i\right) = \sum_{k=0}^{n} \binom{n}{k} a_1^k b^{n-k} h_k(x_1)$$

$$\times \sum_{\substack{\alpha=(0,\alpha_2,\ldots,\alpha_{M+1}) \\ |\alpha|=n-k}} \frac{(n-k)!}{\alpha!} \frac{a^\alpha}{b^{|\alpha|}} h_\alpha(x_2, x_3, \ldots, x_{M+1})$$

$$= \sum_{k=0}^{n} \sum_{\substack{\alpha=(0,\alpha_2,\ldots,\alpha_{M+1}) \\ |\alpha|=n-k}} \frac{n!}{k!\alpha!} a_1^k a_2^{\alpha_2} \cdots a_{M+1}^{\alpha_{M+1}} h_k(x_1) h_\alpha(x_2, \ldots, x_{M+1}).$$

$\square$

Proposition D.3. *Let $a, b_1, b_2, \ldots, b_M, x$ be vectors in $\mathbb{R}^K$. If all the inner products $\langle a, b_i \rangle = 0$ $i = 1, 2, \ldots, M$, then*

$$\sum_{\substack{|\alpha|=n \\ |\beta_1|=1,\ldots,|\beta_M|=1}} \frac{n!}{\alpha!} a^\alpha b_1^{\beta_1} \cdots b_M^{\beta_M} h_{\alpha+\beta_1+\cdots+\beta_M}(x)$$

$$= \sum_{\substack{|\alpha|=n \\ |\beta_1|=1,\ldots,|\beta_M|=1}} \frac{n!}{\alpha!} a^\alpha b_1^{\beta_1} \cdots b_M^{\beta_M} h_\alpha(x) h_{\beta_1+\cdots+\beta_M}(x).$$

Proof By induction again. This time we use induction on the number of β-s. We first consider the case with one β. We use (D.2) to get

$$\sum_{\substack{|\alpha|=n \\ |\beta|=1}} \frac{n!}{\alpha!} a^\alpha b^\beta h_{\alpha+\beta}(x)$$

$$= \sum_{\substack{|\alpha|=n \\ |\beta|=1}} \frac{n!}{\alpha!} a^\alpha b^\beta h_\alpha(x) h_\beta(x) - \sum_{\substack{|\alpha|=n \\ |\beta|=1}} \frac{n!}{(\alpha-\beta)!} a^\alpha b^\beta h_{\alpha-\beta}(x)$$

$$= \sum_{\substack{|\alpha|=n \\ |\beta|=1}} \frac{n!}{\alpha!} a^\alpha b^\beta h_\alpha(x) h_\beta(x) - \sum_{|\beta|=1} a^\beta b^\beta \left(\sum_{|\alpha|=n-1} \frac{n!}{\alpha!} a^\alpha h_\alpha(x) \right).$$

If $\langle a, b \rangle = 0$, the last term vanishes. This proves the case with one β. By induction we will assume that the statement is true on all levels up to M. We use (D.2) again to get

$$\sum_{\substack{|\alpha|=n \\ |\beta_1|=1,\ldots,|\beta_{M+1}|=1}} \frac{n!}{\alpha!} a^\alpha b_1^{\beta_1} \cdots b_{M+1}^{\beta_{M+1}} h_{\alpha+\beta_1+\cdots+\beta_M+\beta_{M+1}}(x)$$

$$= \sum_{\substack{|\alpha|=n \\ |\beta_1|=1,\ldots,|\beta_{M+1}|=1}} \frac{n!}{\alpha!} a^\alpha b_1^{\beta_1} \cdots b_{M+1}^{\beta_{M+1}} h_{\alpha+\beta_1+\cdots+\beta_M}(x) h_{\beta_{M+1}}(x)$$

$$- \sum_{\substack{|\alpha|=n \\ |\beta_1|=1,\ldots,|\beta_{M+1}|=1}} \frac{n!}{\alpha!} a^\alpha b_1^{\beta_1} \cdots b_{M+1}^{\beta_{M+1}} \frac{(\alpha+\beta_1+\cdots+\beta_M)!}{(\alpha+\beta_1+\cdots+\beta_M-\beta_{M+1})!}$$

$$\times h_{\alpha+\beta_1+\cdots+\beta_M-\beta_{M+1}}(x).$$

We now use the induction hypothesis on the first term.

$$= \sum_{\substack{|\alpha|=n \\ |\beta_1|=1,\ldots,|\beta_{M+1}|=1}} \frac{n!}{\alpha!} a^\alpha b_1^{\beta_1} \cdots b_{M+1}^{\beta_{M+1}} h_\alpha(x) h_{\beta_1+\cdots+\beta_M}(x) h_{\beta_{M+1}}(x)$$

$$- \sum_{\substack{|\alpha|=n \\ |\beta_1|=1,\ldots,|\beta_{M+1}|=1}} \frac{n!}{\alpha!} a^\alpha b_1^{\beta_1} \cdots b_{M+1}^{\beta_{M+1}} \frac{(\alpha+\beta_1+\cdots+\beta_M)!}{(\alpha+\beta_1+\cdots+\beta_M-\beta_{M+1})!}$$

$$\times h_{\alpha+\beta_1+\cdots+\beta_M-\beta_{M+1}}(x)$$

Then we use (D.2) backwards in the first expression.

$$= \sum_{\substack{|\alpha|=n \\ |\beta_1|=1,\ldots,|\beta_{M+1}|=1}} \frac{n!}{\alpha!} a^\alpha b_1^{\beta_1} \cdots b_{M+1}^{\beta_{M+1}} h_\alpha(x) h_{\beta_1+\cdots+\beta_M}(x)$$

$$+ \sum_{\substack{|\alpha|=n \\ |\beta_1|=1,\ldots,|\beta_{M+1}|=1}} \left(\frac{n!}{\alpha!} a^\alpha b_1^{\beta_1} \cdots b_{M+1}^{\beta_{M+1}} \frac{(\beta_1 + \cdots + \beta_M)!}{(\beta_1 + \cdots + \beta_M - \beta_{M+1})!} h_\alpha(x) \right.$$

$$\times \left. h_{\beta_1 + \cdots + \beta_M - \beta_{M+1}}(x) \right)$$

$$- \sum_{\substack{|\alpha|=n \\ |\beta_1|=1,\ldots,|\beta_{M+1}|=1}} \left(\frac{n!}{\alpha!} a^\alpha b_1^{\beta_1} \cdots b_{M+1}^{\beta_{M+1}} \frac{(\alpha + \beta_1 + \cdots + \beta_M)!}{(\alpha + \beta_1 + \cdots + \beta_M - \beta_{M+1})!} \right.$$

$$\times \left. h_{\alpha + \beta_1 + \cdots + \beta_M - \beta_{M+1}}(x) \right)$$

$$=: I + II - III$$

Now observe that

$$\frac{(\beta_1 + \cdots + \beta_M)!}{(\beta_1 + \cdots + \beta_M - \beta_{M+1})!} = \#\{\beta_i = \beta_{M+1}, i \leq M\}$$

$$\frac{(\alpha + \beta_1 + \cdots + \beta_M)!}{(\alpha + \beta_1 + \cdots + \beta_M - \beta_{M+1})!} = \frac{\alpha!}{(\alpha - \beta_{M+1})!} + \#\{\beta_i = \beta_{M+1}, i \leq M\}$$

We consider the second term II, and have

$$II = \sum_{\substack{|\alpha|=n \\ |\beta_1|=1,\ldots,|\beta_{M+1}|=1}} \left(\frac{n!}{\alpha!} a^\alpha b_1^{\beta_1} \cdots b_{M+1}^{\beta_{M+1}} h_\alpha(x) h_{\beta_1 + \cdots + \beta_M - \beta_{M+1}}(x) \right.$$

$$\left. \#\{\beta_1 = \beta_{M+1}, i \leq M\} \right)$$

$$= \sum_{|\beta_{M+1}|=1} b_{M+1}^{\beta_{M+1}} b_1^{\beta_{M+1}} \sum_{\substack{|\alpha|=n \\ |\beta_2|=1,\ldots,|\beta_M|=1}} \frac{n!}{\alpha!} a^\alpha b_2^{\beta_2} \cdots b_M^{\beta_M} h_\alpha(x) h_{\beta_2 + \cdots + \beta_M}(x)$$

$$+ \sum_{|\beta_{M+1}|=1} b_{M+1}^{\beta_{M+1}} b_2^{\beta_{M+1}} \sum_{\substack{|\alpha|=n \\ |\beta_1|=1,|\beta_3|=1,\ldots,|\beta_M|=1}} \frac{n!}{\alpha!} a^\alpha b_1^{\beta_1} b_3^{\beta_3} \cdots b_M^{\beta_M} h_\alpha(x) h_{\beta_1 + \beta_3 + \cdots + \beta_M}(x)$$

$$+$$

$$\vdots$$

$$+ \sum_{|\beta_{M+1}|=1} b_{M+1}^{\beta_{M+1}} b_M^{\beta_{M+1}} \sum_{\substack{|\alpha|=n \\ |\beta_1|=1,\ldots,|\beta_{M-1}|=1}} \frac{n!}{\alpha!} a^\alpha b_1^{\beta_1} \cdots b_{M-1}^{\beta_{M-1}} h_\alpha(x) h_{\beta_1 + \cdots + \beta_{M-1}}(x).$$

The induction hypothesis applies (backwards) to all of these, so we get that the second term II is equal to the expression

$$\sum_{\substack{|\alpha|=n \\ |\beta_1|=1,\ldots,|\beta_{M+1}|=1}} \frac{n!}{\alpha!} a^\alpha b_1^{\beta_1} \cdots b_{M+1}^{\beta_{M+1}} h_{\alpha+\beta_1+\cdots+\beta_M-\beta_{M+1}}(x) \#\{\beta_i=\beta_{M+1}, i \le M\}.$$

Now we can finally subtract the third term III from the second II, to obtain

$$II - III = -\sum_{\substack{|\alpha|=n \\ |\beta_1|=1,\ldots,|\beta_{M+1}|=1}} \frac{n!}{\alpha!} a^\alpha b_1^{\beta_1} \cdots b_{M+1}^{\beta_{M+1}} \frac{\alpha!}{(\alpha-\beta_{M+1})!} h_{\alpha+\beta_1+\cdots+\beta_M-\beta_{M+1}}(x)$$

$$= -\sum_{|\beta_{M+1}|=1} a^{\beta_{M+1}} b_{M+1}^{\beta_{M+1}} \left\{ \sum_{\substack{|\alpha|=n \\ |\beta_1|=1,\ldots,|\beta_M|=1}} \frac{n!}{\alpha!} a^\alpha b_1^{\beta_1} \cdots b_{M+1}^{\beta_{M+1}} h_{\alpha+\beta_1+\cdots+\beta_M}(x) \right\} = 0,$$

and this proves the proposition. □

We now proceed to show basis-invariance. We consider two bases, $\{e^{(k)}\}_{k=1}^\infty$ and $\{\hat{e}^{(k)}\}_{k=1}^\infty$ for $\oplus_{k=1}^m L^2(\mathbb{R}^d)$. We let $\theta_k = <\omega, e^{(k)}>$ and $\hat{\theta}_k = <\omega, \hat{e}^{(k)}>$ denote the corresponding first–order integrals, and we let $\diamond$ and $\hat{\diamond}$ denote the Wick products that arise from the two bases. To prove that $\diamond=\hat{\diamond}$, we proceed as follows:

Lemma D.4. *For each pair of integers n and k*

$$\theta_k^{\diamond n} = \theta_k^{\hat{\diamond} n} = h_n(\theta_k).$$

Proof Since $\|\theta_k\|_{L^2(\mu_m)} = 1$, it can be approximated in $L^2(\mu_m)$ by a sum $\sum_{i=1}^M a_i \hat{\theta}_i$ where $\sum_{i=1}^M a_i^2 = 1$. Then by definition of the $\hat{\diamond}$ product,

$$\theta_k^{\hat{\diamond} n} \approx \left(\sum_{i=1}^M a_i \hat{\theta}_i\right)^{\hat{\diamond} n} = \sum_{\substack{\alpha=(\alpha_1,\ldots,\alpha_M) \\ |\alpha|=n}} \frac{n!}{\alpha!} a^\alpha h_\alpha(\hat{\theta}_1, \hat{\theta}_2, \ldots, \hat{\theta}_M)$$

$$= h_n\left(\sum_{i=1}^M a_i \hat{\theta}_i\right) \approx h_n(\theta_k) = \theta_k^{\hat{\diamond} n}.$$

In the third equality we used Proposition D.2. We now let $M \to \infty$, and this proves the lemma. □

Corollary D.5. *If n, m and k are non-negative integers,*

$$h_n(\theta_k) \,\hat{\diamond}\, h_m(\theta_k) = h_n(\theta_k) \diamond h_m(\theta_k) = h_{n+m}(\theta_k).$$

Proof

$$h_n(\theta_k) \,\hat\diamond\, h_m(\theta_k) = \theta_k^{\hat\diamond n} \,\hat\diamond\, \theta_k^{\hat\diamond m} = \theta_k^{\hat\diamond n+m} = h_{n+m}(\theta_k),$$

by Lemma D.4. □

Proposition D.6. *For all finite length multi-indices α and β,*

$$H_\alpha(\omega) \,\hat\diamond\, H_\beta(\omega) = H_\alpha(\omega) \diamond H_\beta(\omega) = H_{\alpha+\beta}(\omega).$$

Proof Because of Corollary D.5 it suffices to prove that for all $n_1, n_2, \ldots, n_K$,

$$h_{n_1}(\theta_1)\hat\diamond h_{n_2}(\theta_2)\hat\diamond \cdot \hat\diamond h_{n_K}(\theta_K) = h_{n_1}(\theta_1) \cdot h_{n_2}(\theta_2) \cdots h_{n_K}(\theta_K).$$

As in the proof of Lemma D.4, we may just as well assume that $\theta_1, \theta_2, \ldots, \theta_K$ is in some finite dimensional subspace generated by the $\hat\theta_k$-s, i.e., we may assume that

$$\theta_1 = \sum_{i=1}^M a_i \hat\theta_i \qquad \theta_2 = \sum_{i=1}^M b_i^{(2)} \hat\theta_i \quad \cdots \quad \theta_K = \sum_{i=1}^M b_i^{(K)} \hat\theta_i,$$

where, in particular, $a = (a_1, a_2, \ldots, a_M)$ is orthogonal to all the $b^{(i)}$-s. By Propositions D.2 and D.3, we get

$$h_{n_1}(\theta_1)\hat\diamond h_{n_2}(\theta_2)\hat\diamond \cdot \hat\diamond h_{n_K}(\theta_K)$$

$$= h_{n_1}\left(\sum_{i=1}^M a_i \hat\theta_i\right) \hat\diamond \underbrace{\left(\sum_{i=1}^M b_i^{(2)} \hat\theta_i\right)\hat\diamond \cdot \hat\diamond \left(\sum_{i=1}^M b_i^{(2)} \hat\theta_i\right)}_{n_2-\text{times}} \hat\diamond \underbrace{\left(\sum_{i=1}^M b_i^{(3)} \hat\theta_i\right)\hat\diamond \cdots}_{n_3-\text{times}}$$

$$= \sum_{\substack{|\alpha|=n_1 \\ |\beta_1|=1,\ldots,|\beta_{n_2+\cdots+n_K}|=1}} \frac{n!}{\alpha!} a^\alpha b_1^{\beta_1} b_2^{\beta_2} \cdot b_{n_2+\cdots+n_K}^{\beta_{n_2+\cdots+n_K}} h_{\alpha+\beta_1+\cdots+\beta_{n_2+\cdots+n_K}}(\hat\theta_1,\ldots,\hat\theta_M)$$

$$= \sum_{\substack{|\alpha|=n_1 \\ |\beta_1|=1,\ldots,|\beta_{n_2+\cdots+n_K}|=1}} \frac{n!}{\alpha!} a^\alpha b_1^{\beta_1} \cdot b_{n_2+\cdots+n_K}^{\beta_{n_2+\cdots+n_K}} h_\alpha(\hat\theta_1,\ldots,\hat\theta_M) h_{\beta_1+\cdots+\beta_{n_2+\cdots+n_K}}(\hat\theta_1,\ldots,\hat\theta_M)$$

$$= h_{n_1}(\theta_1) \cdot \{h_{n_2}(\theta_2)\hat\diamond \cdot \hat\diamond h_{n_K}(\theta_K)\},$$

and the claim follows by repeated use of this argument. □

Proposition D.6 says that the alternative Wick product $\hat\diamond$ is equal to the original Wick product on all elements in a base for $L^2(\mu_m)$. The same then certainly applies to all finite-dimensional linear spans of such elements. This is the result that we used previously in this book. It is certainly possible to extend the result to more generalsituations. Roughly speaking, the above result says that the definition of the Wick product does not depend on any particular choice of base elements. The topological structure of $(\mathcal{S})^{-1}$, however, strongly depends on the choice of the Hermite functions as basis elements. If we want to extend the above result to limits of finite-dimensional linear spans of base elements, a space has to be fixed so that the limit concept is well–defined. The natural space structure is then certainly $(\mathcal{S})^{-1}$. As in the definition of L^1 Wick products in Holden, et al. (1993a), a natural definition of the alternative Wick product $\hat{\diamond}$ is then just to define

$$X \hat{\diamond} Y = \lim_{n \to \infty} X_n \hat{\diamond} Y_n,$$

where X_n and Y_n are finite combinations of base elements converging to X and Y respectively in $(\mathcal{S})^{-1}$ and the limit is taken in this space also. By Proposition D.6, we always have $X_n \hat{\diamond} Y_n = X_n \diamond Y_n$. Hence the limit always exists and is equal to $X \diamond Y$. With the above definition, we easily get

$$X \hat{\diamond} Y = X \diamond Y$$

for all $X, Y \in (\mathcal{S})^{-1}$.

Appendix E
Stochastic Calculus Based on Lévy Processes

This appendix is somewhat analogous to Appendix B except that here we deal with a more general class of processes, the Lévy processes. Nevertheless, it is useful to have seen the special case with Brownian motion first as decribed in Appendix B, rather than proceeding directly to the more general theory in this appendix.

We will not give proofs in the following, but we refer to Applebaum (2004), Bertoin (1996), Jacod and Shiryaev (2003) and Sato (1999) for more details. The summary here is also based on Chapter 1 in Øksendal and Sulem (2007).

Definition E1. Let $(\Omega, \mathcal{F}, P)$ be a probability space. A (1-dimensional) Lévy process is a stochastic process $\eta(t) = \eta(t, \omega) : [0, \infty) \times \Omega \to \mathbb{R}$ with the following properties:

$\eta(0) = 0$ a.s. $\hfill$ (E.1)

η has independent increments (see B.2) $\hfill$ (E.2)

η has *stationary* increments, i.e., for all fixed $h > 0$ the increment process $I(t) := \eta(t + h) - \eta(t); \ t \geq 0$ is a *stationary* process. (A) stochastic process $\theta(t)$ is called *stationary* if $\theta(t + t_0)$ has the same law as $\theta(t)$ for all $t_0 > 0$.
$\hfill$ (E.3)

η is *stochastically continuous*, i.e., for all $t > 0, \epsilon > 0$ we have $\hfill$ (E.4)

$$\lim_{s \to t} P(|\eta(t) - \eta(s)| > \epsilon) = 0,$$

and η has *càdlàg* paths, i.e., the paths of η are continuous from the right (continue à droite) with left-sided limits (limites à gauche).

Note that if we strengthen the condition (E.4) to requiring that η has *continuous* paths, then in fact η is necessarily of the form (see (C.1))

$$\eta(t) = a\,t + \sigma B(t); \quad t \geq 0$$

where a, b are constants and $B(\cdot)$ is a Brownian motion. Thus it is the possible presence of jumps that distinguishes a general Lévy process from a Brownian motion with a constant drift.

The *jump* of η at time t is defined by

$$\Delta\eta(t) = \eta(t) - \eta(t^-)$$

Put $\mathbb{R}_0 = \mathbb{R} \setminus \{0\}$ and let $\mathcal{B}(\mathbb{R}_0)$ be the family of all Borel subsets $U \subset \mathbb{R}$ such that $\overline{U} \subset \mathbb{R}_0$. If $U \in \mathcal{B}(\mathbb{R}_0)$ and $t > 0$ we define

$$N(t, U) = \text{the number of jumps of } \eta(\cdot) \text{ of size } \Delta\eta(s) \in U; \ s \le t$$

Since the paths of η are càdlàg, we see that $N(t, U) < \infty$ for all $t > 0, U \in \mathcal{B}(\mathbb{R}_0)$. It can be proved that for all $\omega \in \Omega$ the function $(a, b) \times U \mapsto N(b, U) - N(a, U); \ 0 \le a < b < \infty, U \in \mathcal{B}(\mathbb{R}_0)$ defines a measure on $\mathcal{B}([0, \infty)) \times \mathcal{B}(\mathbb{R}_0)$, called the *Poisson random measure of* η. The differential form of this measure is denoted by

$$N(dt, dz)$$

The Lévy measure ν of $\eta(\cdot)$ is defined by

$$\nu(U) = \mathrm{E}[N(1, U)]; \ U \in \mathcal{B}(\mathbb{R}_0) \tag{E.5}$$

The Lévy measure need not be finite. In fact, it is even possible that

$$\int_{\mathbb{R}} \min(1, |z|)\nu(dz) = \infty \tag{E.6}$$

On the other hand, we always have

$$\int_{\mathbb{R}} \min(1, z^2)\nu(dz) < \infty \tag{E.7}$$

The Lévy measure ν determines the law of $\eta(\cdot)$. In fact, we have

Theorem E.2 (The Lévy–Khintchine formula). *Let η be a Lévy process with Lévy measure ν. Then*

$$\int_{\mathbb{R}} \min(1, z^2)\nu(dz) < \infty$$

and

$$\mathrm{E}[e^{iu\eta(t)}] = e^{t\Psi(u)}; \ u \in \mathbb{R} \tag{E.8}$$

where

$$\Psi(u) = -\frac{1}{2}\sigma^2 u^2 + i\alpha u \int_{|z|<1} \{e^{iuz} - 1 - iuz\}\nu(dz) + \int_{|z|\ge1} (e^{iuz} - 1)\nu(dz) \tag{E.9}$$

for some constants $\alpha, \sigma \in \mathbb{R}$. Conversely, given constants $\alpha, \sigma \in \mathbb{R}$ and a measure ν on $\mathcal{B}(\mathbb{R}_0)$ such that

$$\int_{\mathbb{R}} \min(1, z^2)\nu(dz) < \infty$$

there exists a Lévy process $\eta(\cdot)$ (unique in law) such that (E.8)-(E.9) hold.

In general one can prove that if we define the *compensated Poisson random measure* $\tilde{N}$ by

$$\tilde{N}(dt, dz) = N(dt, dz) - \nu(dz)dt \tag{E.10}$$

and $\theta(t, z)$ is an $\mathcal{F}_t$-adapted process such that

$$\mathrm{E}\left[\int_0^T \int_{\mathbb{R}} \theta^2(t, z)\nu(dz)dt \right] < \infty$$

then

$$M(t) := \lim_{n \to \infty} \int_0^t \int_{|z| \geq \frac{1}{n}} \theta(t, z)\tilde{N}(dz, dt); \ 0 \leq t \leq T$$

exists as a limit in $L^2(P)$ and it is a *martingale*. If we only know that

$$\int_0^T \int_{\mathbb{R}} \theta^2(t, z)\nu(dz) < \infty \ a.s.$$

then the limit

$$M(t) := \lim_{n \to \infty} \int_0^t \int_{|z| \geq \frac{1}{n}} \theta(t, z)\tilde{N}(dz, dt)$$

exists in probability and it is a *local martingale*. Moreover, the following *Itô isometry* holds:

$$\mathrm{E}\left[\left(\int_0^T \int_{\mathbb{R}} \theta(t, z)\tilde{N}(dt, dz) \right)^2 \right] = \mathrm{E}\left[\int_0^T \int_{\mathbb{R}} \theta^2(t, z)\nu(dz)dt \right] \tag{E.11}$$

A complete description of a Lévy process is given in the following result:

Theorem E.3 (Itô–Lévy decomposition theorem). *Let η be a Lévy process. Then η can be written*

$$\eta(t) = a_1 t + \sigma B(t) + \int_{|z| < 1} z\tilde{N}(t, dz) + \int_{|z| \geq 1} zN(t, dz) \tag{E.12}$$

where a_1, σ are constants and $B(\cdot)$ is a Brownian motion.

In general we have that if, for some $p \geq 1$,

$$E[|\eta(t)|^p] < \infty \text{ for all } t$$

then

$$\int_{|z|\geq 1} |z|^p \nu(dz) < \infty$$

In particular, if

$$E[|\eta(t)|] < \infty \text{ for all } t \tag{E.13}$$

then $\int_{|z|\geq 1} z\nu(dz)$ is well defined and the representation (E.12) simplifies to

$$\eta(t) = at + \sigma B(t) + \int_{\mathbb{R}} z\tilde{N}(t, dz) \tag{E.14}$$

where $a = a_1 + \int_{|z|\geq 1} z\nu(dz)$. For simplicity we will from now on assume that (E.13) (and hence (E.14)) holds. In view of (E.14) it is then natural to consider stochastic processes of the form

$$X(t) = x + \int_0^t \alpha(s, \omega)ds + \int_0^t \beta(s, \omega)dB(s) + \int_0^t \int_{\mathbb{R}} \gamma(s, \omega)\tilde{N}(ds, dz) \tag{E.15}$$

where α, β and γ are predictable processes such that

$$\int_0^T \{|\alpha(s)| + \beta^2(s) + \int_{\mathbb{R}} \gamma^2(s, z)\nu(dz)\}ds < \infty \text{ a.s.} \tag{E.16}$$

We call such processes (1-dimensional) *Itô–Lévy processes*. In analogy with the Brownian motion case we use the short hand differential notation

$$dX(t) = \alpha(t)dt + \beta(t)dB(t) + \int_{\mathbb{R}} \gamma(t, z)\tilde{N}(dt, dz); \quad X(0) = x \tag{E.17}$$

for (E.14).

Theorem E.4 (The 1-dimensional Itô formula). *Let $X(t)$ be the Itô–Lévy process (E.15). Let $f(t, x) : \mathbb{R} \times \mathbb{R} \to \mathbb{R}$ be a function in $C^{1,2}(\mathbb{R} \times \mathbb{R})$ and define*

$$Y(t) = f(t, X(t))$$

Then $Y(t)$ is also an Itô–Lévy process and it is given in differential form by

$$dY(t) = \frac{\partial f}{\partial t}(t, X(t))dt + \frac{\partial f}{\partial x}[\alpha(t)dt + \beta(t)dB(t)] + \frac{1}{2}\frac{\partial^2 f}{\partial x^2}(t, X(t))\beta^2(t)dt$$

$$+ \int_{\mathbb{R}}\left\{ f(t, X(t) + \gamma(t, z)) - f(t, X(t)) - \frac{\partial f}{\partial x}(t, X(t))\gamma(t, z) \right\}\nu(dz)dt$$

$$+ \int_{\mathbb{R}}\{ f(t, X(t) + \gamma(t, z)) - f(t, X(t)) \}\tilde{N}(dt, dz) \qquad \text{(E.18)}$$

In the multidimensional case we are given n Itô–Lévy processes $X_1(t), \ldots, X_n(t)$ driven by m independent 1-dimensional Brownian motions $B_1(t), \ldots, B_m(t)$ and l independent (1-dimensional) compensated Poisson random measures $\tilde{N}_1(dt, dz), \ldots, \tilde{N}_l(dt, dz)$ as follows

$$dX_i(t) = \alpha_i(t)dt + \sum_{j=1}^{m}\beta_{ij}(t)dB_j(t) + \sum_{k=1}^{l}\int_{\mathbb{R}}\gamma_{ik}(t, z_k)\tilde{N}_k(dt, dz_k); \; 1 \le i \le n$$

$$\text{(E.19)}$$

Or, in matrix notation,

$$dX(t) = \alpha(t)dt + \beta(t)dB(t) + \int_{\mathbb{R}}\gamma(t, z)\tilde{N}(dt, dz) \qquad \text{(E.20)}$$

where

$$dX(t) = \begin{bmatrix} dX_1(t) \\ \vdots \\ dX_n(t) \end{bmatrix}, \; \alpha(t) = \begin{bmatrix} \alpha_1(t) \\ \vdots \\ \alpha_n(t) \end{bmatrix},$$

$$\beta(t) = [\beta_{ij}(t)]_{\substack{1 \le i \le n \\ 1 \le j \le m}}, \gamma(t, z) = [\gamma_{ik}(t, z)]_{\substack{1 \le i \le n \\ 1 \le k \le l}}$$

and

$$\tilde{N}(dt, dz) = \begin{bmatrix} \tilde{N}_1(dt, dz_1) \\ \vdots \\ \tilde{N}_l(dt, dz_l) \end{bmatrix}; \; z = (z_1, \ldots, z_l) \in \mathbb{R}^l$$

Then we have the following:

Theorem E.5 (The multidimensional Itô formula). *Let $X(t) \in \mathbb{R}^n$ be as in (E.20). Let $f(t, x) = f(t, x_1, \ldots, x_n) : \mathbb{R} \times \mathbb{R}^n \to \mathbb{R}$ be in $C^{1,2}(\mathbb{R} \times \mathbb{R}^n)$ and define*

$$Y(t) = f(t, X(t))$$

Then

$$dY(t) = \frac{\partial f}{\partial t}(t, X(t))dt + \sum_{i=1}^{n} \frac{\partial f}{\partial x_i}\left[\alpha_i(t)dt + \sum_{j=1}^{m} \beta_{ij}(t)dB_j(t)\right]$$

$$+ \frac{1}{2}\sum_{i,j=1}^{n}(\beta\beta^{\top})_{ij}(t)\frac{\partial^2 f}{\partial x_i \partial x_j}(t, X(t))dt$$

$$+ \sum_{k=1}^{l}\int_{\mathbb{R}}\{f(t, X(t) + \gamma^{(k)}(t, z_k))$$

$$- f(t, X(t)) - \nabla_x f(t, X(t))^{\top}\gamma^{(k)}(t, z_k)\}\nu_k(dz_k)dt$$

$$+ \sum_{k=1}^{l}\int_{\mathbb{R}}\{f(t, X(t) + \gamma^{(k)}(t, z_k)) - f(t, X(t))\}\tilde{N}_k(dt, dz_k)dt \quad \text{(E.21)}$$

where $\gamma^{(k)}(t, z)$ is column number k of the $n \times l$ matrix $\gamma(t, z)$.

One can prove an existence and uniqueness result for stochastic differential equations driven by Lévy processes, analogous to the result stated in Appendix B:

Theorem E.6. *Let $\alpha : \mathbb{R} \times \mathbb{R}^n \to \mathbb{R}^n, \beta : \mathbb{R} \times \mathbb{R}^n \to \mathbb{R}^{n \times m}$ and $\theta : \mathbb{R} \times \mathbb{R}^n \times \mathbb{R}^l \to \mathbb{R}^{n \times l}$ be given functions satisfying the conditions:*
There exists a constant C such that

$$|b(t, x)|^2 + ||\beta(t, x)||^2 + \sum_{k=1}^{l}\int_{\mathbb{R}}|\theta_k(t, x, z_k)|^2\nu_k(dz_k) \leq C(1 + |x|^2) \quad \text{(E.22)}$$

for all $t \in [0, T]; x \in \mathbb{R}^n$.
There exists a constant D such that

$$|b(t, x) - b(t, y)|^2 + ||\beta(t, x) - \beta(t, y)||^2$$

$$+ \sum_{k=1}^{l}\int_{\mathbb{R}}|\theta_k(t, x, z_k) - \theta_k(t, y, z_k)|^2\nu_k(dz_k) \leq D|x - y|^2 \quad \text{(E.23)}$$

for all $t \in [0, T]; x, y \in \mathbb{R}^n$. Then the stochastic differential equation

$$dX(t) = b(t, X(t))dt + \beta(t, X(t))dB(t) + \int_{\mathbb{R}}\theta(t, X(t^-), z)\tilde{N}(dt, dz)$$

$$0 \leq t \leq T, X(0) = x\,(constant) \in \mathbb{R}^n \quad \text{(E.24)}$$

has a unique $\mathcal{F}_t$-adapted càdlàg solution $X(t)$ such that $\mathrm{E}[X^2(t)] < \infty$ for all $t \in [0, T]$.

Example E.6. Consider the stochastic differential equation

$$dY(t) = \mu(t)Y(t)dt + \sigma(t)Y(t)dB(t) + Y(t^-)\int_{\mathbb{R}} \gamma(t,z)\tilde{N}(dt,dz)$$

$$0 \le t, Y(0) = y > 0 \tag{E.25}$$

where $\mu(t), \sigma(t)$ and $\gamma(t,z)$ are deterministic, $\gamma(t,z) \ge -1$. If $\mu(t) = \mu, \sigma(t) = \sigma$ and $\gamma(t,z) = z$ do not depend on t, this may be regarded as a natural jump extension of the classical geometric Brownian motion. We call this process the *geometric Lévy process*. Using the Itô formula we obtain that the solution of (E.25) is

$$Y(t) = y\exp\left[\int_0^T \left\{\mu(s) - \frac{1}{2}\sigma^2(s)\right\}ds + \int_0^t \sigma(s)dB(s)\right.$$

$$+ \int_0^t \int_{\mathbb{R}} \{\ln(1+\gamma(s,z)) - \gamma(s,z)\}\nu(dz)ds$$

$$\left.+ \int_0^t \int_{\mathbb{R}} \ln(1+\gamma(s,z))\tilde{N}(ds,dz)\right] \tag{E.26}$$

under suitable growth conditions on μ, σ and γ.

The Girsanov theorem The Girsanov theorem for Brownian motion was presented in Appendix B. We here just concentrate on what happens in the jump case:

Theorem E.7 (Girsanov theorem for Itô–Lévy processes). *Let $X(t)$ be an Itô–Lévy process in $\mathbb{R}^n$ of the form*

$$dX(t) = \alpha(t)dt + \int_{\mathbb{R}^l} \gamma(t,z)\tilde{N}(dt,dz) \tag{E.27}$$

Assume that there exists a process $\theta(t,z) = (\theta_1(t,z),\ldots,\theta_l(t,z))^\top \in \mathbb{R}^l$ such that $\theta_j(t,z) \le 1$ and

$$\sum_{j=1}^l \int_{\mathbb{R}} \gamma_{ij}(t,z_j)\theta(t,z_j)\nu_j(dz_j) = \alpha_i(t); \quad 1 \le i \le n \tag{E.28}$$

and such that the process

$$Z(t) := \exp\left[\sum_{j=1}^l \int_0^t \int_{\mathbb{R}} \{\ln(1-\theta_j(s,z_j)) + \theta_j(s,z_j)\}\nu_j(dz_j)ds\right.$$

$$+ \sum_{j=1}^{l} \int_0^t \int_{\mathbb{R}} \ln(1 - \theta_j(s, z_j)) \tilde{N}_j(ds, dz_j) \bigg] \qquad (E.29)$$

exists for $0 \leq t \leq T$ and satisfies $\mathrm{E}[Z(T)] = 1$. Define the measure Q on $\mathcal{F}_T$ by

$$dQ(\omega) = Z(T)dP(\omega) \quad \text{on } \mathcal{F}_T \qquad (E.30)$$

Then $X(t)$ is a local martingale with respect to Q. In other words: Q is an equivalent local martingale measure for $X(t)$.

Remark E.8 Note that the condition (E.28) corresponds to the condition (B.28) in the Brownian motion case. However, while equation (B.28) has a unique solution $u(t)$ if $\theta(t)$ is a non-singular quadratic matrix, equation (E.28) will typically have infinitely many solutions $\theta(t, z)$, even if the matrix $\gamma(t, z)$ is quadratic and non-singular. We see this even in the case where $n = l = 1$. Then (E.29) becomes

$$\int_{\mathbb{R}} \gamma(t, z)\theta(t, z)\nu(dz) = \alpha(t)$$

The only case where this has a unique solution is when ν is supported on one point only, say z_0, i.e., when

$$\nu(dz) = \lambda \delta_{z_0}(dz)$$

for some constant $\lambda > 0$. This corresponds to the case when $\eta = \int_{\mathbb{R}} z\tilde{N}(t, dz)$ is a Poisson process with intensity λ and jump size z_0.

The second fundamental theorem of asset pricing states that — under some conditions — a mathematical market is complete if and only if there is *only one* equivalent local martingale measure for the normalized price process of the risky asset. Applied to the Girsanov theorem above, this means that the market consisting of

(i) a safe investment, with price $dS_0(t) = \rho S_0(t)dt$; $S_0(0) = 1$
(ii) a risky investment, with price

$$dS_1(t) = \mu S_1(t)dt + \beta S_1(t^-) \int_{\mathbb{R}} \gamma(t, z)\tilde{N}(dt, dz); \quad S_1(0) > 0$$

where ρ, μ and β are constants, will typically be an *incomplete* market, unless the underlying Lévy process is a Poisson process.

References

K. AASE, B. ØKSENDAL AND J. UBØE (2001): Using the Donsker delta function to compute hedging strategies. *Potential Anal.*, **14**, 351–374.

S. ALBEVERIO, J. KONDRATIEV AND L. STREIT (1993B): How to generalize white noise analysis to non-Gaussian measures. In Ph. Blanchard, L. Streit, M. Sirugue-Collins, D. Testard (editors): *Dynamics of Complex and Irregular Systems*. World Scientific, Singapore, pp. 120–130.

S. ALBEVERIO, YU. L. DALETSKY, YU. L. KONDRATIEV, AND L. STREIT (1996A): Non-Gaussian infinite-dimensional analysis. *J. Funct. Anal.*, **138**, 2, 311–350.

S. ALBEVERIO, Z. HABA AND F. RUSSO (1996B): On nonlinear two-space-dimensional wave equation perturbed by space–time white noise. In *Stochastic Analysis: Random Fields and Measure-valued Processes* (Ramat Gan, 93/95), Israel Math. Conf. Proc., 10, BarIlan Univ., Ramat Gan., pp. 125.

S. ALBEVERIO, S. A. MOLCHANOV, D. SURGAILIS (1996C): Stratified structure of the universe and the Burgers equation: A probabilistic approach. *Prob. Th. Rel. Fields*, **100**, 4, 457–484.

D. APPLEBAUM (2004): *Lévy Processes and Stochastic Calculus*. Cambridge University Press, Cambridge, MA.

F. E. BENTH (1993): Integrals in the Hida distribution space $(S)^*$. In T. Lindstrøm, B. Øksendal and A. S. Üstünel (editors): *Stochastic Analysis and Related Topics*. Gordon and Breach, Philadelphia, pp. 89–99.

F. E. BENTH (1994): A functional process solution to a stochastic partial differential equation with applications to nonlinear filtering. *Stochastics*, **51**, 195–216.

F. E. BENTH (1995): Stochastic partial differential equations and generalized stochastics processes. Dr. Scient. Thesis, University of Oslo.

F. E. BENTH (1996): A note on population growth in a crowded stochastic environment. In H. Körezlioglu, B. Øksendal and A. S. Üstünel (editors): *Stochastic Analysis and Related Topics* 5, Birkhäuser, Boston, pp. 111–119.

F. E. BENTH, T. DECK, J. POTTHOFF AND G. VÅGE (1998): Explicit strong solutions of SPDEs with applications to nonlinear filtering. *Acta Appl. Math.*, **51**, No. 2, 215–242.

F. E. BENTH AND J. GJERDE (1998A): A remark on the equivalence between Poisson and Gaussian stochastic partial differential equations. *Potential Anal.*, **8**, 179–193.

F. E. BENTH AND J. GJERDE (1998B): Convergence rates for finite-element approximations of SPDEs. *Stochastics Stochastics Rep.*, **63**, 313–326.

F. E. BENTH AND H. GJESSING (2000): A nonlinear parabolic equation with noise. A reduction method. *Potential Anal.*, **12**, 4, 385–401.

F. E. BENTH AND J. POTTHOFF (1996): On the martingale property for generalized stochastic processes. *Stochastics Stochastics Rep.*, **58**, 349–367.

F. E. BENTH, B. ØKSENDAL, J. UBØE AND T. ZHANG (1996): Wick products of complex-valued random variables. In H. Körezlioglu, B. Øksendal and A.S. Üstünel (editors): *Stochastic Analysis and Related Topics* 5, Birkhäuser, Boston, pp. 135–155.

Y. M. BEREZANSKY AND Y. KONDRATIEV (1988): Spectral methods in infinite-dimensional analysis. *Naukova Dumka*, Kiev.

C. BERG AND G. FORST (1975): *Potential Theory on Locally Compact Abelian Groups.* Springer, New York.

M. A. BERGER AND V. J. MIZEL (1982): An extension of the stochastic integral. *Ann. Prob.*, **10**, 435–450.

L. BERS, F. JOHN AND M. SCHECHTER (1964): *Partial Differential Equations.* Interscience, Malden, MA.

L. BERTINI, N. CANCRINI AND G. JONA-LASINIO (1994): The stochastic Burgers equation. *Comm. Math. Phys.*, **165**, 211–232.

J. BERTOIN (1996): *Lévy Processes.* Cambridge University Press.

F. BIAGINI, Y. HU, B. ØKSENDAL AND T. ZHANG (2008): *Stochastic Calculus for Fractional Brownian Motion.* Springer, New York.

J. M. BURGERS (1940): Application of a model system to illustrate some points of the statistical theory of free turbulence. *Proc. Roy. Neth. Acad. Sci.* (Amsterdam), **43**, 2–12.

J. M. BURGERS (1974): *The Nonlinear Diffusion Equation,* Reidel, Dordrecht.

E. CARLEN AND P. KREE (1991): L^p estimates on iterated stochastic integrals. *Ann. Prob.*, **19**, 1, 354–368.

P. L. CHOW (1989): Generalized solution of some parabolic equations with a random drift. *J. Appl. Math. Optim.*, **20**, 81–96.

K. L. CHUNG AND R. J. WILLIAMS (1990): *Introduction to Stochastic Integration* (Second Edition). Birkhäuser, Boston.

G. COCHRAN AND J. POTTHOFF (1993): Fixed point principles for stochastic partial differential equations. In P. Blanchard, L. Streit, M. Sirugue-Collins, D. Testard (editors): *Dynamics of Complex and Irregular Systems,* World Scientific, Singapore, pp. 141–148.

W. G. COCHRAN, J.-S. LEE AND J. POTTHOFF (1995): Stochastic Volterra equations with singular kernels. *Stoch. Proc. Appl.*, **56**, 337–349.

J. F. COLOMBEAU (1990): Multiplication of distributions. *Bull. Amer. Math. Soc.*, **23**, 251–268.

G. DA PRATO, A. DEBUSSCHE, R. TEMAM (1994): Stochastic Burgers equation. *Nonlinear Diff. Eq. Appl.*, **1**, 389–402.

G. DA PRATO AND J. ZABZYK (1992): *Stochastic Equations in Infinite Dimensions.* Cambridge University Press, Cambridge, MA.

T. DECK AND J. POTTHOFF (1998): On a class of stochastic partial differential equations related to turbulent transport. *Probability Theory and Related Fields*, **111**, 101–122.

G. DI NUNNO, B. ØKSENDAL AND F. PROSKE (2004): White noise analysis for Lévy processes. *J. Functional Anal.*, **206**, 109–148.

G. DI NUNNO, B. ØKSENDAL AND F. PROSKE (2009): *Malliavin Calculus for Lévy Processes with Applications to Finance.* Springer, New York.

R. L. DOBRUSHIN AND R. A. MINLOS (1977): Polynomials in linear random functions. *Russian Math. Surveys*, **32**(2), 71–127.

R. DURRETT (1984): *Brownian Motion and Martingales in Analysis.* Wadsworth, Belmont, CA.

E. B. DYNKIN (1965): *Markov Processes,* Vols. I, II. Springer, New York.

YU. V. EGOROV AND M. A. SHUBIN (editors) (1991): *Partial Differential Equations III.* Encyclopaedia of Mathematical Sciences 32. Springer, New York.

Yu. V. EGOROV AND M. A. SHUBIN (editors) (1992): *Partial Differential Equations I.* Encyclopaedia of Mathematical Sciences 30. Springer, New York.

A. R. FORSYTH (1906): *Theory of Differential Equations, Part IV. Partial Differential Equations,* Vol. VI; Cambridge University Press, Cambridge, MA.

M. FREIDLIN (1985): *Functional Integration and Partial Differential Equations.* Princeton University Press, Princeton, NJ.

A. FRIEDMAN (1976): *Stochastic Differential Equations and Applications,* Vols. I, II. Academic Press. Reprint, Dover Publications, New York, 2006.

I. M. GELFAND AND N. Y. VILENKIN (1964): *Generalized Functions, Vol. 4: Applications of Harmonic Analysis.* Academic Press. (English translation.)

J. GJERDE (1996a): Two classes of stochastic Dirichlet equations which admit explicit solution formulas. In H. Körezlioglu, B. Øksendal and A.S. Üstünel (editors): *Stochastic Analysis and Related Topics,* Vol. 5. Birkhäuser, Boston, pp. 157–181.

J. GJERDE (1996b): An equation modelling Poisson distributed pollution in a stochastic medium: a white noise approach. Manuscript, University of Oslo.

J. GJERDE (1998): Existence and uniqueness theorems for some stochastic parabolic partial differential equations, *Stoch. Anal. Appl.,* **16,** 261–289.

J. GJERDE, H. HOLDEN, B. ØKSENDAL, J. UBØE AND T. ZHANG (1995): An equation modelling transport of a substance in a stochastic medium. In E. Bolthausen, M. Dozzi and F. Russo (editors): *Seminar on Stochastic Analysis, Random Fields and Applications.* Birkhäuser, Boston, pp. 123–134.

H. GJESSING (1993): A note on the Wick product. Preprint, University of Bergen No. 23.

H. GJESSING (1994): Wick calculus with applications to anticipating stochastic differential equations. Manuscript, University of Bergen.

H. GJESSING, H. HOLDEN, T. LINDSTRØM, J. UBØE AND T. ZHANG (1993): The Wick product. In H. Niemi, G. Högnäs, A. N. Shiryaev and A. Melnikov (editors): *Frontiers in Pure and Applied Probability,* Vol. 1. TVP Publishers, Moscow, pp. 29–67.

J. GLIMM AND A. JAFFE (1987): *Quantum Physics* (Second Edition). Springer, New York.

G. GRIPENBERG, S.-O. LONDEN AND O. STAFFANS (1990): *Volterra Integral and Functional Equations.* Cambridge University Press, Cambridge, MA.

J. GRUE AND B. ØKSENDAL (1997): A stochastic oscillator with time-dependent damping. *Stochastic Processes and Their Applications,* **68,** 113–131.

S. GURBATOV, A. MALAKHOV AND A. SAICHES (1991): *Nonlinear Random Waves and Turbulence in Nondispersive Media: Waves, Rays and Particles.* Manchester University Press, Manchester, UK.

H. HANCHE-OLSEN (1992): The Bochner–Minlos theorem. A pedestrian approach. Manuscript, Norwegian Institute of Technology.

T. HIDA (1980): *Brownian Motion.* Springer, New York.

T. HIDA AND N. IKEDA (1965): Analysis on Hilbert space with reproducing kernel arising from multiple Wiener integral. *Proc. Fifth Berkeley Symp. Math. Stat. Probab. II,* part 1, pp. 117–143.

T. HIDA AND J. POTTHOFF (1990): White noise analysis: an overview. In T. Hida, H.-H. Kuo, J. Potthoff and L. Streit (editors): *White Noise Analysis.* World Scientific, Singapore, pp. 140–165.

T. HIDA, H.-H. KUO, J. POTTHOFF AND L. STREIT (1993): *White Noise: An Infinite Dimensional Calculus.* Springer, New York.

E. HILLE AND R. S. PHILLIPS (1957): Functional Analysis and Semigroups. *Amer. Math. Soc. Colloq. Publ.,* **31.**

H. HOLDEN AND Y. HU (1996): Finite difference approximation of the pressure equation for fluid flow in a stochastic medium: a probabilistic approach. *Comm. Partial Diff. Eq.,* **21,** 9–10, 1367–1388.

H. HOLDEN AND N.H. RISEBRO (1991): Stochastic properties of the scalar Buckley-Leverett equation. *SIAM J. Appl. Math.*, **51**, 1472–1488.

H. HOLDEN AND N.H. RISEBRO (1997): Conservation laws with a random source. *Appl. Math. Optim.*, **36**, 2, 229–241.

H. HOLDEN, T. LINDSTRØM, B. ØKSENDAL AND J. UBØE (1992): Discrete Wick calculus and stochastic functional equations. *Potential Anal.*, **1**, 291–306.

H. HOLDEN, T. LINDSTRØM, B. ØKSENDAL AND J. UBØE (1993): Discrete Wick products. In T. Lindstrøm, B. Øksendal and A.S. Üstünel (editors:) *Stochastic Analysis and Related Topics*. Gordon & Breach, New York, pp. 123–148.

H. HOLDEN, T. LINDSTRØM, B. ØKSENDAL, J. UBØE AND T. ZHANG (1993A): Stochastic boundary value problems: a white noise functional approach. *Prob. Th. Rel. Fields*, **95**, 391–419.

H. HOLDEN, T. LINDSTRØM, B. ØKSENDAL, J. UBØE AND T. ZHANG (1993B): A comparison experiment for Wick multiplication and ordinary multiplication. In T. Lindstrøm, B. Øksendal and A. S. Üstünel (editors): *Stochastic Analysis and Related Topics*. Gordon & Breach, New York, pp. 149–159.

H. HOLDEN, T. LINDSTRØM, B. ØKSENDAL, J. UBØE AND T. ZHANG (1994): The Burgers equation with a noisy force. *Comm. PDE*, **19**, 119–141.

H. HOLDEN, T. LINDSTRØM, B. ØKSENDAL, J. UBØE AND T. ZHANG (1995): The pressure equation for fluid flow in a stochastic medium. *Potential Anal.*, **4**, 655–674.

H. HOLDEN, T. LINDSTRØM, B. ØKSENDAL, J. UBØE AND T. ZHANG (1995B): The stochastic Wick-type Burgers equation. In A. Etheridge (editor): *Stochastic Partial Differential Equations*. Cambridge University Press, Cambridge, MA, pp. 141–161.

Y. HU AND B. ØKSENDAL (1996): Wick approximation of quasilinear stochastic differential equations. In H. Körezlioglu, B. Øksendal and A.S. Üstünel (editors): *Stochastic Analysis and Related Topics*, Vol. 5. Birkhäuser, Boston, pp. 203–231.

Y. HU, T. LINDSTRØM, B. ØKSENDAL, J. UBØE AND T. ZHANG (1995): Inverse powers of white noise. In M.G. Cranston and M. Pinsky (editors): *Stochastic Anal.* American Mathematical Society., pp. 439–456.

N. IKEDA AND S. WATANABE (1989): *Stochastic Differential Equations and Diffusion Processes* (Second Edition). North-Holland/Kodansha.

E. ISOBE AND SH. SATO(1993): Wiener–Hermite expansion of a process generated by an Itô stochastic differential equation *J. Appl. Prob.*, **20**, 754–765.

K. ITÔ (1944): Stochastic integral. *Proc. Imp. Acad. Tokyo*, **20**, 519–524.

K. ITÔ (1951): Multiple Wiener integral. *J. Math. Soc. Jpn*, **3**, 157–169.

Y. ITÔ (1988): Generalized Poisson functionals. *Prob. Th. Rel. Fields*, **77**, 1–28.

Y. ITÔ AND I. KUBO (1988): Calculus on Gaussian and Poisson white noises. *Nagoya Math. J.*, **111**, 41–84.

J. JACOD AND A. SHIRYAEV (2003): *Limit Theorems for Stochastic Processes* (Second Edition). Springer, New York.

F. JOHN (1986): *Partial Differential Equations* (Fourth Edition). Springer-Verlag, New York.

Y. KABANOV (1975): On extended stochastic integrals. *Theory Probab. Appl.*, **20**, 710–722.

G. KALLIANPUR AND J. XIONG (1994): Stochastic models of environmental pollution. *Adv. Appl. Prob.*, **266**, 377–403.

I. KARATZAS AND S. E. SHREVE (1991): *Brownian Motion and Stochastic Calculus* (Second Edition). Springer, New York.

M. KARDAR, G. PARINI AND Y.-C. ZHANG (1986): Dynamic scaling of growing interfaces. *Phys. Rev. Lett.*, **56**, 889–892.

Y. KONDRATIEV (1978): Generalized functions in problems of infinite-dimensional analysis. Ph.D. thesis, Kiev University.

Y. KONDRATIEV, P. LEUKERT AND L. STREIT (1994): Wick calculus in Gaussian analysis. *Acta Appl. Math.*, **44** , 3, 269–294.

Y. KONDRATIEV, J. L. DA SILVA AND L. STREIT (1997): Generalized Appel systems. *Math. Funct. Anal. Topology*, **3**, 28–61.

Y. KONDRATIEV, L. STREIT AND W. WESTERKAMP (1995A): A note on positive distributions in Gaussian analysis. *Ukrainian Math. J.*, **47:5**, 649–659.

Y. KONDRATIEV, L. STREIT, W. WESTERKAMP AND J. YAN (1998): Generalized functions in infinite-dimensional analysis. *Hiroshima Math. J.*, **28**, 2, 213–260.

H. KÖREZLIOGLU, B. ØKSENDAL AND A. S. ÜSTÜNEL (EDITORS) (1996): *Stochastic Analysis and Related Topics*, Vol. 5. Birkhäuser, Boston.

H. KUNITA (1984): Stochastic differential equations and stochastic flows of diffeomorphisms. In P. L. Hennequin (editor): École d'Été de Probabilités de Saint-Flour XII. Springer LNM 1097, New York, pp. 143–303.

H. H. KUO (1996): *White Noise Distribution Theory*. CRC Press, Boca Raton, FL.

J. LAMPERTI (1966): *Probability*. W. A. Benjamin. Reprint, Dover Publications, New York, 2007.

A. LANCONELLI AND F. PROSKE (2004): On explicit strong solution of Itô SDE's and the Donsker delta function of a diffusion. *Inf. Dim. Anal. Quant. Probab. Rel. Top.*, **7**, 437–447.

T. LINDSTRØM, B. ØKSENDAL AND J. UBØE (1991A): Stochastic differential equations involving positive noise. In M. Barlow and N. Bingham (editors): *Stochastic Analysis*. Cambridge University Press, Cambridge, MA, pp. 261–303.

T. LINDSTRØM, B. ØKSENDAL AND J. UBØE (1991B): Stochastic modelling of fluid flow in porous media. In S. Chen and J. Yong (editors): *Control Theory, Stochastic Analysis and Applications*. World Scientific, Singapore, pp. 156–172.

T. LINDSTRØM, B. ØKSENDAL AND J. UBØE (1992): Wick multiplication and Itô–Skorohod stochastic differential equations. In S. Albeverio, J. E. Fenstad, H. Holden and T. Lindstrøm (editors): *Ideas and Methods in Mathematical Analysis, Stochastics, and Applications*. Cambridge University Press, Cambridge, MA, pp. 183–206.

T. LINDSTRØM, B. ØKSENDAL AND A.S. ÜSTÜNEL (editors)(1993): *Stochastic Analysis and Related Topics*. Gordon & Breach, New York.

T. LINDSTRØM, B. ØKSENDAL, J. UBØE AND T. ZHANG (1995): Stability properties of stochastic partial differential equations. *Stochas. Anal. Appl.*, **13**, 177–204.

E. W. LYTVYNOV AND G. F. US (1996): Dual Appell systems in non-Gaussian white noise calculus. *Meth. Func. Anal. Topol.*, **2**, 2, 70–85.

A. LØKKA, B. ØKSENDAL AND F. PROSKE (2004): Stochastic partial differential equations driven by Lévy space–time white noise. *Ann. App. Prob.*, **14**, 1506–1528.

A. LØKKA AND F. PROSKE (2006): Infinite-dimensional analysis of pure jump Lévy processes on the Poisson space. *Math. Scand.*, **98**, 237–261.

E. LUNGU AND B. ØKSENDAL (1997): Optimal harvesting from a population in a stochastic, crowded environment. *Math. Biosci.*, **145**, 47–75.

S. MATARAMVURA, B. ØKSENDAL AND F. PROSKE (2004): The Donsker delta function of a Lévy process with application to chaos expansion of local time. *Ann. Inst. Henri Poincaré*, **40**, 553–567.

K. J. MÅLØY, J. FEDER AND T. JØSSANG (1985): Viscous fingering fractals in porous media. *Phys. Rev. Lett.*, **55**, 2688–2691.

E. MEDINA, T. HWA, M. KARDAR AND T.-C. ZHANG (1989): Burgers equation with correlated noise: Renormalization group analysis and applications to directed polymers and interface growth, *Phys. Rev.*, **39A**, 3053–3075.

P. A. MEYER AND J. A. YAN (1989): Distributions sur l'espace de Wiener (suite). *Sém. de Probabilités XXIII*, J. Azéma, P. A. Meyer and M. Yor (editors). Springer LNM 1372, Paris, pp. 382–392.

R. MIKULEVICIUS AND B. ROZOVSKI (1998): Linear parabolic stochastic PDEs and Wiener chaos. *SIAM J. Math. Anal.*, **29**, 452–480.

D. NUALART AND M. ZAKAI (1986): Generalized stochastic integrals and the Malliavin calculus. *Prob. Th. Rel. Fields*, **73**, 255–280.

D. NUALART AND M. ZAKAI (1989): Generalized Brownian functionals and the solution to a stochastic partial differential equation. *J. Func. Anal.*, **84**, 279–296.

N. OBATA (1994): *White Noise Calculus and Fock Space*. LNM 1577. Springer, New York.

M. OBERGUGGENBERGER (1992): *Multiplication of Distributions and Applications to Partial Differential Equations*. Longman Scientific and Technical, Wiley, Hoboken, NJ.

M. OBERGUGGENBERGER (1995): Generalized functions and stochastic processes. In E. Bolthausen, M. Dozzi and F. Russo (editors): *Seminar on Stochastic Analysis, Random Fields and Applications*. Birkhäuser, Boston, pp. 215–229.

S. OGAWA (1986): On the stochastic integral equation of Fredholm type. *Studies in Mathematics and Its Applications* **18**; Patterns and Waves: Qualitative Analysis of Nonlinear Differential Equations, 597–605.

B. ØKSENDAL (2003): *Stochastic Differential Equations* (Sixth Edition). Springer, New York.

B. ØKSENDAL (2008): Stochastic partial differential equations driven by multiparameter white noise of Lévy processes. *Quart. Appl. Math.*, **66**, 521–537.

B. ØKSENDAL, F. PROSKE AND M. SIGNAHL (2006): The Cauchy problem for the wave equation with Lévy noise initial data. *Inf. Dim. Anal., Quantum Prob. Rel. Topics*, **9**, 249–270.

B. ØKSENDAL AND F. PROSKE (2004): White noise of Poisson random measures. *Potential Anal.*, **21**, 375–403.

B. ØKSENDAL AND A. SULEM (2007): *Applied Stochastic Control of Jump Diffusions* (Second Edition). Springer, New York.

B. ØKSENDAL AND G. VÅGE (1999): A moving boundary problem in a stochastic medium. *Inf. Dim. Anal. Quant. Prob. Rel. Top.*, **2**, 179–202.

B. ØKSENDAL AND T. ZHANG (1993): The stochastic Volterra equation. In D. Nualart and M. Sanz Solé (editors): *Barcelona Seminar on Stochastic Analysis*. Birkhäuser, Boston, pp. 168–202.

B. ØKSENDAL AND T. ZHANG (1996): The general linear stochastic Volterra equation with anticipating coefficients. In I. M. Davis, A. Truman and K. D. Elworthy (editors): *Stochastic Analysis and Applications*. World Scientific, Singapore, 1996, pp. 343–366.

U. OXAAL, M. MURAT, F. BOGER, A. AHARONY, J. FEDER AND T. JØSSANG (1987): Viscous fingering on percolation clusters. *Nature*, **329**, 32–37.

E. PARDOUX (1990): Applications of anticipating stochastic calculus to stochastic differential equations. In H. Korezlioglu and A. S. Üstünel (editors): *Stochastic Analysis and Related Topics II*. Springer LNM 1444, New York, pp. 63–105.

E. PARDOUX AND P. PROTTER (1990): Stochastic Volterra equations with anticipating coefficients. *Annals Prob.*, **18**, 1635–1655.

S. C. PORT AND C. J. STONE (1978): *Brownian Motion and Classical Potential Theory*. Academic Press, New York.

J. POTTHOFF (1992): White noise methods for stochastic partial differential equations. In B.L. Rozovskii and R.B. Sowers (editors): *Stochastic Partial Differential Equations and Their Applications*, Springer, New York, 238–251.

J. POTTHOFF (1994): White noise approach to parabolic stochastic partial differential equations. In A. J. Cardoso, M. de Faria, J. Potthoff, R. Sénéor and L. Streit (editors): *Stochastic Analysis and Applications in Physics*, Kluwer, Norwell, MA, pp. 307–327.

J. POTTHOFF AND P. SUNDAR (1996): Law of large numbers and central limit theorem for Donsker's delta function of diffusion. *Potential Anal.*, **5**, 487–504.

J. POTTHOFF AND M. TIMPEL (1995): On a dual pair of smooth and generalized random variables. *Potential Anal.*, **4**, 637–654.

J. POTTHOFF, G. VÅGE AND H. WATANABE (1998): Generalized solutions of linear parabolic stochastic partial differential equations. *Appl. Mat. Optim.*, **38**, 95–107.

M. REED AND B. SIMON (1975): *Methods of Modern Mathematical Physics*, Vol. 2. Academic Press, New York.

M. REED AND B. SIMON (1980): *Methods of Modern Mathematical Physics*, Vol. 1. Academic Press, New York.

D. REVUZ AND M. YOR (1991): *Continuous Martingales and Brownian Motion.* Springer, New York.

W. RUDIN (1973): *Functional Analysis.* McGraw-Hill, New York.

F. RUSSO (1994): Colombeau generalized functions and stochastic analysis. In A. J. Cardoso, M. de Faria, J. Potthoff, R. Sénéor and L. Streit (editors): *Stochastic Analysis and Applications in Physics*, Kluwer, Norwell, MA, pp. 329–349.

K. SATO (1999): *Lévy Processes and Infinitely Divisible Distributions.* Cambridge University Press, Cambridge, MA.

S. F. SHANDARIN AND YA. B. ZELDOVICH (1989): The large-scale structure of the universe: turbulence, intermittency, structures in a self-gravitating medium. *Rev. Mod. Phys.*, **61**, 185–220.

B. SIMON (1974): *The $P(\phi)_2$ Euclidean (Quantum) Field Theory.* Princeton University Press, Princeton, NJ.

B. SIMON (1979): *Functional Integration and Quantum Physics.* Academic Press, New York.

D. W. STROOCK AND S. R. S. VARADHAN (1979): *Multidimensional Diffusion Processes.* Springer, New York.

A. TRUMAN, H. Z. ZHAO (1996): On stochastic diffusion equations and stochastic Burgers equations. *J. Math. Phys.*, **37**, 1, 283–307.

G. F. US (1995): Dual Appel systems in Poisson analysis. *Methods Funct. Anal. Topology*, **1**, 1, 93–108.

G. VÅGE (1995a): Stochastic differential equations and Kondratiev spaces. Dr. Ing. Thesis, University of Trondheim.

G. VÅGE (1996a): Hilbert space methods applied to stochastic partial differential equations. In H. Körezlioglu, B. Øksendal and A. S. Üstünel (editors): *Stochastic Analysis and Related Topics*, Vol. 5. Birkhäuser, Boston, pp. 281–294.

G. VÅGE (1996b): A general existence and uniqueness theorem for Wick-SDEs in $(S)^n_{-1,k}$. *Stochastics Stochastics Rep.*, **58**, 3-4, 259–284.

A. YU VERETENNIKOV AND N. V. KRYLOV (1976): On explicit formulas for solutions of stochastic differential equations. *Math. USSR Sbornik*, **29**, 229–256.

J. B. WALSH (1986): An introduction to stochastic partial differential equations. In R. Carmona, H. Kesten and J.B. Walsh (editors): *École d'Été de Probabilités de Saint-Flour XIV-1984.* Springer LNM 1180, New York, pp. 265–437.

S. WATANABE (1983): Malliavin's calculus in terms of generalized Wiener functionals. In G. Kallianpur (editor): *Theory and Applications of Random Fields.* Springer, New York.

G. C. WICK (1950): The evaluation of the collinear matrix. *Phys. Rev.*, **80**, 268–272.

N. WIENER (1924): Differential space. *J. Math. Phys.*, **3**, 127–146.

E. WONG AND M. ZAKAI (1965): On the relation between ordinary and stochastic differential equations. *Intern. J. Engr. Sci.*, **3**, 213–229.

T. ZHANG (1992): Characterizations of white noise test functions and Hida distributions. *Stochastics*, **41**, 71–87.

List of frequently used notation and symbols

$\mathbb{N}$ = the natural numbers $1, 2, 3, \dots$

$\mathbb{N}_0 = \mathbb{N} \cup \{0\}$

$\mathbb{Z}$ = the integers

$\mathbb{Q}$ = the rational numbers

$\mathbb{R}$ = the real numbers

$\mathbb{R}^+ = [0, \infty)$ the nonnegative real numbers

$\mathbb{R}_0 = \mathbb{R} \setminus \{0\}$

$\mathbb{C}$ = the complex numbers

$\mathbb{R}^n = \mathbb{R} \times \cdots \times \mathbb{R}$ (n times)= the n-dimensional Euclidean space

$\mathbb{R}^n_+ = \mathbb{R}_+ \times \cdots \times \mathbb{R}_+$ (n times)

$\mathbb{R}^{m \times n}$ = the set of all $m \times n$ matrices with real entries

I_n = the $n \times n$ identity matrix

$\mathbb{C}^{\mathbb{N}}$ = the set of all sequences $z = (z_1, z_2, \dots)$ with $z_k \in \mathbb{C}$

$(\mathbb{C}^{\mathbb{N}})_c$ = the set of all *finite* sequences in $\mathbb{C}^{\mathbb{N}}$ (a finite sequence $(z_1, \dots, z_k)$ is identified with the sequence $(z_1, \dots, z_k, 0, 0, \dots) \in \mathbb{C}^{\mathbb{N}}$)

$(\mathbb{R}^{\mathbb{N}})_c$ = the set of all *finite* sequences in $\mathbb{R}^{\mathbb{N}}$ (a finite sequence $(x_1, \dots, x_k)$ is identified with the sequence $(x_1, \dots, x_k, 0, 0, \dots) \in \mathbb{R}^{\mathbb{N}}$)

$\mathcal{J} = (\mathbb{N}_0^{\mathbb{N}})_c$ the set of all finite sequences (multi-indices) $\alpha = (\alpha_1, \alpha_2, \dots, \alpha_k)$ where $\alpha_i \in \mathbb{N}_0$, $k = 1, 2, \dots$ (see equation (2.2.9))

Index $\alpha = \max\{j; \alpha_j \neq 0\}$ if $\alpha = (\alpha_1, \alpha_2, \dots) \in \mathcal{J}$

$l(\alpha)$ = the length of α = the number of nonzero elements of $\alpha \in \mathcal{J}$

$\epsilon^{(k)} = (0, 0, \dots, 0, 1, 0, \dots) \in \mathcal{J}$, with 1 on entry number k, $k = 1, 2, \dots$

$\Gamma_n = \{\alpha \in \mathcal{J}; \text{Index } \alpha \leq n \text{ and } \alpha_j \in \{0, 1, \dots, n\} \text{ for all } j\}$

$z^\gamma = z_1^{\gamma_1} z_2^{\gamma_2} \cdots$ if $z \in \mathbb{C}^N$ and $\gamma \in (\mathbb{R}^N)_c$ (where 0^0 is interpreted as 1)

$(2\mathbb{N})^\gamma = \prod_{j=1}^{\infty}(2j)^{\gamma_j}$ for $\gamma \in (\mathbb{R}^N)_c$ (see equation (2.3.8))

$\mathbb{K} = \{z = (z_1, \ldots, z_n) \in \mathbb{C}^n \; ; \; c_1|z_1|^2 + \cdots + c_n|z_n|^2 \leq \delta^2\}$

$\mathbb{K}_q(R) = \{\zeta = (\zeta_1, \zeta_2, \ldots) \in \mathbb{C}^N; \sum_{\alpha \neq 0} |\zeta^\alpha|^2 (2\mathbb{N})^{q\alpha} < R^2\}$ (Definition 2.6.4)

$h_n(x) =$ the Hermite polynomials; $n = 0, 1, 2, \ldots$ (see equation (2.2.1))

$\xi_n(x) =$ the Hermite functions; $n = 1, 2, \ldots$ (see equation (2.2.2)) (they constitute a basis for $L^2(\mathbb{R})$)

$\xi_\delta = \xi_{\delta_1} \otimes \cdots \otimes \xi_{\delta_d} \; ; \; \delta = (\delta_1, \ldots, \delta_d) \in \mathbb{N}^d$ (the tensor product of $\xi_{\delta_1}, \ldots, \xi_{\delta_d}$, i.e. $\xi_\delta(x_1, \ldots, x_d) = \xi_{\delta_1}(x_1) \cdots \xi_{\delta_d}(x_d) \; ; \; (x_1, \ldots, x_d) \in \mathbb{R}^d)$ (see equation (2.2.6))

$\eta_j = \xi_{\delta^{(j)}}$, where $\delta^{(1)}, \delta^{(2)}, \ldots$ is a fixed ordering according to size of all d-dimensional multiindices $\delta = (\delta_1, \ldots, \delta_d) \in \mathbb{N}^d$ (see equation (2.2.8)). The family $\{\eta_j\}_{j=1}^{\infty}$ constitutes a basis for $L^2(\mathbb{R}^d)$.

$H_\alpha(\omega) = \prod_{i=1}^{\infty} h_{\alpha_i}(\langle \omega, \eta_i \rangle)$ (see equation (2.2.10))

$K_\alpha(\omega) = I_{|\alpha|}(\delta^{\hat\otimes\alpha})(\omega)$ (see equations (5.3.22) and (5.4.42))

$e^{(k)} = \eta_j \epsilon^{(i)}$ if $k = i + (j-1)m; i \in \{l, \ldots, m\}, j \in \mathbb{N}$. The family $\{e^{(k)}\}_{k=1}^{\infty}$ constitutes a basis for $\mathcal{K} = \bigoplus_{k=1}^{m} L^2(\mathbb{R}^d)$. (See equation (2.2.11)).

$(\mathcal{S})_\rho^N = (\mathcal{S})_\rho^{m;N}; -1 \leq \rho \leq 1 =$ the Kondratiev spaces (Definition 2.3.2 and Definition 5.3.6)

(m is the dimension of the white noise vector and N is the state space dimension)

$(\mathcal{S}), (\mathcal{S})^* =$ the Hida test function space and the Hida distribution space, respectively (see Proposition 2.3.7 and Definition 5.3.6)

$(\mathcal{S})_{\rho,r}^{m;N} =$ the Kondratiev Hilbert spaces (Definition 2.7.1 and Definition 5.3.6)

$\diamond =$ the Wick product (Section 2.4 and Definition 5.3.9)

$\mathcal{H}(F)(z) = \widetilde{F}(z) =$ the Hermite transform of $F \in (\mathcal{S})_{-1}$ (Definition 2.6.1 and Definition 5.4.15)

$(\mathcal{S}F)(\lambda\phi) =$ the $\mathcal{S}$-transform of $F \in (\mathcal{S})_{-1}$ at $\lambda\phi$ (Definition 2.7.5)

$\mathcal{S}(\mathbb{R}^d) =$ the Schwartz space of rapidly decreasing smooth functions (d is called the parameter dimension)

$\mathcal{S}'(\mathbb{R}^d) =$ the space of tempered distributions on $\mathbb{R}^d$

$\mathcal{S} = \prod_{i=1}^{m} \mathcal{S}(\mathbb{R}^d), \mathcal{S}' = \prod_{i=1}^{m} \mathcal{S}'(\mathbb{R}^d)$ (see equation (2.1.33))

$\mu_1 =$ the white noise probability measure on $\mathcal{S}'(\mathbb{R}^d)$ (Theorem 2.1.1)

$\mu_m = \mu_1 \times \cdots \times \mu_1$ d-parameter, m-dimensional white noise probability measure on $\mathcal{S}'$ (see equation (2.1.24)). If the value of m is clear from the context, we sometimes write μ for μ_m

$E_\mu =$ the expectation with respect to μ

$B(t); t \in \mathbb{R} =$ 1-parameter Brownian motion

$B(x); x \in \mathbb{R}^d =$ d-parameter Brownian motion/field (see equations (2.1.11)–(2.1.13))

$W(x); x \in \mathbb{R}^d =$ d-parameter singular white noise (Definition 2.3.9)

$\mathbf{W}(x) = (W_1(x), \ldots, W_m(x)) =$ m-dimensional, d-parameter singular white noise (Definition 2.3.10) $x \in \mathbb{R}^d$

$\langle \omega, \phi \rangle =$ the action of $\omega \in \mathcal{S}'(\mathbb{R}^d)$ on $\phi \in \mathcal{S}(\mathbb{R}^d)$ (see Theorem 2.1.1)

$w(\phi); \phi \in \mathcal{S}(\mathbb{R}^d) =$ 1-dimensional, smoothed white noise (Definition 2.1.4)

$\mathbf{w}(\phi); \phi \in \mathcal{S} =$ m-dimensional, smoothed white noise (see equation (2.1.27))

$\phi_x(y) = \phi(y - x); x, y \in \mathbb{R}^d =$ the x-shift of $\phi \in \mathcal{S}$ (see equation (2.1.31))

$W_\phi(x) = w(\phi_x), \mathbf{W}_\phi(x) = \mathbf{w}(\phi_x) =$ the 1-dimensional and the m-dimensional smoothed white noise process, respectively. (See equation (2.1.30))

$\eta(t); t \in \mathbb{R} = 1 -$ parameter Lévy process

$\eta(x); x \in \mathbb{R}^d = d -$ parameter Lévy process/field (see Definition E1 and Theorem 5.4.3)

$N(dt, dz) =$ the jump measure of a Lévy process (Definition E1)

$\nu(dz)$ = the Lévy measure of a Lévy process (see equation (E.5))

$\tilde{N}(dt, dz)$ = the compensated jump measure/Poisson random measure (see equation (E.10))

$\dot{\eta}(t); t \in \mathbb{R} = 1 -$ parameter Lévy white noise (Definition 5.3.7)

$\dot{\eta}(x); x \in \mathbb{R}^d = d -$ parameter Lévy white noise (see equation (5.4.50))

$\overset{\cdot}{\tilde{N}}(t, z); (t, z) \in \mathbb{R} \times \mathbb{R}_0 = (1 -$ parameter) white noise of $\tilde{N}(dt, dz)$ (Definition 5.3.8)

supp f = the support of the function f

$dx, dy, dt, \ldots$ = Lebesgue measure on $\mathbb{R}^n$

$L^p_{\mathrm{loc}}(\mathbb{R}^d)$ = the functions on $\mathbb{R}^d$ that are locally in L^p with respect to the Lebesgue measure

$C^k(U)$ = the real functions on $U \subset \mathbb{R}^n$ that are k times continuously differentiable ($k \in \mathbb{N}_0$)

$C(U) = C^0(U)$

$C_0^k(U)$ = the functions in $C^k(U)$ with compact support in U

$C^\infty(U) = \bigcap_{k=0}^\infty C^k(U)$

$C_0^\infty(U) = \bigcap_{k=0}^\infty C_0^k(U)$

$C^k = C^k(\mathbb{R}^n)$

$C^{0+\lambda}(U)$ = the functions g in $C(U)$ that are *Hölder continuous of order* λ, i.e., that satisfy

$$\|g\|_{C^{0+\lambda}(U)} := \sup_{\substack{x,y \in U \\ x \neq y}} \frac{|g(x) - g(y)|}{|x - y|^\lambda} < \infty \quad (0 < \lambda \leq 1)$$

$C^{k+\lambda}(U)$ = the functions in $C^k(U)$ with whole partial derivatives up to order k are Hölder continuous of order $\lambda \in (0, 1)$

$$\|g\|_{C^{k+\lambda}(U)} = \sum_{0 \leq |\alpha| \leq k} \|\partial^\alpha g\|_{C^\lambda(U)}$$

If $g = g(t, x)$; $t \in [0, T]$, $x \in D \subset \mathbb{R}^d$, then

$$\|g\|_{C^{(\lambda)}} = \|g\|_{C^{(\lambda)}((0,T) \times D)} = \sup_{\substack{t \in (0,T) \\ x \in D}} |g(t, x)| + \sup_{\substack{y_1, y_2 \\ \in (0,T) \times D}} \frac{|g(y_1) - g(y_2)|}{d(y_1, y_2)^\lambda}$$

where

$$d(y_1, y_2) = (|x_1 - x_2|^2 + |t_1 - t_2|)^{\frac{1}{2}} \quad \text{when} \quad y_i = (t_i, x_i) \in (0, T) \times D.$$

$$\|g\|_{C^{(2+\lambda)}} = \|g\|_{C^{(\lambda)}} + \left\|\frac{\partial g}{\partial t}\right\|_{C^{(\lambda)}} + \sum_{j=1}^{d} \left\|\frac{\partial g}{\partial x_j}\right\|_{C^{(\lambda)}} + \sum_{i,j=1}^{d} \left\|\frac{\partial^2 g}{\partial x_i \partial x_j}\right\|_{C^{(\lambda)}}$$

$(\cdot, \cdot)$ = the inner product in $L^2(\mathbb{R}^d)$ or the inner product in $\mathbb{R}^n$ (depending on the context)

$\Delta = \sum_{k=1}^{n} \frac{\partial^2}{\partial x_k^2}$ (the n-dimensional Laplace operator)

$\Delta_j = 2^d \delta_1^{(j)} \delta_2^{(j)} \cdots \delta_d^{(j)}$; $j = 1, 2, \ldots$ (see equation (2.3.19))

$\Delta^\alpha = \Delta_1^{\alpha_1} \Delta_2^{\alpha_2} \cdots \Delta_j^{\alpha_j} \cdots = \prod_{j=1}^{\infty} (2^d \delta_1^{(j)} \delta_2^{(j)} \cdots \delta_d^{(j)})^{\alpha_j}$ (see equation (2.3.20))

∂D = the (topological) boundary of the set D in $\mathbb{R}^n$

$\partial_R D$ = the regular boundary points of D (with respect to the given process)

$\overline{D}$ = the closure of D

$V \subset\subset D$ means that $\overline{V} \subset D$ and that $\overline{V}$ is compact

$\delta_0(x)$ = Dirac measure at 0

$\delta_{ij} = 1$ if $i = j$, 0 otherwise

$E[X] = E_\nu[X]$ = expectation of a random variable X with respect to a measure ν

$\text{Var}[X] = E[(X - E[X])^2]$ = the variance of X

$:=$ by definition equal to

C = a constant (its value may change from place to place)

$\square$ = end of proof.

Index

Der skal et par dumheder
med i en bog ...
for at også de dumme
skal syns, den er klog.

True wisdom knows
it must comprise
some nonsense
as a comprise,
lest fools should fail
to find it wise.

Piet Hein